Anna Zieglerin
and the Lion's Blood

Anna Zieglerin and the Lion's Blood

Alchemy and End Times in Reformation Germany

Tara Nummedal

PENN

UNIVERSITY OF PENNSYLVANIA PRESS

PHILADELPHIA

A volume in the Haney Foundation Series,
established in 1961 with the generous support of
Dr. John Louis Haney.

Published by
University of Pennsylvania Press
Philadelphia, Pennsylvania 19104-4112
www.upenn.edu/pennpress

Printed in the United States of America on acid-free paper
1 3 5 7 9 10 8 6 4 2

Library of Congress Cataloging-in-Publication Data

Names: Nummedal, Tara E., author.
Title: Anna Zieglerin and the lion's blood: alchemy and end times in
    Reformation Germany / Tara Nummedal.
Other titles: Haney Foundation series.
Description: 1st edition. | Philadelphia: University of Pennsylvania Press,
    [2019] | Series: Haney Foundation series | Includes bibliographical
    references and index.
Identifiers: LCCN 2018045210 | ISBN 9780812250893 (hardcover)
Subjects: LCSH: Zieglerin, Anna Maria, approximately 1545–1575. |
    Alchemists—Germany—Biography. | Alchemy—Germany—
    History—16th century. | Alchemy—Religious aspects—
    Christianity—History—16th century. | Religion and science—
    Germany—History—16th century. | Germany—History—1517–1648.
Classification: LCC QD24.Z54 N86 2019 | DDC 540.1/12092 [B]—dc23
LC record available at https://lccn.loc.gov/2018045210

For Seth and Mila

CONTENTS

Anna Maria Zieglerin (ca. 1545–1575), a.k.a. Anna Maria Ziegler or Schombachin, Schlüterliese, or Schlüter Ilsche (see "A Note on Names"). Saxon alchemist, prophet, and courtier. Married to Heinrich Schombach. Executed in 1575 in Wolfenbüttel.

Anna, Electress of Saxony (1532–1585). Born a Danish princess, later Electress of Saxony. Renowned for her medical and botanical skill. Married to August of Saxony.

August, Elector of Saxony (1526–1586). Adversary of his cousin Duke Johann Friedrich II of Saxony during the "Grumbach Feud" and its aftermath. Interested in mining, alchemy, and agriculture. Married to Anna of Saxony.

Hänschen, a.k.a. Hans, Hänsen, or Hänsel Müller, Henkel, or Tausendschön (ca. 1550–1567). Peasant boy from Sundhausen, a village near Gotha. "Angel seer" for Duke Johann Friedrich II of Saxony. Executed in 1567 in Gotha.

Hedwig, Duchess of Braunschweig-Lüneburg (1540–1602). Anna's nemesis in Wolfenbüttel, also known for her medical and botanical skill. Married to Julius of Braunschweig-Lüneburg.

Heinrich Julius, Duke of Braunschweig-Lüneburg (1564–1613). Julius and Hedwig's eldest son, known for his interest in the law and baroque theater, as well as witch-hunting and the expulsion of the Jews from the duchy. Presided over Anna, Philipp, and Heinrich's sentencing in 1575 at age ten.

Heinrich Schombach (d. 1575), a.k.a. Schielheinz (Cross-Eyed Heinz). Valet and court jester at the court of Duke Johann Friedrich II of Saxony until 1567, then alchemical assistant and courtier in Wolfenbüttel. Married to Anna Zieglerin. Executed in 1575 in Wolfenbüttel.

Jobst Kettwig (d. 1575). Mercenary captain from Holstein, then courtier to Duke Julius of Braunschweig-Lüneburg. Executed in 1575 in Wolfenbüttel.

Johann Friedrich II, Duke of Saxony (1529–1595). Lutheran polemicist, possible "Last World Emperor," and Heinrich Schombach and Philipp Sömmering's patron in the 1560s. Imprisoned in Austria after the "Grumbach Feud" and his defeat during the siege of Gotha in 1567.

Johann Georg, Elector of Brandenburg (1525–1598). Carried out a pogrom and expelled Berlin's Jews after executing Lippold in 1573. Hedwig's brother, Julius's mentor.

Julius, Duke of Braunschweig-Lüneburg and Prince of Braunschweig-Wolfenbüttel (1528–1589). Anna's patron in Wolfenbüttel, married to Hedwig of Braunschweig-Lüneburg.

Katharina, Margravess of Brandenburg-Küstrin (1518–1574). Anna Zieglerin's foe. Julius's sister, married to Hedwig's uncle, Margrave Johann of Küstrin.

Lippold, or Lipman ben (Judel) Chluchim (before 1530–1573). Treasurer (*Schatullenverwalter*) for Elector Joachim Hector II of Brandenburg, administrator (*Oberältester*) of the Jewish community, and master of the mint in Brandenburg. Executed in 1573 in Berlin.

Margarete, Duchess of Münsterberg-Oels (1516–1580). Julius's eldest sister. Married to Johann of Münsterberg-Oels.

Philipp Sömmering (d. 1575). Lutheran pastor in Thuringia until 1567, then alchemist, mining, and church adviser in Wolfenbüttel. Married to Catharina Sömmering. Executed in 1575 in Wolfenbüttel.

Silvester Schulverman (d. 1575). Mercenary soldier from Lübeck, then alchemical assistant and courtier to Duke Julius of Braunschweig-Lüneburg. Executed in 1575 in Wolfenbüttel.

Wilhelm von Grumbach (1503–1567). Frankish imperial knight, adviser to Duke Johann Friedrich II of Saxony. Executed in 1567 in Gotha.

A NOTE ON NAMES

Early modern German names were fluid, particularly for women, both in spelling and in form. Noblewomen, for example, took on the names or titles of their husbands, changing their names over the course of their lives, which makes them sometimes confusing to track. Hedwig was born as Margravess Hedwig of Brandenburg, but once she married Duke Julius, she became Duchess Hedwig of Braunschweig-Lüneburg and Princess of Braunschweig-Wolfenbüttel. Anna's surname was also unstable in the sixteenth century, and she signed her own name variously as Anna Maria Zigler/Ziegler, Ziglerin/Zieglerin, or more rarely (taking her husband's name) Schombach/Schombachin. The suffix *-in* was added irregularly to Anna's surname in the sixteenth century and should be understood simply as an indicator of the female form of her family name, Ziegler. Because the suffix *-in* has been abandoned in modern German, it is tempting to drop it in Anna's name today. Moreover, the *-in* suffix can sometimes have a pejorative ring to modern German ears, as in "that Zieglerin woman." Because Anna seems to have referred to herself most often as Anna Maria Zieglerin, however, I have chosen to follow her own usage.

FIGURE I. "Schlüterliese's witch's chair." Braunschweigisches Landesmuseum, ZG 3905. Photo credit: Braunschweigisches Landesmuseum, I. Simon.

A Witch's Chair?

In October 2011, the State Museum in Braunschweig, Germany, opened an exhibit to celebrate its 120th anniversary. On display were 120 objects from the museum's collections, each meant to spark a dialogue between the viewer and a particular moment in the region's past. Among the objects was an ornate iron chair said to be from the sixteenth century (see Figure 1). With a curved back, decorative ironwork, a swivel, pedestal stem, and five wheels (only one of which remains), the chair's elegant design hints at lofty origins, perhaps in a patrician's home or at a princely court. It may have been a work chair, originally a kind of Renaissance office chair. The 2011 exhibition's wall text, however, suggested a far more unsettling history for this object. It is "Purportedly A Witch's Chair," ostensibly the chair in which a woman named Anna Maria Zieglerin was executed in nearby Wolfenbüttel on February 7, 1575, for numerous crimes, including sorcery, poison, fraud, theft, adultery, and infanticide. For these offenses, according to local lore, Zieglerin was led to the public site of execution, her skin was torn six times with red-hot tongs, and then she was executed by fire, strapped to this iron chair and suspended over the flames by chains attached to the loops under the armrests.[1]

What sort of connection are we in the twenty-first century to make to this object? According to the museum's exhibition catalog, Zieglerin's execution in 1575 was "one of the most gruesome and sensational cases in Wolfenbüttel," and indeed, the chair certainly prompts revulsion.[2] Whatever Zieglerin's crimes may have been, this was a ghastly way to die, and it is all the more chilling that she was executed legally at the hands of the state after a rigorous and procedurally correct trial. Nevertheless, the museum tells us, during her trial, Anna Zieglerin stated that "she confessed to everything, but didn't do everything; she said these things out of fear."[3] The chair, then, is first and foremost a disturbing artifact of an early modern legal system that not only sanctioned torture but also depended on it in exceptionally serious cases to produce confessions. The chair vividly reminds us, too, of the carefully choreographed and public nature of early modern state-sponsored punishment, which was meant simultaneously to punish and dishonor criminals' bodies, to restore the communities they had violated, and to communicate the precise nature of their offenses to onlookers. Although each of the crimes to which Zieglerin confessed warranted its own punishment, in the end, the chair suggests, state authorities chose to execute her by fire, the punishment reserved for the most heinous crimes: heresy, sorcery, poison, and sodomy, as well as crimes related to fire, such as arson. Fire was meant to completely obliterate the early modern criminal's body, preventing a burial and functioning almost as a purifying ritual to purge the community of his or her offenses.[4] By choosing this particular form of execution, then, the Wolfenbüttel officials would have intended to send a very clear message: Anna Zieglerin had violated not only the laws, but also the spiritual community of the Duchy of Braunschweig-Lüneburg. As a relic of execution by fire, finally, the chair also evokes the witch hunts that proliferated in Europe during this same period, condemning tens of thousands of men and women to their deaths.[5] As we gaze upon this chair today, we are faced first and foremost with a spectacular death and the brutality of early modern justice, leaving us unsettled but perhaps also oddly comforted, glad that we live in a different era.

And yet, in the exhibition wall text next to the chair, the State Museum encouraged viewers to ponder Zieglerin's life as well as her death. A Lutheran woman who spent her entire life in proximity to three German princely courts, Anna Zieglerin (ca. 1545–1575) was wife, courtier, and alchemist before her arrest and conviction in 1574. For a time, in fact, her alchemy earned her the support and patronage of Duke Julius of the northern German territory of Braunschweig-Lüneburg (1528–1589). Julius was drawn to Anna's recipes for a golden oil she called the lion's blood, which was said to possess the extraordinary ability to stimulate the growth of plants, make gemstones, and even transform lead into the coveted philosophers' stone, a precious secret in its own right that could transmute metals and promote health. Equally intriguing was the way Anna framed her alchemy as a tool for addressing a matter that weighed heavily on the minds of pious Christians in sixteenth-century Europe: the rapidly approaching end of time. Anna claimed that she and an enigmatic adept named Count Carl von Oettingen were to fulfill a prophecy, using the lion's blood to conceive children whose alchemical bodies would help prepare the world for its end. Anna and those close to her drew parallels to the Virgin Mary, likening Anna's extraordinary generation of life, as well as its profound implications for sacred history, with the virgin birth of Christ. The real promise of the lion's blood was not simply gold and gemstones, but nothing less than the redemption of the world in its final moments. As riveting as it is, therefore, the iron chair on display in Braunschweig turns out to be a red herring. It misdirects our attention toward early modern witchcraft, a brutal justice system, and a narrative of modern progress away from that pitiless era, obscuring a far more interesting story about the intersection of alchemy, gender, and belief in the era of the Reformation.[6]

The iron chair is a red herring in another sense as well, for upon closer scrutiny it begins to fall apart as a stable historical artifact. The sixteenth-century documents related to Zieglerin's case are oddly silent about this detail of her execution, in fact—a surprising omission, given how unusually extensive the records surrounding her trial otherwise are.[7] The Lower Saxony Regional Archive in Wolfenbüttel contains thirty-one folders pertaining to the case, including letters, recipes, supply orders, and various other items written before her arrest, as well as transcripts of interrogations and other legal documents generated during the trial itself. The record of the final judgment (*Urteil*) has not survived, unfortunately, but a summary of the legal judgment and sentences recommended by two panels of jurists (*Schöffensprüche*) has, and we can assume that these guided Duke Julius's court in issuing the final sentence.[8] The jurists recommended that Zieglerin be punished "with fire," but we find nothing here about a chair, nor do any contemporary sixteenth-century sources bear witness to Zieglerin's execution in this manner.[9] This silence is telling, because such a punishment would have been highly unusual in the sixteenth century and certainly would have sparked comment.[10] Nor does the chair as a material object offer any conclusive evidence of either its use in Zieglerin's execution or its origins in the sixteenth century. The catalog entry from the Braunschweig State Museum notes that the chair bears white "blooms" on its surface, suggesting that at some point it may have been in a fire, but when the marks were created is unknown. The design and condition of the chair, meanwhile, are equally puzzling. Both its capacity to swivel and the small wheels at its base are striking, seemingly out of place for a sixteenth-century chair. And yet, swivel chairs were already in use in the fifteenth century in the German lands in work or study settings; nor were early modern medical wheelchairs unheard of (most famously, a wheeled, reclining chair that King Philipp II of Spain used because of his gout).[11] Such hints suggest that this iron chair was possible to imagine in the sixteenth century, but even the language of the catalog entry reveals how little is known about this specific chair's origins. It is "purportedly" Zieglerin's chair. Prior to her execution, it "could have belonged" to her patron Duke Julius, who was disabled and who might have used it as either a wheelchair or (an extremely heavy, it must be said) sedan chair before repurposing it for Zieglerin's execution.[12] These conclusions, while intriguing, remain speculative. In the end, there is no contemporary evidence confirming that Zieglerin was executed in any chair, let alone this chair, nor even that this chair was made in the sixteenth century.

The fact that this iron chair stands in a museum today as an artifact of Anna Zieglerin's execution, despite the uncertainty about its origins, offers a powerful reminder of both the influence still exerted by nineteenth-century historical narratives about early modern Europe and how careful we need to be in accepting them uncritically. Modern Braunschweigers' interest in Zieglerin is rooted in the nineteenth century, when a handful of articles and one major book on the case, the jurist and historian Albert Rhamm's 1883 *The Fraudulent Goldmakers at the Court of Julius of Braunschweig, According to the Trial Records*, appeared alongside a painting of Anna Zieglerin's execution, and, of course, the identification of "her" chair at the Wolfenbüttel castle.[13] The story had evidently circulated locally for centuries, but in the nineteenth century historians were drawn to the episode anew for a variety of reasons, ranging from the compelling triad of alchemy, witchcraft, and torture and the titillation of a good scandal to, in Rhamm's case, the opportunity to demonstrate rigorous, modern, professional historical research.[14] The modern scholars who first wrote about Zieglerin and her fellow alchemists were, in part, interested in what their story had to say about their patron, Duke Julius, who was a crucial figure in Braunschweig's history. Not only did Duke Julius stabilize the territory after decades of confessional warfare, but he also declared the state officially Protestant, founded the University of Helmstedt, and launched numerous projects to develop and modernize the infrastructure, commercial economy, and mining industry in the region.[15] Julius was, in other words, the founding father of the modern Duchy of Braunschweig, which in the nineteenth century made his pursuit of alchemy particularly complicated to explain. In grappling with this issue, some authors resorted to age-old polemics, caricaturing alchemy as doubly tainted by folly and fraud.[16] As one scholar wrote in 1857, for example, in the sixteenth century, the study of "chemistry and physics . . . was the shoals on which common sense foundered very easily. Even Julius did not know how to avoid these shoals. He became an adept."[17] Others offered a more nuanced view, recognizing that alchemy was a respected pursuit among scholars, princes, and artisans alike in the sixteenth century. As the legal scholar Johannes Merkel put it in 1896, "It would have been a wonder if a prince drawn to the study of mineralogy in that period did not possess his alchemists. No court, not even the imperial court, was without them at that time."[18] These historians also differed in their views of Julius's support for Zieglerin's fellow alchemist Philipp Sömmering, who also became an influential adviser on church matters at Julius's court. Some saw Sömmering as the consummate duplicitous adviser, who used fraudulent promises of alchemical success to captivate and take advantage of Duke Julius. Others presented Sömmering as initially well intentioned, but eventually forced to resort to desperate measures as his alchemical projects inevitably failed.[19]

Despite differences of interpretation regarding Duke Julius's support for alchemy or Philipp Sömmering's motivations, one thing that all of the nineteenth-century historians who wrote about this case agreed on was Anna Zieglerin. She was "a crafty, shrewd, and seductive woman"; "radiant and robust, wanton and cunning," an "uneducated courtesan" able to seduce Julius and thereby deafen him to the warnings of his "lovable, pious, and prudent wife" and skeptical advisers; "a wily schemer with a disarming manner."[20] Even Albert Rhamm, generally the most evenhanded and historically sensitive of this generation of historians to write about the episode, compared Zieglerin to a spider: "She lured poor Philipp [Sömmering] deeper and deeper into her web, no less with her promises than with her personal charm, until he saw that he was caught in her threads and recognized, too late, that he was a cheat who had been cheated."[21] The nineteenth-century historians who wrote about Zieglerin, in other words, could only describe her as a dangerous blend of sexuality, intelligence, charisma, and ambition. In doing so, they drew on both age-old classical and Christian stereotypes that had long associated the acquisition of knowledge by women with unbridled sexuality, and the image of Zieglerin produced during her own trial, as we shall see. To the nineteenth-century historians who wrote about her at all, Zieglerin was terrifying, sexually voracious, and powerful—a perfect confirmation of the witch stereotype; beyond this, however, she was uninteresting.

Much has changed since the nineteenth century, of course. Anna Zieglerin has largely (although not entirely) faded from view and from historical memory in Braunschweig, even as the iron chair still sits in the Braunschweig State Museum. Meanwhile, more than a century of scholarship on the history of alchemy, gender, and Reformation Germany makes it possible to move beyond her execution and the distorted image of Zieglerin as witch and, instead, use her life to raise a new set of questions. Above all, Zieglerin refocuses attention on the religious, gendered, and political meanings of early modern European alchemy in the wake of a generation of revisionist scholarship. In the past two decades, the "new historiography of alchemy" has explored the contributions of alchemy to science and medicine, firmly and importantly establishing its central, if complicated, role in the development of new ways of understanding nature in the sixteenth and seventeenth centuries.[22] The history of alchemy is not coterminous with the history of science and medicine, however, for it also extends into the history of Christianity, gender, politics, commerce and political economy, material culture, art, literature, and even music in early modern Europe. Following Zieglerin's life and ideas makes many of these connections visible in new and surprising ways.

From the time of alchemy's arrival from the Islamic world, its promoters in Europe had attempted to forge connections between alchemy and Christianity.[23] Alchemists articulated an analogical relationship between the production of the philosophers' stone and the life of Christ, for example, likening the dissolution and reconstruction of matter in the furnace and flask to Christ's suffering on the cross; both stone and Christ emerged from their ordeals with the power to redeem.[24] Alchemy had also long been a resource for imagining and understanding the peculiar qualities of biblical bodies, whether of Adam and the long-lived patriarchs, or of the millennial or resurrected bodies at the end of time.[25] Anna Zieglerin pulled on all of these threads in her engagement with alchemical tradition, but she came to focus especially on how alchemy could help her explore, if not manipulate, the vexed relationship between matter and spirit in early modern Christianity. Zieglerin understood alchemy primarily as a powerful technology for generating unusually subtle matter, whether gemstones, plants, gold, or human bodies. This view led her to draw analogies not between the philosophers' stone and the crucifixion, as had so many alchemists before her, but rather between the alchemist and the Virgin Mary, whose own paradoxical body made possible the incarnation of Christ. Ultimately, Zieglerin proposed that alchemy could help her enact another pivotal event in sacred history, using the alchemical lion's blood and her own body to transcend ordinary generation and produce subtle alchemical bodies, thereby bringing about the end of time. The fact that Zieglerin, who seems to have encountered only a few alchemical texts, could nonetheless draw on alchemical ideas and practices to forge such a creative and sophisticated view of Christian eschatology is striking, underscoring the fact that alchemy was as entangled with the Reformation as it was with science or medicine during the turbulent sixteenth century in Europe.

Zieglerin's preoccupations also point to the importance of the body in the early modern period.[26] The fact that she situated her alchemical process in her own body, ultimately conflating the blood that ran through her veins and the generative space of her womb with the elixirs and vessels at the center of the alchemist's art, was in some ways idiosyncratic and deeply personal. In another sense, however, Zieglerin merely typified much broader understandings of the body as a site of both knowledge and piety in medieval and early modern Europe. Pamela H. Smith, for example, has drawn attention to the ways that the "body of the artisan" was central to the ability to know nature in the early modern period; it was an instrument for knowing, manipulating, and transforming matter.[27] At the same time, the body was also a site of piety. As Caroline Walker Bynum's work has demonstrated, it was a highly charged "locus of the sacred," particularly for medieval holy women, who often demonstrated their piety through bodies that "displayed unusual changes, closures, openings, and exudings," or who attributed "religious significance" to pain and illness.[28] Zieglerin's body, too, bore signs of her spiritual purity, and was crucial to her articulation of her own piety and authority. The fact that she was a Lutheran might make this surprising; in some ways she calls to mind medieval Catholic holy women more than sober Lutheran *Hausmütter,* and the fact that she chose to model herself on the Virgin Mary at the court of a recently converted Lutheran prince underscores the longevity of medieval modes of piety even for all of the dramatic ruptures of the Reformation.[29] Although we often imagine alchemists producing precious metals in a laboratory, Zieglerin highlights the fact that the human body was also a primary object of alchemical technologies. Alchemically produced elixirs, tinctures, quintessences, and balsams, whether ingested or applied externally, held out the promise of transforming the body to make it healthier, more productive, and long-lived. Some alchemists also extended their art far beyond medicine, aiming to produce especially subtle bodies that straddled the divide between the worldly and divine, continuing to exist in this world, while accessing qualities and gifts that were beyond it. Anna Zieglerin's alchemical practice sat at the intersection of all of these corporeal engagements with her world. She used her own body to demonstrate her piety, authority, and expertise as a holy alchemist, promising to manipulate the entanglements of body and spirit that vexed and inspired early modern Europeans.[30]

Zieglerin also brings the history of women and gender to the fore in this period. Historians continue to add to our understanding of the many ways in which women produced, consumed, and contributed to knowledge of nature in the early modern world. Like their male counterparts, women were patrons and practitioners, and engaged with science and medicine in the court, as well as in the household, library, laboratory, workshop, and Republic of Letters. We should not be surprised, therefore, to find Anna Zieglerin among the ranks of women who drew, read, and wrote about nature and tested, collected, sold, and exchanged their knowledge.[31] Studies of these women have expanded, deepened, and also complicated older narratives about what was at stake in understanding and manipulating nature in early modern Europe, as well as where and how knowledge of nature was produced and consumed. They have drawn attention to the household as a site of knowledge production, to the gendered dimensions of publicity and credit, and forced us to redraw the social boundaries of science and medicine in the early modern period.

The records of Zieglerin's successes and failures in Wolfenbüttel are unusually voluminous, offering a rare opportunity to examine how, and on what terms, a woman could claim plausibly to be an alchemist in the sixteenth century. Most practitioners of alchemy still had to rely on wealthy patrons to support their work in this period, and this plunged many alchemists into the dangerous world of the early modern court.[32] Zieglerin's career in Wolfenbüttel demonstrates the peril and the opportunity that courts could offer any alchemist, but it also highlights the particular challenges that she faced as a female courtier. In attempting to secure a place at the ducal court in Wolfenbüttel and to assert her own authority as an alchemist, Zieglerin creatively navigated extremely dangerous terrain, including the formidable opposition of the ranking woman at court, Duchess Hedwig. In the end, Zieglerin (like her male companions but for different reasons) was unsuccessful. Even her downfall is instructive, however, for it makes visible a set of associations between alchemy and poison, sorcery, and political sabotage that pulsed through the courts of the Holy Roman Empire, revealing the instability and fear lurking at the heart of the nascent centralizing state.[33]

When Anna Zieglerin imagined herself to be a new Virgin Mary, with the alchemical lion's blood in hand, she made a surprisingly bold claim not only to being a court alchemist, but also to playing a central role in bringing about the Last Days. Zieglerin wove together her biography, her body, her piety, and her alchemy in sometimes startling ways, suggesting that alchemy was, for her, deeply personal and constitutive of self.[34] More than just a good story, however, Zieglerin's life and ideas reveal how tightly alchemy was entwined with the political, religious, social, and intellectual culture of Reformation Europe. In her encounters with networks of Lutheran pastors searching for the philosophers' stone or with princes trying to pry secrets out of angels and alchemists, as well as in her own extraordinarily creative explorations of the relationship between matter and spirit, Anna Zieglerin found in alchemy a powerful way to engage, manipulate, and interrogate her world.

How Zieglerin came to construct, believe, and convince others of her ideas makes for a dramatic narrative, not least because some of them seem to defy belief today. She was not the only person in her circle of companions, patrons, and foes to put forward claims that some modern readers may find outrageous, implausible, or simply ridiculous, however. Such stories and contentions challenge us to take Zieglerin and her contemporaries on their own terms, to explore the cultural logic of their assertions. We must seek to understand why seemingly preposterous claims might have seemed reasonable to early modern ears, as well as the standards that *they* used to evaluate them and to determine whether they were plausible.[35] At the same time, Anna Zieglerin asks us to acknowledge and consider the productive role of fantasy, aspiration, and invention in the sixteenth century (and in subsequent centuries as well). In one way or another, most of the figures in this book were engaged in spinning tales, both about themselves and others. Zieglerin, too, may have reinvented herself and her past repeatedly, both in response to very real events and to less tangible rumors and matters of reputation, and it is sometimes difficult to know how to make sense of her claims. We might wonder whether her tales were improvised or strategic, knowing dissimulation or self-fashioning.[36] *Anna Zieglerin and the Lion's Blood* begins with a proposal, however—namely, that it is more productive to set aside the question of what really happened and to ask not only how Zieglerin and her contemporaries negotiated the middle ground between self-narration and outright lying, but also how fantasy both propelled and limited politics, faith, and knowledge in Reformation Germany.[37]

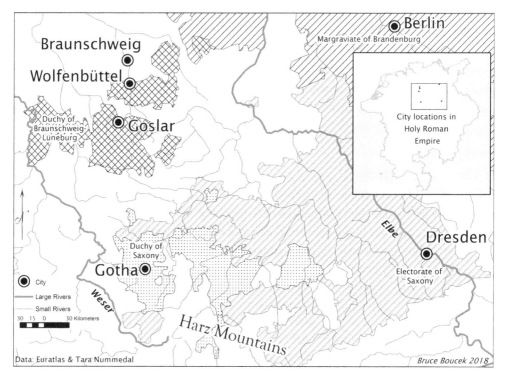

MAP 1. Map of Electoral Saxony, Ducal Saxony, and the Duchy of
Braunschweig-Lüneburg, late sixteenth century. The boundaries displayed are
approximate, as the exact boundaries of these territories (particularly in
Ducal Saxony) were in flux during the 1560s and 1570s. Bruce Boucek, Brown
University Library. Map drawn with the Euratlas Georeferenced Historical
Vector Data © Copyright 2008, Christos Nüssli, Euratlas.

CHAPTER I

The Shadow of Gotha

Anna Zieglerin's tale unfolded mostly in the small northern German city of Wolfenbüttel, at the court of Duke Julius (1528–1589) and Duchess Hedwig (1540–1602) of Braunschweig-Lüneburg. This is where Anna arrived in the fall of 1571, accompanied by her husband Heinrich Schombach and their companion Philipp Sömmering, and where, for a time, they flourished as alchemists and courtiers. Wolfenbüttel is also where Anna, Heinrich, and Philipp would face their downfall three years later, as controversy around them evolved into a legal process and, ultimately, condemnation at the scaffold. Although these events in Wolfenbüttel would prove decisive for Anna and her companions, their lives there were overshadowed by the events that had propelled them to Julius and Hedwig's court in the first place: the 1567 siege of Gotha and the downfall of their first patron, Duke Johann Friedrich II of Saxony (1529–1595). Johann Friedrich had been led into this ill-advised military confrontation by his adviser Wilhelm von Grumbach (1503–1567) and by a host of angelic spirits, who had been communicating with the ducal court for years via a peasant boy named Hänschen (or "Little Hans"). These angels turned out to be quite bellicose, exacerbating long-simmering tensions between duke and emperor, which eventually erupted into a three month siege and Duke Johann Friedrich II's overthrow. In the wake of the political and military defeat, rumors that Grumbach used sorcery to manipulate the duke circulated, adding to the disturbing implications of this episode.

Heinrich, Philipp, and especially Anna were not directly involved in the "Grumbach Feud" or the 1567 "Gotha Rebellion," as these events came to be known. Nevertheless, they were affiliated with Duke Johann Friedrich II's court in the 1560s and thus were forced to flee ducal Saxony in the ensuing political fallout. Cast adrift in the Holy Roman Empire and in search of the security of a new patron, they eventually found refuge in Wolfenbüttel, where they hoped to begin life anew in the 1570s. Much to their dismay and frustration, however, the heady blend of angelic prophecy and sorcery associated with Duke Johann Friedrich II's demise continued to hover over Anna, Heinrich, and Philipp, shaping their new lives in ways they could not always control. The story of Anna's years in Wolfenbüttel, therefore, must begin nearly a decade before in Gotha.

The Grumbach Feud and Hänschen's Little Men

When Anna first came to ducal Saxony in 1562, the Grumbach Feud was already well under way. The episode began years earlier with a shared sense of injustice and an assassination. The Franconian knight Wilhelm von Grumbach arrived at Duke Johann Friedrich II's court in 1557, seeking employment following the death of his previous patron, Margrave Albrecht Alcibiades of Brandenburg-Kumbach (1522–1557). Both the knight and the duke recently had been dispossessed of their lands. Duke Johann Friedrich II's misfortunes had to do with an old rivalry between two branches of the Saxon Wettin dynasty: the Albertines, centered in Dresden, and the Ernestines in Weimar. As the eldest son of the Ernestine Saxon Elector Johann Friedrich I, Johann Friedrich II was groomed as a child to inherit the prestigious title Elector of Saxony. This presumed future was entirely scuttled at the 1547 Battle of Mühlberg, however, when the seventeen-year-old Johann Friedrich II, fighting alongside his father, was defeated by his Albertine cousin Moritz and subsequently stripped of the electoral title and his lands. Almost in an instant, Johann Friedrich II saw his political fortunes plummet, his father taken off to prison, and, in his view, the ideals of the evangelical movement betrayed by political expediency, as the Protestant Moritz had joined Catholic imperial troops to seize his Protestant cousin's lands. Grumbach's woes, meanwhile, were rooted in a decade-long feud with the Catholic prince-bishops of Würzburg, during which Prince-Bishop of Würzburg Melchior Zobel von Giebelstadt confiscated Grumbach's fiefs in the area. In retaliation, Grumbach's men shot and killed the bishop in 1558. The murdered bishop's successor, Prince-Bishop Friedrich von Wirsberg, pressed the issue with imperial authorities, launching a legal process that continued for nearly a decade. In the meantime, Grumbach and Duke Johann Friedrich II grew closer. As one modern scholar has put it, "The two men, prince and knight, soon became one in a desperate desire to recover their heritages and take vengeance on those who had deprived them of what was rightfully theirs."[1]

When Anna and Heinrich first arrived at Duke Johann Friedrich II's court in the early 1560s, therefore, they stepped into a tense political situation. Grumbach and the duke were feeling increasingly embattled as they nursed their grievances against Elector August of Saxony, on the one hand, and the prince-bishop of Würzburg, on the other. Meanwhile, they anxiously waited to hear whether Holy Roman Emperor Ferdinand I would place Grumbach under an imperial ban for his role in Zobel von Giebelstadt's murder (thereby nullifying all of Grumbach's legal and property rights). The duke and the knight soon found a new focus for their desires and anxieties, however, in the form of another young person who arrived at Duke Johann Friedrich II's court around the same time as Anna Zieglerin.[2]

In the fall of 1562, Grumbach urged the duke to come to the Grimmenstein fortress in Gotha to meet a thirteen-year-old illiterate peasant boy named Hänschen from Sundhausen, a nearby village.[3] For three or so years, Grumbach explained, Hänschen had been receiving regular visits from benevolent spirits, who protected Hänschen's family from demonic harassment, issued prophecies, and divulged numerous other useful secrets. These "Little Men," as Hänschen called them, were about the size of two- or three-year-old children, wore white garments with black caps, and each carried a burning torch and a little white rod, with which they opened up the earth to return to their underground dwelling.[4] Grumbach had learned of Hänschen and his Little Men when Grumbach's secretary, Mortiz Hausner, met the boy's father at a baptism in the Thuringian countryside, and Grumbach immediately arranged the urgent meeting in Gotha with Duke Johann Friedrich II, the boy, and his father.

Eventually, Johann Friedrich II would come to believe that the Little Men were in fact angels sent by God to communicate with him through Hänschen. The duke did not arrive at this conclusion easily, however, and he was initially skeptical of the angelic nature of Hänschen's apparitions and their eschatological meaning. He knew that he had to be cautious, recognizing that Hänschen's visions could easily be the product of sorcery, diabolic deception, melancholy, or fraud.[5] Well informed about the dangers of demonic activity and sorcery, Johann Friedrich II understood that his role as a magistrate was to root them out in his lands; indeed, during the years before Hänschen arrived at his court, he had launched fifteen trials against

sorcerers.[6] Even Grumbach was skeptical when Hausner first reported Hänschen's visions, reportedly expressing his belief that Hänschen's visions "were
a product of sorcery or demonic delusions."[7] Duke Johann Friedrich II sent
officials to Hänschen's village to interview the boy's family, friends, and pastor in order to determine whether his Little Men might be the result of
sorcery. Perhaps someone had bewitched Hänschen in order to make him
believe he saw the Little Men, or Hänschen had conjured them up himself.[8]

Concluding from this first inquiry that no sorcery was involved and
that Hänschen's Little Men were real, the duke then conducted a rigorous
investigation to determine whether the spirits were demons or angels. This
was no easy task. Johann Friedrich II himself interrogated the boy for a few
days, during which time the duke ordered Hänschen to ask the Little Men a
battery of theological questions, including "whether they believed that the
eternal word of God had become flesh, and whether Jesus Christ was both
true God and true Man." The duke was impressed with the answers he
received, but even so both he and Grumbach remained skeptical. Johann
Friedrich II then asked Hänschen's father whether the apparitions had ever
lied (for angels, theologians agreed, were incapable of falsehood), and the
duke also ordered Moritz Hausner to continue to interrogate the boy and the
Little Men. They never contradicted scripture or blasphemed, and Hänschen
never demonstrated fear when they appeared, as demonologists agreed he
would have had they been demons. After a week of rigorous questioning
deeply informed by numerous authorities on demons and angels, the duke
finally reached his conclusion: Hänschen's spirits were indeed angels.[9]

From that point on, the duke never wavered in his confidence in the
Little Men. Not all of the duke's theological advisers agreed, however. A later
observer claimed that all of the good pastors and theologians in Thuringia
spoke out against this "magical business" and "these things known only to
the devil."[10] Indeed, in early 1566, two prominent theologians in Jena, the
superintendent Johann Stößel and the theology professor Nikolaus Selnecker,
who later joined Duke Julius's court in Wolfenbüttel, denounced Hänschen,
Grumbach, and the "angels of darkness."[11] When Stößel asked the duke why
he was so certain that the Little Men were not demons, Johann Friedrich
replied, "Because Satan never states the truth. But this spirit [i.e., the Little
Men] has spoken the truth more than a hundred times, therefore it is not an
evil spirit."[12] This argument—that accurate revelations were themselves proof
of the angelic nature of Hänschen's spirits—failed to convince Stößel and
Selnecker. Elector August of Saxony, too, agreed that his cousin Johann

Friedrich was horribly mistaken, and that the only explanation for the duke's stubborn conviction and the rebellious actions it inspired was that he *himself* had been bewitched.

Despite these objections, Duke Johann Friedrich II was convinced that the Little Men were neither demonic nor the product of sorcery, but divine messengers. Moreover, he believed that they were sent to signal that the Last Days had arrived.[13] This conclusion testifies to the climate of eschatological expectation that pervaded Lutheran circles in this period. In the Revelation of John, angels played a pivotal role, not only in dispensing God's wrath, but also heralding the various stages of the end times to come and dispatching the chosen to prophesy on earth. Moreover, to Duke Johann Friedrich II, it seemed reasonable that angels would seek him out as one of God's chosen prophets to reform the world, for it accorded well with his belief that his family was on earth to serve as God's instruments.[14] Johann Friedrich II's forefathers had been among Luther's earliest supporters and had played an important role in the promulgation and preservation of the reformer's message. The inheritor of this legacy, the duke held the deep conviction that he had been, as one historian has put it, "called upon by God to preserve the True Word and Luther's legacy, hold fast to the Augsburg Confession, and defend the righteous from the grips of the Antichrist 'until the grave or the end of the earth.' "[15] To the pious duke, the angels' attempts to communicate with him through the peasant boy Hänschen merely confirmed what he already knew: God had special plans for Johann Friedrich II and his realm.

For the next four and a half years (during which time Anna's husband Heinrich also served in Johann Friedrich II's court as a valet and perhaps also court fool),[16] Hänschen lived in Grumbach's household, disguised as a servant boy and under the immediate supervision of Grumbach's secretary, Moritz Hausner.[17] During these years, the duke and his closest advisers consulted almost daily via Hänschen with the Little Men, who carried Johann Friedrich II and Grumbach's inquiries to God and returned with answers about matters of political, economic, medical, and alchemical import. Some of what they revealed was "everyday" knowledge that had practical benefits for the duke and his court. They consulted on the duke's constipation, for instance, the duchess's toothache, the maladies that afflicted the duke and duchess's son, Friedrich IV, and Grumbach's gout, as well as the illnesses that plagued various other members of the court.[18] The Little Men also served as guardian angels, protecting the duke and his court from plague and other diseases, sorcery, and the perils of travel.[19] The angels' talents extended beyond the health of the court as well, as they helped Duke Johann Friedrich II address his significant financial woes. By the 1560s, he was in desperate need of money, as his father's debts, decades of warfare, the ensuing loss of territory, and the expenses of the ducal court had depleted his coffers. He addressed this problem in a variety of ways, including reducing the number of people he fed at court each day and seeking out loans from the French king and the duke of Lorraine, among others. Johann Friedrich also looked for new sources of wealth in his lands, seeking buried treasure, new mines, and a magical "spring-root" (*Springwurtzel*), which supposedly could help them locate both by "springing open" locked doors, exposing buried treasure, and revealing hidden veins of precious metals.[20] The angels contributed to the duke's financial agenda, helping him locate buried treasures and a new mine hidden in the Thuringian Forest.[21]

The cluster of related things—finance, treasure, mines, and magic—that promised to secure new inflows of cash to the court also led Duke Johann Friedrich II to alchemy. He owned two books on the subject, both of which were recently printed compendia of alchemical texts: *De alchimia opuscula complura veterum philosophorum* compiled by Cyriacus Jacob in 1550 and Guglielmo Gratarolo's 1561 anthology, *Verae alchemiae artisque metallica*. These texts offered a nice introduction to the study and practice of alchemy, but certainly not a deep scholarly engagement with the vast (and growing) corpus of alchemical texts available to readers by the 1560s, suggesting that the duke's interest in alchemy was not particularly scholarly or extensive.[22] Rather, it seems to have emerged largely in the context of his search for new sources of wealth. To pursue these entrepreneurial alchemical goals, the duke established a laboratory about ten miles outside of Gotha in Reinhardsbrunn, the site of the former Reinhardsbrunn Abbey, which the Saxon dukes had been using as a hunting lodge since the mid-sixteenth century. There the duke employed a handful of alchemists, including a man named Matthes (or Mattheus Friedrich), another named Karl (or Carol), and a third named Hans Rudolf Blumenecker (or Plumenecker).[23] All of this, reportedly, amounted to the duke's total expenditure on alchemy of 10,000 gulden, a modest sum in the context of princely alchemical projects.[24] Still, Johann Friedrich II's support for these alchemists was not uncontroversial. While the duke himself was interested in alchemy, and several of his own councillors were practitioners of the art, a number of skeptics also resided at court. Anna Zieglerin would later report, for example, that even Wilhelm von Grumbach was an "enemy of the alchemists," as was the ducal secretary, Hans Rudolf.[25]

In this context, the angels made themselves useful by weighing in on proposals for new alchemical projects presented to the duke. Some came from his own councillors, for example, the ducal envoy to France and legal scholar Dr. Lukas Thangel and the Augsburg patrician and financier David Baumgartner.[26] Other proposals came from alchemists outside the court who sought the duke's support. In 1566, for example, two pastors, Abel Scherdinger and the man who would soon become Anna Zieglerin and Heinrich Schombach's collaborator, Philipp Sömmering, approached the duke with a proposal to produce the philosophers' stone in exchange for 750 thaler.[27] Johann Friedrich decided to support the project, but the angels made their objections known, reporting that while God would indeed reveal the true art of alchemy, it would not be via Scherdinger and Sömmering. Furthermore, the angels said, even if the pastor-alchemists accomplished something, it would not produce a profit, costing nearly as much to carry out the recipe as it would yield in the end. The angels also rejected the two alchemical proposals from Johann Friedrich II's own councillor, Lukas Thangel, warning that his alchemy was not to be trusted either.[28]

The angels did not stop at consulting on practical issues like toothaches, treasures, and alchemy. They also dictated ducal political policy, including how Johann Friedrich II should proceed in the escalating conflict with the emperor over the Grumbach Feud. On such matters, the angels were extremely aggressive, ordering the duke and the knight to undertake a series of increasingly violent acts against their enemies. From the moment they arrived, in fact, the angels advocated war. In December 1562, shortly after Hänschen's arrival, they commanded Grumbach to send a servant boy to shoot "the great lord who is not of the right belief and who is leading his people away from God's word."[29] The target of this intended assassination, the angels then revealed, was none other than Holy Roman Emperor Ferdinand I. Astoundingly, Grumbach did not hesitate, revealing his steely resolve in dealing with the emperor. Reminding the duke "how miraculously God directs all of our affairs from day to day [and] that He alone guides our deeds and wants to use no one else other than Your Princely Grace," Grumbach informed Johann Friedrich II that, "therefore, on orders from God, early today I armed the boy with his gun, and he awaits further orders from the angels about what he should do next."[30] The angels ordered the boy to walk out into a field and fire a shot into the sky; they would then guide the bullet through the air nearly eighty miles to Speyer, where the emperor was hunting, leading it into his body and killing him. On December 26, 1562, the boy actually fired off his shot, and Grumbach reported that the emperor was dead. In fact, Emperor Ferdinand I would continue to live for another year and a half, but Grumbach and the duke's confidence in the angels' pronouncements was strong enough that they continued to believe that their assassination attempt had been successful, waffling in this belief only a couple of years later, when evidence to the contrary made it difficult to sustain.[31]

With this attempt at the angelic assassination of an emperor, Duke Johann Friedrich II and his councillor Grumbach stepped onto a truly radical path. Despite decades of warfare between Protestants and Catholics in the sixteenth century and sustained debates about imperial authority in religious affairs, assassinating the emperor remained unthinkable to the empire's princes. This was the stuff of the radical peasant revolts of the 1520s, not the rhetoric of princes forty years later.[32] The angels, however, convinced Duke Johann Friedrich II that imperial authority itself was illegitimate, and this allowed the duke to think the unthinkable. This delegitimization applied equally to Emperor Ferdinand I and to his heir, Maximilian II, who would ascend the imperial throne in 1564. Asserting (incorrectly) that Ferdinand was dead, the angels told Grumbach that the emperor's successor would be even worse: "This newly elected king [Maximilian II] is an even greater blasphemer and more worthless man than his father is," they said.[33] With the conviction that they were fulfilling divine commands, Johann Friedrich and Grumbach went on to prepare for war against a group of powerful enemies that included not only the emperor, but also a long list of others: the prominent Protestant princes Elector August of Saxony and Duke Christoph of Württemberg, the free imperial city of Nuremberg, the Catholic Duke Albrecht of Bavaria, and the Catholic prince-bishops of Würzburg and Bamberg, among others.[34] Increasingly embattled and with no meaningful allies, Johann Friedrich and Grumbach nevertheless faced their enemies with full faith that God would ensure their victory.

With the emperor supposedly dead, the angels next targeted Grumbach's old nemesis, the prince-bishop of Würzburg. "God shall give the Junker [Grumbach] His grace that he should torture everyone from the least to the greatest [in Würzburg]," the angels prophesied; the priests in Würzburg "shall be exterminated, struck dead, and none of them shall remain in the land any longer."[35] In the fall of 1563, therefore, Grumbach led the hundreds of horsemen and soldiers he had assembled in an attack on the prince-bishopric and, accompanied by Hänschen, brought Würzburg to surrender within a single day. The soldiers plundered the city and seized Grumbach's former lands. The success of this campaign, of course, merely proved to Grumbach that the angels spoke the truth. As Grumbach put it to Johann Friedrich II a few days after the attack on Würzburg, "the whole business went off just as the lad, Hänschen, stated it would beforehand."[36] Shortly thereafter, Emperor Ferdinand I—very much still alive, despite beliefs to the contrary in Gotha—responded by finally declaring Grumbach an imperial outlaw and breaker of the peace, warning Duke Johann Friedrich II that he, too, would be placed under an imperial ban if he did not turn over Grumbach. Grumbach, however, was not particularly troubled. "I won't let it worry me," he wrote; "I have an honorable . . . case and a just God who lives in heaven, and who has thus far miraculously protected me from my enemies, and who has given me recently my victory in Würzburg. The same God still lives, and is able to give me even more."[37]

Indeed, the angels prepared for war again, this time against Johann Friedrich II's cousin, Elector August of Saxony. They vacillated about whether Johann Friedrich II should go on the attack or merely prepare to defend himself, but the angels consistently promised that the duke would regain his lands and electoral title. They warned him that the battle would be horrific, but they also assured him that ultimately he would be victorious. In January of 1564, the angels urged the duke to prepare for another attack from "a great people whom others would shudder in fear upon seeing. And therefore there will be such an immense outpouring of blood that one will be wading in the blood of the papists and other godless people until it goes over the tops of the shoes."[38] Rumors circulated at Johann Friedrich's court that Elector August of Saxony was preparing to attack the duchy, while the angels continued their bellicose prophesies and Duke Johann Friedrich II began to assemble what forces he could, and prepared the Grimmenstein fortress adjacent to the Gotha town walls.[39] Under ordinary circumstances, Duke Johann Friedrich II could hardly hope to resist his enemies with these meager troops and provisions, but the angels assured him that God would provide. "A great war will begin," they said in January of 1564, "and everyone will say that it will not be possible to defend against such military forces. But these large forces will be defeated by very few people. And it will be a great sign from God about which everyone in the whole world will write and talk, as nothing like it has ever been heard of before. God will give this victory with thirty-two horses. He will order the people who will ride them."[40] Persuaded by these promises of divine aid, Johann Friedrich II forged ahead, readying for war.[41]

As Johann Friedrich II became increasingly isolated and embattled in the face of immense imperial pressure to give up Grumbach, the apocalyptic tenor of the angelic missives escalated. The angels began to draw more explicitly on apocalyptic imagery from scripture, making clear that this was to be no ordinary battle, but something far more epic. They instructed Johann Friedrich II to make a set of six flags, for example, two each with imperial, electoral Saxon, and knightly insignia (that is, the titles that Johann Friedrich II and Grumbach hoped to gain or regain), explaining that "when [Johann Friedrich II] captures a castle or a city and plants one of these flags inside it, because of God no person will be able to recapture it."[42] The angels promised Johann Friedrich II that he would become king of France, as well as of Spain or Bohemia. They showed Hänschen another vision of horses, one of which bore an angel with a letter "written by the hand of God and the blood of Jesus Christ." This heavenly lettering commanded that "all priests will be silenced. Whoever does not allow them to be silenced will have to answer for it on Judgment Day. Whoever are Christians will see that it is not a human script."[43] The angels declared that Elector August would be killed in battle, and that Johann Friedrich II would become Holy Roman Emperor "because religious faith in the world has never been in such a bad way as right now."[44]

Meanwhile, when Maximilian II finally donned the imperial crown in the summer of 1564, the angels issued yet another stunning prophesy: Maximilian "is not an emperor and will not become an emperor," they declared. "God will not allow it. Another shall become [emperor] who will be God-fearing. God shall elect him with His holy cross and he [God's emperor] shall allow His clear and pure word to be preached."[45] Shortly thereafter, the angels promised that, "in fifty-two weeks, God will arrange for the duke [Johann Friedrich II] to assume the [imperial] office. But he must win it with the sword. And God shall give him His success and blessings."[46] They then ordered Johann Friedrich II to send Hänschen, along with Grumbach's secretary, Moritz Hausner, to Torgau, where they were to oversee the forging of four swords to be used in the upcoming battles.[47] Underscoring his warlike mood, in November of 1564 Johann Friedrich II also transferred his ducal seat from its traditional location in Weimar to the Grimmenstein fortress at Gotha.[48]

A year and a half later, at the Imperial Diet in Augsburg in March 1566, the imperial estates confirmed the ban against Grumbach once again. Emperor Maximilian II followed up a few months later by stripping Johann Friedrich II of his ducal title, awarding it to his younger brother, Johann

Wilhelm, instead, and then by making it clear that Johann Friedrich II had three months either to comply with the ban or to be placed under the ban himself. Johann Friedrich II, meanwhile, received pleas from his younger brother, local clergy, faculty at the local University of Jena, imperial envoys, and fellow princes begging him to comply with imperial orders. He continued to ignore these entreaties.[49] As embattled as he was, Johann Friedrich II remained certain that God was on his side; the angels told him so.

In May of 1566, the angels upped the ante even further, predicting that Margrave Johann of Brandenburg-Küstrin, Duke Eric II of Braunschweig-Lüneburg, and King Erik XIV of Sweden "will join with the duke [Johann Friedrich II] and his people and help crush all the estates. . . . Thereupon, because of God, in this and next year an overturning of the world will happen, [and] the four lords . . . shall turn the world upside down." After these four lords emerged victorious, the angels continued, "then the Christians will have peace and rest . . . and [they] will multiply," enjoying an earthly millennium before the final battles of the apocalypse. "Also, in this and the next year," the angels continued, "the Jewish prophecy that tells of four gods will be fulfilled.[50] These [lords] shall be the four who turn the world upside down, and [who] also rush all of the Jews and the priests, [who], for God's sake, shall [themselves] become Christians. [All of this shall occur] as soon as the duke receives his lands [i.e., the lands granted to August of Saxony], on account of God."[51] Finally, the angels continued, with divine aid Johann Friedrich II would even defeat the Ottoman Empire, preventing it from "crush[ing] and exterminat[ing] the Christian religion."[52] According to the angels' steady stream of increasingly apocalyptic prophecies, in other words, Johann Friedrich II would vanquish Elector August of Saxony and the Catholic emperor in battle and take the electoral and imperial thrones himself. Then he would persecute the priests, restore the gospel, destroy the Ottoman Empire, and convert the Jews before laying down his crown and ushering in the final events of the apocalypse.[53]

With this stunning prophecy, Hänschen's angels identified Duke Johann Friedrich II with the Last World Emperor, who, it was widely believed, would bring about peace and prepare the way for the return of the Messiah. According to an enormously influential apocalypse first recorded in Syriac and attributed in the seventh century to the fourth-century bishop Methodius of Patara, the Last World Emperor would arise to restrain Antichrist, defeat Christ's enemies (including Gog and Magog), reform the church, and usher in a reign of peace and prosperity before finally laying down his crown in

Jerusalem, thus releasing Antichrist and launching the final battles of the end times.[54] In medieval central Europe, the Last World Emperor came to be identified with a messianic German emperor, a "third Friedrich," successor to the twelfth- and thirteenth-century Hohenstaufen emperors, Friedrich I, or Barbarossa, and Friedrich II. (One legend claimed that one of these emperors had in fact never died, and was still sleeping in the Thuringian Kyffhäuser Mountains, awaiting the proper moment to restore the glory of the German empire as the Last World Emperor.) By the sixteenth century, the legend that the Last World Emperor would be a "third Friedrich" was particularly strong in Saxony and Thuringia, where he was identified variously with Fredrick the Wise in Saxony and Johann Friedrich II's own father, Elector Johann Friedrich I of Saxony.[55] Bolstered by this tradition, Duke Johann Friedrich II was prepared to accept the full eschatological implications of the angels' prophecy—namely, that he was the Last World Emperor, and his upcoming confrontation with Elector August and Emperor Maximilian II was only the beginning of his campaign to reform the world before its end. In the event, the angels proved to be horribly mistaken, of course, although Duke Johann Friedrich II's faith in their prophecies never wavered.

The Siege of Gotha and Its Aftermath

Having failed in his attempts to persuade Duke Johann Friedrich II to hand over Grumbach, Emperor Maximilian II finally turned to force. On December 12, 1566, Emperor Maximilian II placed the duke himself under imperial ban, divesting him of ducal title and lands, for his refusal to recognize the long-standing imperial order to surrender Grumbach. In order to enforce the bans against Grumbach and now Johann Friedrich II, Maximillian activated the old intra-Saxon rivalry once again, authorizing the duke's cousin and rival, Elector August of Saxony, to carry out the imperial order by force.[56] With August's aid, imperial troops besieged Gotha and the Grimmenstein fortress on New Year's Day, 1567. Johann Friedrich II, his wife, Duchess Elisabeth, and their children, along with Grumbach and Hänschen, hunkered down in the fortress, while a relatively miniscule Gothaner militia skirmished with about 18,500 imperial cavalry, infantry, and trench diggers (Figure 2). With their water supply cut off, food supplies rapidly depleting, and unusually cold weather, the Gothaners were vulnerable to the propaganda campaign against Grumbach, which Elector August delivered in the

FIGURE 2. The siege of Gotha, 1566–1567, depicting the Grimmenstein fortress
on the right, containing the chancellery, as well as the duke and duchess's
chambers. *GOTHA/ belagert den 30. Dec. Anno 1566 und eingenommen den 13.
April Anno 1567* (undated). © Deutsches Historisches
Museum Berlin, Inventar-Nr. Gr 64/2147.

form of pamphlets wrapped around arrows and shot into town. If Johann
Friedrich II and Grumbach viewed the conflict as an apocalyptic holy war
for the preservation of the gospel, August's missives proposed a much more
disturbing interpretation: Grumbach was a deceitful, murderous, godless sorcerer who had bewitched the duke and willingly led Johann Friedrich II's
subjects into needless war just for his own preservation.[57]

August's soldiers, meanwhile, made it clear that they would cease their
assault as soon as the Gothaners gave up Grumbach, reportedly by shouting

from the trenches up and over the city walls.[58] According to one witness to the siege, it was also widely known among the Gothaners that the emperor had declared that they were no longer subject to Johann Friedrich II's authority.[59] By early April, as the winter siege entered its fourth month, conditions inside the city walls had grown miserable. As supplies dwindled, thousands of people died of hunger and disease.[60] Finally, on April 4, the Gotha militia mutinied, taking the duke, Grumbach, and several other advisers prisoner. The mutineers then negotiated a surrender with the imperial and electoral forces, handing the duke and his entourage over to the custody of Emperor Maximillan II.

Elector August quickly appointed a commission to investigate the events that had led to the siege.[61] After a brief trial, Grumbach, Hänschen, and several other ducal advisers were sentenced to death. On April 18, 1567, Grumbach and three others were executed; Grumbach was carried to the gallows on a stool because he suffered from gout and made his confession before the crowd before being hanged. The executioner then cut out his heart and hurled it at Grumbach's mouth, proclaiming, "Behold, Grumbach, your false heart!" Grumbach was then dismembered, his body parts dispersed and hung on the edges of town. Hänschen, by then seventeen years old, followed the rebels to the gallows the next week (Figure 3).[62] Duke Johann Friedrich II, meanwhile, was spared the sword, but he spent the rest of his life in prison. Even the Grimmenstein fortress was razed, obliterating any trace of the rebellion. As Elector August himself put it on April 23, "If the rebellious murderers' nest is not razed to the ground and is thus endowed with the qualities of eternal memorial, then the victorious, glorious execution [of Grumbach and the others] would not praise, but rather displease God, give the rebellious mobs cause for further mutiny, and embolden them."[63] This was an unequivocal defeat for Duke Johann Friedrich II's vision of the world and his place in it.

The Grumbach Feud and dramatic events in Gotha that culminated in six executions and Duke Johann Friedrich II's lifelong imprisonment reverberated throughout the Holy Roman Empire for years to come. The local repercussions were clear: the Gothaners had a new duke, Johann Friedrich II's younger brother, Johann Wilhelm.[64] With this new regime came a reorientation of both church and state, as allies of the former duke were called to account for their roles in the disaster and replaced with supporters of the new duke, Johann Wilhelm.

More generally, the victors began to craft a narrative about what had happened, apportioning blame in the hopes of reestablishing stability in the

FIGURE 3. The execution of Wilhelm von Grumbach in the center of Gotha,
1567, as imagined in 1717, a century and a half later. The image depicts
Grumbach being carried to the site of execution, quartered alive, and shown his
own heart as the executioner exclaims, "Behold, Grumbach, your false ♥ !"
Grumbach's and his coconspirator Chancellor Brück's severed limbs are
depicted hanging at each of the four gates to the city. Friderich Rudolphi,
*Gotha Diplomatica, Oder Ausführliche Historische Beschreibung des Fürstentums
Sachsen-Gotha* (Frankfurt am Main; Leipzig, 1717), vol. 2, image between
pages 144 and 145. Bayerische Staatsbibliothek München, 2 Germ.sp.
133–1/2, Abschnitt 144.

Saxon lands. The lessons of Gotha offered themselves up for any number of
different interpretations, all of which vied for dominance during and after
the siege. The crimes for which Grumbach and the others were legally convicted were straightforward, all having to do with Grumbach's original assassination of the prince-bishop of Würzburg and Duke Johann Friedrich II's
refusal to comply with imperial decrees; on the surface, then, the Grumbach
Feud and the Gotha Rebellion that it sparked were a dispute about imperial
authority. And yet, the entire episode raised far more complicated and

troubling issues, which participants and observers struggled to resolve in its wake. Johann Friedrich II's assertion that he was the sole protector of the "true religion," for example, had the potential to be explosive in an era in which the outlines of a stable, institutionalized "Lutheranism" were still almost as hotly contested as disputes between Catholics and Protestants. After all, Johann Friedrich II's campaign sought to destroy not only Catholics, but also other Lutherans, who were, in his view, dangerously willing to compromise the gospel for the sake of political expediency. The fact that the Protestant August of Saxony had allied himself with the Catholic Emperor Maximillian II to defeat the Protestant Johann Friedrich II, of course, seemed only to prove the duke's point. Worse, the siege of Gotha in 1567 echoed the events of 1547, when August's Protestant brother, Moritz, joined the Catholic emperor to defeat Johann Friedrich II's father at the Battle of Mühlberg. Elector August, in other words, potentially had a public relations problem, even as he emerged victorious from the siege of Gotha. If Johann Friedrich's narrative prevailed, and observers came to understand the Gotha Rebellion as a battle to defend the pure gospel in the face of both the emperor's Catholicism and August's corrupted Lutheranism, then August risked his credibility as a leader among Lutheran princes. He might have won the battle, in other words, but he was well aware that he still faced losing the war.

In the aftermath of the Gotha Rebellion, therefore, Elector August of Saxony not only purged ducal Saxony of Johann Friedrich II's political supporters and razed the Grimmenstein fortress; he also turned to print to ensure that observers drew the proper lessons from the catastrophe. To facilitate this, August's envoy to France, the Protestant exile Hubert Languet (1518–1581), wrote a history of the Gotha siege, published anonymously in Latin in 1568 and translated into German the following year.[65] The "Letter to the Reader" explained that the text hoped to correct the many "lies and false allegations" that were circulating, particularly the claim that "His Electoral Grace [August] allied himself with Papist princes and bishops and was willing to suppress or even to split up the true Christian religion," and even that August had "those devious and heretical knaves, the Jesuits," among his councillors. Such rumors, Languet averred, were pernicious lies.[66] While Languet could not deny the fact that the Protestant August had, indeed, sided with the Catholic emperor against his Protestant cousin, he argued that the elector had taken such extreme measures reluctantly and only as a last resort when Grumbach, the true "instigator of war," repeatedly provoked August and the emperor. Having defended August's motivations, Languet then turned to discredit the elector's enemies: Grumbach and his allies. Above all, Languet disputed the Grumbach party's claim to be the "protectors and defenders of the true religion," portraying them instead as proponents of disorder and war, assassins, and agents of "all kinds of immoral and evil deeds" from the safety of their "nest" at Johann Friedrich's court.[67] In short, Languet's pamphlet pinned the disastrous events of 1567 on Grumbach and those who supported his violent, manipulative campaign to upend political order and pit Luther's followers against one another. The executions of these dangerous men, he maintained, were well deserved.

It was easy enough to demonize Grumbach, already executed, and blame him for all that had gone wrong at Gotha. Duke Johann Friedrich II's support for the rebellion and his belief in Hänschen, however, demanded a more careful explanation, and his status as a prince of the empire made it more complicated to condemn him outright. To be sure, the duke's political aspirations had drawn him to Grumbach. As Languet put it, Johann Friedrich II was "deceived by his own good will and understanding, or, even more, deceived and led astray by his futile hopes" that he could regain the electorship of Saxony.[68] And yet, Hänschen's presence at the ducal court suggested that something more than political strategy had motivated Johann Friedrich II's actions. Seeking explanations for his cousin's behavior, after the siege Elector August sent his agents to go through Johann Friedrich II's papers, where they found the voluminous records of Hänschen's "angelic pronouncements" while at the ducal court.[69] These and other documents uncovered in Gotha confirmed August's suspicion that Johann Friedrich had been the victim of an ingenious plot involving not just political persuasion, but also sorcery. According to this scenario, which both August and Johann Friedrich II's wife, Elisabeth, embraced, both Hänschen and the duke had been deceived. "Either Grumbach or someone in his circle" had discovered the youthful Hänschen and instructed him in "magic and heretical arts" so that he came to believe he could converse with angels and prophesy the future.[70] Meanwhile, Grumbach "bewitched" the duke with a potion of red wine and berries, which made Johann Friedrich II believe the boy's prophecies and allowed Grumbach to manipulate the duke.[71] August and Languet, in other words, promoted the view that a dangerous mixture of sorcery, prophecy, and political ambition had turned Johann Friedrich II into Grumbach's dupe, motivating the duke, against his better judgment, to protect the knight while he and his allies plotted ever more outrageous rebellions.[72] By reframing the angelic conversations conducted at Johann Friedrich II's court, and the

duke's actions, as the products of magical deception, rather than divine revelation, Languet located an explanation for Johann Friedrich's behavior that,
in a sense, let him off the hook. This was, essentially, a "sorcery defense," a
claim that the duke literally was not in his right mind. In the long run,
August's propaganda campaign against what he contemptuously called the
"spirit of Gotha" was successful, and his version of events prevailed, thanks,
in part, to Languet's pamphlet. The result was a powerful and enduring
narrative that conflated everything at Johann Friedrich II's court, even
alchemy, with Grumbach's politically motivated sorcery.

Collateral Damage

Alchemy had very little to do with the events that eventually led to the siege
of Gotha. Certainly, it was marginal to the prophecies that the angels revealed
to Johann Friedrich and his advisers, and both Grumbach and the angels
generally opposed the duke's investments in alchemy, favoring mining projects and the search for buried treasure instead. Nevertheless, many observers
of and participants in the events at Gotha during the winter of 1566–1567
drew connections between Johann Friedrich II's alchemical projects and his
political maneuverings, angelic prophecies, and fearless preparation for apocalyptic holy war, in the process associating alchemy generally with a set of
extremely dangerous activities.

Anna Zieglerin's collaborator Philipp Sömmering insisted that the duke's
alchemical projects had nothing to do with Grumbach. Writing to the new
Saxon duke, Johann Wilhelm, on the day of the mutiny that finally brought
an end to the Gotha siege, Philipp and his fellow pastor-alchemist Abel
Scherdinger described at length the "grisly, un-Christian, pagan, and Turkish
manner" in which Elector August of Saxony's troops had descended upon
Thuringia several months earlier. After detailing "all kinds of evils and harassment [that were] inflicted upon both the common man and a number of
pastors at that time," the two pastor-alchemists underscored their own heroic
efforts to perform Easter services for their congregations in the face of bands
of horsemen who literally were chasing them all over ducal Saxony, breaking
down church doors, storming Scherdinger's rectorate, and pursuing them
from village to village.[73] They attributed the fury of the electoral troops to
the fact that "the Elector, or several of His Electoral Grace's highest [minis-
ters], held us to be suspect because [we were] humble clients of our Princely

Grace and Lord, Duke Johann Friedrich, to whom we came recently and were remunerated [i.e., for alchemical work], and this identified us with the Grumbach Feud, *which was not our office and about which we knew too little to take part.*"[74] To the contrary, the pastors reminded the new duke, Johann Wilhelm, that they had helped the now-deposed Johann Friedrich assess which alchemists were legitimate and which were incompetent or, worse, frauds. More importantly (at least to one reader, likely the duke, who underlined this section of their letter), the pastors had signed a contract to pursue their own alchemical projects at ducal expense, although they lamented the fact that their work "had been cut off and hindered by the sudden siege."[75] "This," they concluded, "we take to be the single reason that we were pursued [by August's troops] in such a way. Otherwise, we have had nothing to do with princely or related business."[76]

Scherdinger and Sömmering's attempt to draw a firm line between their alchemical work for Duke Johann Friedrich II and the treasonous Grumbach Feud was never entirely successful. To make matters worse, their involvement in the ducal alchemical projects also became a reason to question their fitness as pastors. Immediately after the siege, the two pastors faced sanction for their flight from August's electoral troops during the battle. After all, they had left their congregations without a pastor for the most important time in the liturgical calendar, Easter, which fell at the end of March that year. The pastors apparently fled their parishes, "perhaps out of fear," according to one observer, just as the siege got under way in January. By February, Scherdinger's and Sömmering's congregations requested replacements because, as one witness put it, they "were associated with Grumbach's alchemists" and had been missing for a month.[77] The pastors later defended their actions, claiming that they had tried valiantly to reach their congregations for the important Easter services, even though the war was still under way, and that they arranged for pastors in neighboring parishes to administer communion and lead services in their absence. The charge that they had abandoned their congregations persisted, however.

For Gotha City-Superintendent Melchior Weidemann,[78] the pastors' absence during Easter was unforgivable, but this was only the latest in a string of condemnable acts, including their practice of alchemy. After the siege, Weidemann wrote to the new duke, Johann Wilhelm, to issue a passionate condemnation of the two "runaway" pastors. Even before the siege, Weidemann wrote, the "runaway pastor" Scherdinger had "led a life that was scandalous and entirely unbridled for a man of his vocation and estate; he was accompanied by the runaway pastor of Schonaw [i.e., Sömmering], as his consort." Weidemann left vague the exact nature of Scherdinger and Sömmering's "scandalous" lifestyle, but clearly he viewed alchemy as part of it. "With punishable neglect of their office, to the great disgrace of the holy ministry," he continued, "they fled to the futile, fraudulent, and false art called alchemy and goldmaking, whereby they first racked up a great debt, yet repaid no one, and then they lived luxuriously in and around Gotha with one Heinz Schumbach [Heinrich Schombach], who was close to the pastor of Hohenkirchen [i.e., Scherdinger]." Weidemann underscored the fact that he had long disapproved of the pastors' alchemical activities and their "disorderly and un-Christian lifestyles." Two years before, he maintained, he had even tried to have the men removed from their positions as pastors, but Grumbach and Schombach had intervened to prevent him from holding a hearing.[79] Clearly Weidemann had been biding his time until he could oust Scherdinger and Sömmering, and Johann Friedrich II and Grumbach's downfall in April of 1567 offered him his chance. The pastors were replaced immediately, their property confiscated in order to repay the salary they had collected in absentia, and they would spend the rest of their lives in exile.[80]

Philipp Sömmering and Abel Scherdinger attributed Weidemann's animus to the Lutheran doctrinal battles that had consumed ducal Saxony in the 1550s and early 1560s, before Grumbach arrived.[81] The pastor-alchemists may or may not have been correct about Weidemann's real motivations; regardless, the fact that Superintendent Weidemann chose to frame his denunciation of the pastors in terms of alchemy, rather than theology, is revealing. By the summer of 1567, Grumbach was dead, and his and Johann Friedrich II's cause was discredited, while Duke Johann Wilhelm sought to rebuild what was left of his shattered lands. The best way to condemn one's enemies in this climate was to link them to the disgraced court, thus tainting them with a whiff of sorcery and political danger. Weidemann and his allies took advantage of this strategy to condemn Scherdinger and Sömmering as Grumbach's alchemists, rather than to attack their doctrinal positions, as they might have done five years earlier. Weidemann's tactic was effective. Scherdinger never returned to his position in Hohenkirchen, although he managed to secure a position at the court of Count Georg Ernst of nearby Henneberg, who defended Scherdinger to Duke Johann Wilhelm and helped the pastor work out a plan to repay his debts and clear his record.[82] Sömmering, however, never managed to settle with the new ducal administration after the siege. He steadfastly denied any involvement with Grumbach, maintaining for years that he had been condemned without a proper hearing and was a victim of injustice; he even signed a letter to Johann Wilhelm's councillors, "Philippus Summering . . . expelled [*verjagter*] pastor of Schonau," directly countering Weidemann and others' claim that he had "run away."[83] More concretely, he lost not only his position as a pastor, but his property in his Thuringian homeland, all of which was confiscated in order to pay the salary Weidemann claimed he collected without performing his duties. Not even Sömmering's wife, Catharina, was able to win Johann Wilhelm's sympathy with her last-ditch effort to plead Philipp's case, for the sake of their children, in 1569.[84]

In the aftermath of Gotha, figures such as Weidemann, seeking simultaneously to demonstrate loyalty to a new sovereign and to marginalize old enemies, conflated the disgraced Duke Johann Friedrich II's military, political, religious, and alchemical projects into one nefarious tangle. Those who had worked for the duke in any capacity, however distant from Grumbach's machinations, fell under suspicion, the collateral damage of the Grumbach Feud and the Gotha Rebellion. Philipp Sömmering firmly rejected any connection between alchemy and the duke's catastrophic angelic holy war, but the association, once forged, became difficult to dissolve. No matter how many times Philipp tried to put the events of 1566–1567 behind him, much to his frustration, they kept cropping up, even several years later in Wolfenbüttel.[85]

*　*　*

In the wake of the 1567 Gotha Rebellion and resulting regime change, Philipp and Heinrich lost their posts at court, while Philipp also lost his post as a pastor in Schönau, as well as his property. He immediately went into exile, joined by his wife, Catharina, and their children, as well as Anna Zieglerin and Heinrich Schombach, whom he had befriended at the ducal court in Gotha before the siege. They all made their way through the Thuringian Forest and over the ducal Saxon border to the city of Schmalkalden, where Philipp had lived for three years as a youth, just after he finished school. The Sömmerings, and possibly also Anna and Heinrich, lived off of the money that Duke Johann Friedrich II had advanced Philipp for his alchemical work.[86] Over the next few years, to no avail, Philipp and Catharina Sömmering issued multiple pleas to the new Saxon duke, Johann Wilhelm, for the restoration of Philipp's property and parish post in Schönau.[87] Meanwhile,

Elector August's campaign to eliminate the "spirit of Gotha" moved beyond the realm of print; he launched a legal campaign against Grumbach's remaining collaborators, as well as others associated with the ducal court whom he suspected of criminal activity.[88] In May of 1568, roughly a year after the siege, August targeted Anna's husband, Heinrich Schombach, and sought his arrest for his affiliation with Johann Friedrich II and Grumbach. Heinrich was captured in Schmalkalden, but was able to avoid extradition to electoral Saxony thanks to the protection of Landgrave Wilhelm IV of Hessen, who held joint sovereignty over Schmalkalden and refused to hand over Heinrich. Anna and her husband managed to escape this legal wrangle. Nevertheless, the hint of legal trouble continued to hang over them for years to come.[89]

Anna, Heinrich, and the Sömmerings endured exile in Schmalkalden for a couple of years. Eventually, Philipp met Johannes Rhenanus, a Lutheran pastor and salt expert (*Salinist*) who oversaw the Hessian saltworks in Allendorf an der Werra. Philipp claimed that it was his reputation as a skilled man that brought him to Rhenanus's attention, but it seems more likely that Philipp sought out Rhenanus instead. Desperate for employment and exiled from his homeland, Philipp hoped that Rhenanus might be able to find him a position in the Hessian saltworks, but Rhenanus suggested that the pastor-alchemist seek employment in the Duchy of Braunschweig-Lüneburg instead, where the new duke, Julius, was investing in the industry.[90] Soon enough, Philipp landed a job in Juliushall, Julius's new saltworks in Bad Harzburg, where he met Julius's former court physician, the Hamburg doctor Jodokus Pellitius. This new acquaintance served as a broker for Philipp to obtain an audience with Duke Julius to pitch some ideas about increasing fuel efficiency in the Juliushall saltworks. After gaining the duke's attention with this new method in salt production, in 1571 Philipp parlayed it into an opportunity to place his many other skills, both alchemical and theological, at Julius's service. As a new duke with an ambitious agenda for economic and religious reform, Julius saw much of value in the combination of talents that Philipp possessed; he promised Philipp a new position as a court alchemist, and Heinrich could be his laboratory assistant. The duke provided Philipp with some "English cloth" (perhaps his clothes were a bit worse for the wear after several years on the road?), money, and the use of a horse (*Klopfer*)[91] to collect Anna Zieglerin and Heinrich Schombach, who had in the meantime moved to the Hessian town of Eschwege, near the saltworks that Rhenanus oversaw in Allendorf.[92]

As Anna, Heinrich, and Philipp prepared to begin their lives in Wolfenbüttel, they finally seemed to be putting the 1567 siege of Gotha behind them.
More broadly, too, the finely graded political and religious distinctions that
had galvanized factions in ducal Saxony under Johann Friedrich II were gradually losing meaning, especially for outsiders who had never been particularly
immersed in the intricacies of Saxon politics. What lingered, however, were
terrifying stories about sorcery and betrayal, alchemy and apocalypse, politics
and magical potions, all culminating in the tragedy of the Gotha siege and a
duke's downfall. As the councillors and courtiers who had been affiliated with
Duke Johann Friedrich II's court in the 1560s, including Anna, Heinrich,
and Philipp, scattered into neighboring territories and tried to begin new
lives, they struggled to shake these associations and allegations that they may
have been somehow involved in Grumbach's plot, simply by virtue of their
affiliation with Duke Johann Friedrich II's court.

CHAPTER 2

The Road to Wolfenbüttel

Anna, Heinrich, and Philipp's journey to their new positions in Wolfenbüttel in the summer of 1571 should have been the start of something new and peaceful. The nearly one-hundred-mile trip was overshadowed, however, by a set of frightening omens that surely raised new worries. Along the way, they saw "dreadful signs in the heavens and apparitions, and they had ghoulish dreams, which they themselves interpreted as meaning that they were in for great misfortune and that they would all be killed." Anna even thought that one of the specters they encountered, a woman on a bridge, was the devil. Indeed, when they asked in the local inn about this particular apparition, they were told that the female figure appeared there often to kill people. Despite these ominous portents, however, the group pressed on toward their new lives at the court in Wolfenbüttel (Figure 4).[1]

When Anna, Heinrich, and Philipp arrived at Wolfenbüttel Palace, they brought with them all of the baggage of their pasts, even as they hoped to move on to something new. Their patrons, Duke Julius and Duchess Hedwig, were also at the beginning of a new chapter of their lives, for they had only recently ascended the ducal throne. Like Anna, Heinrich, and Philipp, Julius and his wife, Hedwig, also had ambitious plans. They were still in the early years of their reign as Duke and Duchess of Braunschweig-Lüneburg, however, so whether they would be able to realize those plans remained to be seen. Obviously, the sovereigns and their courtiers were separated by a tremendous gap in status, and yet, in a sense, they all faced the uncertainty of beginnings. They envisioned successful lives and careers in Wolfenbüttel, but were also keenly aware of how fragile those futures were. This tension between ambition and uncertainty had a profound effect on the dynamic among this group of patrons and alchemists, men and women, ruler and

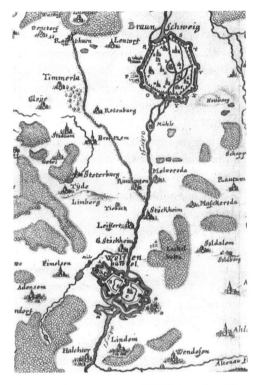

FIGURE 4. Seventeenth-century map of Wolfenbüttel and Braunschweig.
Martin Zeiller and Matthäus Merian the Elder, *Topographia und Eigentliche
Beschreibung der Vornembsten Stäte, Schlösser auch anderer Plätze und Örter in
denen Hertzogthümer[n] Braunschweig und Lüneburg, und denen dazu
gehörende[n] Grafschaffien Herrschaffien und Landen* (Franckfurt: Merian,
1654), copperplate engraving before page 15. © SLUB Dresden, Hist.Sax.inf.4,
http://digital.slub-dresden.de/id404350887/30.

ruled, as their lives became increasingly entangled over the next several years
in Wolfenbüttel.

Anna and Heinrich

Perhaps Anna Zieglerin never expected to be so far from home. Although
Wolfenbüttel was only about two hundred miles from her birthplace near
Dresden, it was across the border in the Duchy of Braunschweig-Lüneburg,
which is to say that it was ruled by another duke and was located beyond the

reach of her family and their immediate networks of political influence. Born
sometime around 1545,[2] Anna claimed to be the child of Caspar von Ziegler
and Clara von Schönberg, thus descended from two respected noble families
in electoral Saxony. Caspar Ziegler (d. 1547) came from a venerable family
that first made its fortune in silver mining in the fourteenth century and by
the sixteenth century possessed numerous properties around Meissen and
Dresden.[3] Anna's mother, Clara, hailed from another influential family, the
Schönbergs, who had long been in the service of the church, court, and
mining administration in Saxony.[4] Anna reported that she was born at
Pillnitz, an estate that had been in her father's family since the mid-fifteenth
century, and that, as befit her noble station, she was baptized in Dresden at
the electoral palace and had "princes and others as godparents."[5] Caspar and
Clara had at least two sons, Christoph and Hans Ziegler, neither of whom
went on to distinguish themselves particularly. The younger of the two, Hans
Ziegler, was among a group of nobles and others who were accused of committing a highway robbery in electoral Saxony in 1566, resulting in both the
Holy Roman Emperor and Elector August of Saxony issuing warrants for his
arrest.[6] The elder brother, Christoph Ziegler, meanwhile, did not produce an
heir and was eventually reduced to debt, which was so great that he was
forced to sell the family estate in Pillnitz in 1569.[7]

At least this is what Anna claimed was her lineage. In fact, nothing is
known about Anna's family or early years beyond what she and her husband
Heinrich later told Duke Julius's court, raising the possibility that this part
of her life is at least in part a fabrication.[8] Perhaps she was not of noble lineage
after all, but developed this aspirational autobiography in Wolfenbüttel in
order to elevate her social status by associating herself with the court of Elector August of Saxony and his wife Anna, a Danish princess. Such a narrative
would have situated Anna at the cultural and political center of an electoral
Saxony on the rise, rather than on the margins of Johann Friedrich II's court,
a ducal Saxony on the wane and destroyed in the wake of Gotha. In the
1550s, Elector August was busy consolidating and expanding his lands, but
the heart and symbol of electoral Saxony was Dresden, which numerous
building projects transformed during these years from an average city of middling size into a glorious and permanent seat for the ascendant electoral court.
The old palace was expanded and renovated during August and Anna's reign
into a palace suitably elegant for princely festivities, and Renaissance architectural forms began to replace older medieval buildings in the city. A number
of new buildings—including a mint and "Goldhouse," where August's

alchemists and assayers tested and prepared metals (1556), chambers for August's councillors (1565–1567), and a court apothecary (1581)—sprang up around the palace to house and represent the expanding electoral bureaucracy of the Saxon state. This was indeed the beginning of Dresden's transformation into the elegant residential city—"Florence on the Elbe"—that tourists flock to today. At the center of this activity were Elector August and his wife, Anna, who ruled electoral Saxony together for over thirty years.[9]

If Anna Zieglerin did indeed grow up around Dresden, she likely would have been there only for the first decade or so of August and Anna's reign; moreover, her claims to close familiarity with the Dresden court may well have been an exaggeration. Nevertheless, even if she observed the electoral couple only from afar, she could have learned a great deal from August and Anna about the value of nature at Renaissance courts. Over Elector August's long rule, he devoted particular attention to expanding the borders of his territory (at the expense of his cousins in ducal Saxony, as we have seen, as well as local bishoprics) and developing the Saxon economy. In addition to repaying the massive debts his brother and predecessor Moritz had accrued, August took a special interest in two key industries in Saxony: mining and agriculture. He established the Dresden Goldhouse in 1556, for example, as a site of experimentation with alchemical and other techniques for processing and multiplying Saxony's rich mineral resources. He also set up model farms, where he could undertake his own experiments with new methods for the cultivation of new trees and grains, and livestock, and even authored a treatise on grafting and the propagation of fruit trees.[10] In most of these things, his wife, Anna, was his partner. If the paternalistic rhetoric of the time figured August as the "father" of his lands, then the Danish princess Anna became the "mother," with her own responsibilities for the economic, medical, and spiritual well-being of her subjects. She shared August's interest in developing Saxon agriculture, taking a particular interest in fruit cultivation and dairy works, which earned her the nicknames "Danish Cheese Woman" and "The Apple and Pear Woman." Like Elector August, Electress Anna was a powerful patron in her own right, with a well-developed network of clients. Above all, however, she was known for her medical expertise. Sharing August's interest in experimenting with ways to improve nature, Anna spent a lifetime creating and maintaining recipe collections, gardens, distilling houses, and stores of medicaments. She not only took responsibility for the health of her courtly household and local poor, as was commonly expected of noblewomen in this period, but also offered medical advice and expertise to a vast network of other European nobles, physicians, and even Holy Roman Emperor Maximilian II and Empress Maria. To someone like Anna Zieglerin, Electress Anna offered a powerful example of how German noblewomen could make skill and knowledge in medicine a centerpiece of their place at court.[11]

When Elector August ascended the throne in 1553, he was twenty-six, Electress Anna was twenty-one, and Anna Zieglerin was just a girl. Moreover, Zieglerin would have been an orphan, for her purported father, Caspar von Ziegler, died shortly after she was born, and, as Anna later claimed, her mother died in childbirth. It is not clear whom she lived with as a child, but as Anna came of age, she might have looked forward to marriage.[12] As she entered her teens, however, her life seems to have been thrown inalterably off course by a series of difficult events that would eventually lead her on that uncertain road to Wolfenbüttel with Heinrich Schombach and Philipp Sömmering. When Anna was not yet fourteen years old, she later recalled, a nobleman named Nikolaus von Handorf raped her somewhere near Meissen in the Saxon countryside.[13] Whatever Anna's social status, this horrific event would have damaged her reputation, as well as her hopes of securing a favorable marriage. Nevertheless, she claimed, a Silesian nobleman named Rottenburg agreed to marry her anyway, and the couple celebrated the wedding in Königswartha, about forty miles from Dresden. Unfortunately, only nine weeks later, Rottenburg was thrown from a horse and died, leaving the ill-fated Anna a teenage widow.[14]

What are we to make of Anna's story of sexual assault and early widowhood? Perhaps the sexual encounter she described was consensual, if conducted out of wedlock. On the other hand, if it was an assault, as she maintained, then it would make some sense to invent a subsequent marriage (and a quick death for the husband to explain his absence). This would have restored Anna's honor in some sense; it was better to be a widow than the victim of a rape or a fornicator. Indeed, widowhood was a common fate for women in sixteenth-century Europe.[15] This was true for noblewomen, artisans, and peasants alike, in part because grooms tended to be older than brides, which meant that wives frequently outlived their husbands.[16] Although many older widows did not remarry, either because they could not find a new spouse or because they chose to remain independent, young widows, particularly those who had not yet had children, typically sought (or were pressured to seek) new husbands.[17] Whether Anna was in fact a widow or had never married at all, she was still young and childless in the 1560s. Moreover, her putative parents were no longer alive, and although her brother Christoph was still in Saxony, he was moving toward financial ruin and was perhaps unable to support young Anna. A new marriage may have seemed her best chance for a stable future.

Before long, Anna found her chance. While Anna was staying to the northeast of Dresden in the Silesian city of Bautzen (with her brother-in-law, she said),[18] she met Heinrich Schombach, also known as "Cross-Eyed Heinz" (Schielheinz). Heinrich was not a particularly appealing marriage prospect, to be sure. Employed in the service of Duke Johann Friedrich II of Saxony[19] but not noble himself, Heinrich was unquestionably a step down from the elite social status Anna imagined for herself and attributed to her first husband, Rottenburg. Heinrich may have been disagreeable in other ways as well. Duke Johann Friedrich II's councillor Grumbach, for one, found him to be rather sycophantic, always keeping an ear out for bits of information that might be useful in some way later (although this characterization could describe many an early modern courtier).[20] When Anna met Heinrich, in any case, he apparently did not impress, and she certainly did not want to marry him. It is probably a sign of how limited Anna's options were, because of either her own social status or damaged reputation, that eventually she consented.

Like many of the details of Anna's early life, the circumstances of her marriage to Heinrich are a bit murky. The only extant descriptions came several years later, from Anna and Heinrich, and each party told a very different tale of how they came to be wed. Heinrich simply explained that "he took her [in marriage] faithfully, as she did him. He went to a pastor in a church and prayed for her to be given to him, and she took a solemn oath to give herself to him in marriage."[21] Anna's version of the story, however, was much less sanguine. As she recalled, when she met him she "had no thought of having [i.e., marrying] Heinrich, but was joking around with him." Heinrich, however, claimed that her teasing was serious, and "said that she had consented [to marry him] and wrote this to Duke Hans Friedrich [i.e., his patron Duke Johann Friedrich II of Saxony]." Although Anna was insistent that she "did not want to have him," all apparently claimed that she had already agreed to the marriage. Duke Johann Friedrich II, meanwhile, supposedly used his power to pressure Anna to marry Heinrich. According to Anna, the duke declared his support for the union, decreed that the wedding go forward, enlisting "two Lords" to ensure that it did, and made Heinrich promise to look after and support Anna. Despite her objections, the wedding occurred in Bautzen in 1563 "as the entire city knows," Anna later said. "A marriage contract was not drawn up," she explained, but "there were a number of people there, including Joachim Haubnitz and other Silesian Junckers."[22] From our perspective, neither Anna's nor Heinrich's story seems quite right. Heinrich's is far too harmonious, given how much Anna seemed to dislike him, yet Anna's claim that a duke was so invested in her welfare that he would personally become involved in matchmaking rings a bit untrue as well. Perhaps the truth lay somewhere in the middle: either Anna was desperate, or her brothers coerced her into a marriage with Heinrich, hoping that he would take responsibility for her welfare.

However it came about, Anna's marriage to Heinrich led her away from her homeland in electoral Saxony to the court of Johann Friedrich II of ducal Saxony, where her husband Heinrich was employed. This move took her from Dresden, a city on the rise in economic, cultural, and political terms, to Gotha, the seat of a rival court on the wane, in political disarray after a series of military defeats, and heading toward the disaster of the Grumbach Feud and the siege of Gotha. Surely Anna would have seen it as something of an exile or a failure to live up to her aspirations, mirroring, perhaps even confirming, the drop in status represented by her marriage to Heinrich.[23] On the other hand, Anna's time in ducal Saxony would expose her to many new things as well. Indeed, her move opened a second chapter of her life, one that included a new court, marriage to Heinrich, and a fateful introduction to a local pastor-alchemist named Philipp Sömmering.

Philipp

The son of a pastor named Johann Sömmering, Philipp was born in the Thuringian Forest, in Tambach, about fourteen miles south of Gotha and forty miles west of Weimar, probably in the 1530s. After an itinerant youth as a student, schoolmaster, and then chaplain in the region around Saxony and Thuringia, in 1555 Philipp took up a position as a pastor in the Thuringian town of Schönau, where he remained until 1567. He was, in other words, a small-town pastor who followed in his father's footsteps and stayed close to home, hardly a denizen of princely courts. The pastor's life in 1560s Germany was not quite as humdrum as it sounds, however, for these were raucous years in Saxon Lutheran circles. Theologians, pastors, and princes struggled to control Luther's legacy and define the evangelical movement's relationship to political power after the reformer's death in 1546. Until 1580, when a majority of Lutheran leaders finally endorsed the *Book of Concord* as a hard-won statement of doctrinal unity, intense debates, both in person and in print, polarized Luther's followers.[24] Particularly fierce were the disputes between the orthodox "Gnesio" (or "Real") Lutherans,[25] on the one hand, who advocated a strict adherence to Luther's teachings, and the more flexible "Philippists," on the other, who accepted the authority and theology of Luther's disciple Philipp Melanchthon as well and were more inclined to adapt Luther's teachings to changing political realities. Doctrinal differences between these two groups were heightened by their political disagreements about earthly affairs, especially about the degree to which Luther's followers should compromise with the Catholic emperor.[26] By the 1560s and early 1570s, additional theological controversies arose over justification, good works, free will, Christology, and the Lord's Supper, among other things, creating what Irene Dingel has called a "culture of conflict" among the Lutheran leadership. These debates often had real-world consequences. Theologians lost appointments at courts and universities, sometimes even landing in prison, as a result of their doctrinal positions.[27] Sömmering specifically cited one of these conflicts, along with the encouragement of two friends from his "patria," as the reasons that he began to consider another line of work.[28] And when he did, he turned to mining and alchemy.

Given the importance of mining in Saxony and Thuringia, it is not entirely surprising that metals would come to interest the native son Sömmering. Nor is it surprising that what Philipp called the "business of mining" (*Berghandel*) included alchemy, for the intellectual and practical overlaps between these two metallurgical traditions were numerous in the sixteenth century.[29] It is also fitting that Philipp—who, as a student, schoolmaster, and then pastor, had long been immersed in the world of texts—would first encounter alchemy through a book given to him by friends. When a pastor in the nearby Harz mining region gave Philipp another book, this time containing information about the *Lunaria* plant, reading about alchemy quickly turned to practicing it. As Philipp would later explain, this abundant plant had a yellow stem and flowers, and it released sap when one picked it. He and another pastor-alchemist "obtained the sap from it, and with that they began to do alchemy according to the process [in the book]."[30] Philipp was likely referring to one of the many "moonworts" (*Mondraute*), plants with crescent-shaped leaves or seedpods that evoked the moon, fascinating naturalists and alchemists alike in the sixteenth century.[31] The prominent Swiss naturalist Conrad Gessner, for instance, devoted an entire treatise to moonworts, and they also appeared in the Neapolitan polymath Giambattista della Porta's *Phytognomonica*, which developed the notion that the medical uses of plants could be detected from their forms.[32] In a 1551 herbal, German naturalist Hieronymus Bock reported that the herb should be plucked by moonlight and, playing on the astrological connection between the moon and silver, mentioned that alchemists were known to use it.[33] A 1599 collection of alchemical texts, *Aureum Vellus*, also discussed this plant, noting that the ancient alchemist Salenius "says of the herb Lunatica or Bera: 'Its root is a metallic earth; it has a red stem, overlaid or rather stained with black, and fades easily. It also acquires or grows short-lived blossoms. If one puts it for three days into mercury, it changes into perfect silver; and if one boils it further, it turns into gold. This same gold turns 100 parts [of mercury] into the very finest gold.' "[34] Sömmering, unfortunately, was unsuccessful with the *Lunaria*, even after learning about distillation and sublimation from a woodcutter, consulting a philosopher about mercury, and procuring several more books. Undeterred, however, Philipp continued to seek out alchemical knowledge from a network of people that included a merchant and his wife, as well as other pastors.[35]

Philipp then returned to books, hoping that they might supplement the knowledge he had gained from other practitioners. He first acquired the "*Book of Isaac*, which contained many good things," perhaps one of the many manuscripts in circulation attributed to the legendary fifteenth-century alchemist Isaac Hollandus.[36] Given that Philipp was first and foremost a pastor, it was perhaps natural that he also would find scripture to be a useful resource for thinking about alchemical creation (and vice-versa). Convinced that a book called the *Hexameron of Bernard* or the *Book of Creation* would give him some insight into alchemy, Sömmering traveled to Erfurt to purchase it; at 1,000 thaler, the beautiful book was too expensive, but when he fell on his knees and begged to purchase the book for only 400 thaler, he later recounted, the seller gave him the reduced price, and Philipp got the book.[37] Although it is difficult to determine precisely what book Philipp purchased, the title makes it clear that it was some kind of commentary on the first six days of creation in Genesis. Philipp was not alone in finding parallels between scripture and alchemy. Just as Lutheran physicians found material for thinking about the generation of infants in the biblical account of the creation of the world, so, too, did some alchemists understand a connection between biblical descriptions of the creation of the world and the alchemist's creation of matter.[38]

Eventually Philipp Sömmering decided to team up with Abel Scherdinger, a pastor who held a post in the nearby town of Hohenkirchen and with whom he would eventually collaborate on alchemical projects for Duke Johann Friedrich II in Gotha.[39] Scherdinger already had been pursuing alchemy on his own, perhaps as unsuccessfully as Sömmering, and so the two pastors decided to pool their resources and split any resulting profits between them. They initially worked with a third pastor named Martin Gerlach, who was convinced he "knew a secret that no one else did,"[40] but Scherdinger and Sömmering parted ways with Gerlach before long. They believed that they had the correct process for producing gold, yet they still could not get it to work properly. In 1565 a man who claimed to be an alchemist and to have successfully calcined tartar visited Scherdinger and Sömmering. He, in turn, claimed to have discovered the secret of alchemy from yet another pastor in nearby Altenstein, Nicolaus Solea.[41] Sömmering rode to Altenstein, where Solea showed him a quintessence of arsenic[42] but would not reveal how it was made. Eventually, however, Solea changed his mind, and told Sömmering that he would reveal the secret. Emboldened by this promise, Sömmering decided to approach the local duke, Johann Friedrich II of Saxony.[43]

This was no small decision on Sömmering's part. Not only did it mark a more serious commitment to alchemy, which until then had been more of a hobby than anything; it also had the potential to transform Sömmering from a small-town Thuringian pastor into a courtier working in much closer proximity to the ducal court. This move promised to be quite rewarding, of course, in terms of both riches and influence, but it also carried risks, as Sömmering would learn, in that it linked his fate to the vagaries of court politics. As Sömmering undoubtedly knew, Johann Friedrich II was increasingly occupied with alchemical projects in the 1560s, as were many of his fellow German princes, who in the second half of the sixteenth century built alchemical laboratories, collected books and recipes about the art, and hired practitioners to try out techniques and develop new ones.[44] Princely patrons were not naïve, however, and they recognized that alchemy attracted practitioners with a range of motives and abilities. Some alchemists were scholars, others were artisans. Some pursued alchemy as a leisure activity; others sought to make a living practicing the art for profit. Finally, some engaged with alchemy out of genuine curiosity about nature's workings, while a few opportunists (although far fewer than contemporary satires and critiques assumed) merely pretended to be alchemists in order to swindle wealthy investors out of their money. Early modern patrons knew that they would have to carefully evaluate any alchemists who came their way, and they employed a number of measures, including test demonstrations, contracts, and the scrutiny of expert advisers, to try to weed out the true from the false alchemists. All of these techniques, not to mention the discernment of angels, came into play when Sömmering and Scherdinger first approached Duke Johann Friedrich II in 1566.[45]

The way the two pastors made their case offers useful insight into not only Philipp's early alchemical career, but also alchemists' strategies for securing patronage in general. In one of their letters, Sömmering presented himself as both a practitioner in his own right and a broker who could bring Nicolas Solea, whom he described as a particularly knowledgeable alchemist, to Johann Friedrich II's court. Philipp told Johann Friedrich II that he had long pursued alchemy (but only insofar as his position as pastor afforded him the time to do so, he was careful to note) at great cost. Not only had he consulted many books, he explained, but he had also sought out people who could teach him the practical skills necessary to actually work with metals and other alchemical materials. Moreover, he was in contact with a group of other learned men who were also interested in alchemy, all of whom shared their secrets with one another. Sömmering had consulted often with Nicolas Solea, who, he explained, was particularly trustworthy, and together they had tried to sort through the confusing alchemical corpus of texts, which had been corrupted by the inexpert people who had compiled, and, in some cases—notably, in texts attributed to the renowned physician and author Paracelsus—attached the names of famous authors to texts they had written themselves. This man Solea, Sömmering continued, had written several lovely little books about *chymia*. Although they had not yet appeared in print, Sömmering included the manuscripts as a gift with his letter to the duke. As a loyal servant of the ducal Saxon crown, he emphasized, Sömmering had begged Solea not to take his skills to Johann Friedrich II's archrivals: Holy Roman Emperor Maximilian II; the emperor's brother, Archduke of Austria Ferdinand II; and Elector August of Saxony. Rather, Sömmering persuaded Solea to share his knowledge with Johann Friedrich II himself. Solea required no money or expenses in exchange for his alchemical secrets, moreover—only a place to work.[46]

Duke Johann Friedrich II replied favorably to the proposal from Sömmering, who prepared to bring Solea to court.[47] Eventually, for reasons that are unclear, Nicolaus Solea backed out of this arrangement, but Sömmering and Scherdinger went on to sign a contract of their own with the duke on November 6, 1566. No longer were Scherdinger and Sömmering alchemical brokers, bringing other, more experienced practitioners to the attention of the court. They were now officially court alchemists in their own right. The contract stipulated that the duke pay the alchemists 750 thaler so that they could work on the production of the philosophers' stone. In return, Scherdinger and Sömmering promised to provide the duke with a full written report of their work, to work honestly and without fraud as long as it pleased the duke, and to maintain strict secrecy about their knowledge; they and their families would also receive the duke's protection (*Schutz und Schirm*) for the duration of their lives, and should they be successful, they would receive 10 percent of any profits remaining after the cost of supplies.[48] Not everyone at court was enthusiastic about this deal, as we have seen. Apparently Grumbach was skeptical about the arrangement, and he warned the duke that Sömmering's proposal was mere roguery (*Schelmerei*), calling into question later claims that Sömmering was one of "Grumbach's alchemists." And, of course, the angels were skeptical as well. Indeed, it is stunning to think that, on the eve of the siege, as Johann Friedrich II prepared for holy war against his Saxon cousin August and Holy Roman Emperor Maximilian II, he and the angels spent any time at all thinking about alchemy, let alone the terms of a contract with Philipp Sömmering. And yet they did. Perhaps the duke was wondering how he might finance the coming war, although, given the angels' constant reassurance that all would be provided, it seems unlikely that Johann Friedrich II was worrying too much about his bills. Duke Johann Friedrich II initially waffled about what to do, but eventually, at the encouragement of one of his courtiers, decided at least to explore what the alchemists might know. Sömmering later claimed to have admitted to Johann Friedrich II that he was not entirely sure about the art but suggested that if the duke would pay any expenses, then Sömmering would do his best and hand over the greater part of the profits should they succeed. Sömmering promised to work at Abel Scherdinger's house, and, once Grumbach's secretary Moritz Hausner gave them each 450 thaler, the two alchemists commenced work in the late fall of 1566.[49]

The courtier whom Philipp Sömmering had to thank for encouraging the duke to sign this contract was Anna Zieglerin's husband, Heinrich Schombach. Whether Philipp and Heinrich already knew each other or whether Heinrich simply saw some advantage in bringing alchemists to Johann Friedrich II's court is unclear. In any case, Philipp's contract linked the fates of these two men, and, by extension, Anna Zieglerin's as well. Perhaps Heinrich and Philipp decided that they made a good team, between Heinrich's skills as a courtier and Philipp's skills as an alchemist. Philipp might also have felt that he was in Heinrich's debt for brokering the contract, or perhaps the two men became genuine friends. In any case, Philipp and Heinrich continued to work together from this point onward, and several years later, after the debacle in Gotha, Philipp would return the favor by extending an invitation to Heinrich to work as his laboratory assistant in Wolfenbüttel.

The association between Heinrich and Philipp, meanwhile, also drew Anna Zieglerin more closely into alchemical circles at Duke Johann Friedrich II's court. It is difficult to know precisely how much contact she had with the duke's alchemists during this period, but as Heinrich's connection with the pastor-alchemist Sömmering grew, so, too, must have Anna's awareness of the promise, pitfalls, and profits of court alchemy. This was not yet the moment for Anna Zieglerin to observe Sömmering's alchemical operations firsthand, however, for far more urgent matters intervened before Scherdinger and Sömmering could begin their work. Shortly after Duke Johann Friedrich II signed his 1566 contract with Philipp, the siege of Gotha began, making alchemical work impossible in ducal Saxony, and eventually sending Philipp, Heinrich, and Anna into exile and on the road to Wolfenbüttel.

Wolfenbüttel, 1571

After braving apparitions and portents on the harrowing journey that followed, Anna, Heinrich, and Philipp finally arrived in Wolfenbüttel in the summer of 1571, four years after leaving Gotha. They encountered a burgeoning city, its fate tied to the rising fortunes of the Braunschweig dukes. Before the sixteenth century, a visitor to this region would not have found a proper town at all where Wolfenbüttel is today, in fact[50]—only Wolfenbüttel Palace, which had been the dukes' administrative seat since the fifteenth century, and a small settlement, known as "At Our Lady" (Zu unser lieben Frawn), clustered around St. Stephan's Church along the eastern bank of the Oker River opposite the palace.[51] Julius's father, Heinrich the Younger (1489–1568), began Wolfenbüttel's transformation from medieval fortress town into a Renaissance city that would eventually rival nearby Braunschweig for regional cultural and economic power.[52] He expanded the settlement across the Oker River, renaming it Newstadt and establishing a set of governing laws in 1540. Although this settlement suffered significant damage as a result of the Catholic duke Heinrich's battles with the Protestant Schmalkaldic League in the

1540s, it was rebuilt again in the 1560s with the addition of a new parish church, the Marienkapelle (1561), which replaced the destroyed St. Stephan's, and the establishment of two *Bürgermeister* and a town council (1567). Heinrich also built up embankments to protect the settlement from encroaching marshland as well as to serve as fortifications to protect Wolfenbüttlers from their enemies. Finally, Heinrich initiated the renovation of the medieval castle, enlisting the Italian architect Francesco Chiaramella de Gandino to repair, update, and expand the structure, transforming it into a Renaissance palace.[53] Heinrich's son and successor, Julius, continued the transformation of the palace and city upon inheriting his father's throne in 1568. He began with the palace itself, which not only served as the residence for the duke and his family, but also housed the ducal collections of art and wonders, laboratories, an apothecary, and, until 1588, when a new chancellery building (*Kanzlei*) adjacent to the palace was completed, the ducal administrative offices. In August of 1570, Julius elevated Newstadt's status, officially designating it a town, renaming it Heinrichsstadt in honor of his father, and granting it important market and heraldic rights. The following spring, Julius announced an ambitious plan for the city. He hoped "to extend, build up, and fortify Heinrichsstadt somewhat, in order to place streets and houses so that more people might take up residence there, as time and opportunity allows. . . ." He continued, "indeed, we decree that all the dilapidated houses and nasty little firetraps that can be moved must be reorganized . . . so that Heinrichsstadt can be built with straight, spacious streets and uniform houses of the same width and height, so that such streets may all lead to our Wolfenbüttel Palace and basically be in view of it."[54]

This strikingly modern plan to rationalize Wolfenbüttel, which sounds more like the vision of an eighteenth-century urban planner (or even a twentieth-century housing developer) than that of a Renaissance prince, was accompanied by a series of equally ambitious projects for the duchy more broadly (Figure 5). In the mid-1570s, for instance, Julius brought in a series of Dutch engineers to make the various rivers in his lands (the Oker, Radau, and Altenau) navigable by ship so that he could transport wood and metals from the mines in the Harz Mountains to Wolfenbüttel. He also developed a plan to create a system of canals that would link the Oker and Elbe Rivers, thus connecting the duchy to broader transportation networks. Not surprisingly, perhaps, Julius would never completely realize all of these grand plans for Wolfenbüttel and the duchy. Nevertheless, urban redevelopment and expansion were certainly under way in the duchy, at the palace, and in

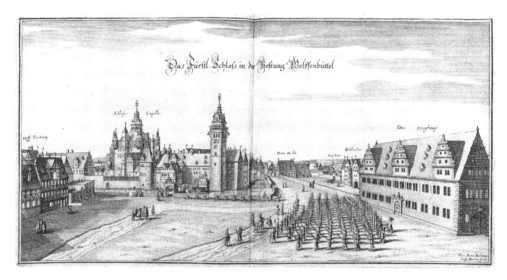

FIGURE 5. Seventeenth-century view of Wolfenbüttel Palace. Martin Zeiller
and Matthäus Merian the Elder, *Topographia und Eigentliche Beschreibung der
Vornembsten Stäte, Schlösser auch anderer Plätze und Örter in denen
Hertzogthümer[n] Braunschweig und Lüneburg, und denen dazu gehörende[n]
Grafschafften Herrschafften und Landen* (Franckfurt: Merian, 1654),
copperplate engraving between pages 208 and 209. © Herzog August
Bibliothek Wolfenbüttel.

Heinrichsstadt when Anna, Heinrich, and Philipp arrived in the summer of
1571. One historian has estimated that the population of the city quadrupled
in the twenty years before Julius took the throne, and that in 1571 Heinrichsstadt alone (i.e., excluding the palace and its inhabitants) had about one hundred houses filled with artisans and peasants. Anna and her companions, in
other words, would have found a modest but thriving town, and also likely
would have seen numerous ongoing building projects in the city, suggesting
that the future there was bright.[55]

Even as Wolfenbüttel was coming into its own as a city, though, the
palace still dominated the cityscape, a reminder that Duke Julius was
the heart, soul, and economic engine of its expansion and transformation.[56]
The youngest son of one of the last Catholic princes of northern Germany,
Julius was never supposed to take the throne. While he was still a child, his
father, Duke Heinrich the Younger, determined that the boy would have a
career in the church, rather than on the battlefield. This was a conventional

strategy in the sixteenth century for princes' younger sons, who were unlikely to inherit their father's positions, but in Julius's case, an additional factor may have determined his father's choice. Julius's legs had been disabled since childhood, and he had difficulty riding a horse, making him ill-suited for a military career.[57] Heinrich, therefore, secured a position for the young Julius as canon (*Domherr*) of the Cologne Cathedral, hoping that his youngest son would find a career in the Catholic Church instead. The fact that Julius would take a different path from the other men in his family became particularly clear in 1542, when the empire-wide war between the Catholic and Protestant princes came to Wolfenbüttel. The Protestant Schmalkaldic League, a military alliance ostensibly formed to defend Protestantism in the Holy Roman Empire, intervened when two prosperous cities in the duchy, Goslar and Braunschweig, declared their intent to implement the Reformation formally, thus directly defying the Catholic Heinrich's political and religious authority. In support of these two cities, the Schmalkaldic League attacked Heinrich's seat in Wolfenbüttel in 1542, destroying the palace and temporarily ousting Heinrich from power. Julius was sent to Cologne. His father and brothers, on the other hand, waged war against the Schmalkalders for the next five years, part of which Heinrich spent as a prisoner of the Protestant Landgrave of Hessen when an attempt to recapture his throne failed. Only when the Schmalkaldic League was defeated at the Battle of Mühlberg in 1547 did Heinrich return to the throne and Julius rejoin his family in Wolfenbüttel. After two years, however, Heinrich sent Julius away again, this time to Bourges, Paris, and then Louvain, where he stayed for three years to study law. Perhaps damning Julius's years of study with faint praise, Julius's sixteenth-century biographer, Franz Algermann (1548–1613), noted that the future duke "learned so much [in Louvain], that he could understand some Latin."[58]

While he was in Louvain, Julius reportedly also decided to undergo a risky medical treatment on his legs. Although a prominent doctor named N. Haerdael appears to have developed and arranged the treatment, he was not present for the procedure itself, which was carried out by an executioner (*Meister*[59]) and his sons. According to Algermann's remarkable description (recorded a few decades after the fact, although Algermann claimed Duke Julius relayed the story to him personally), the executioner placed his sons, Julius, and "all kinds of instruments, comfort measures, and medicaments" in a coach "so that no cries of pain and screams could be heard, and they rode from Louvain to Antwerp." While the coach made the forty-mile trip, the executioner used the kinds of leg presses ordinarily used in the context of judicial torture to crush Julius's legs (presumably so that the bones could then be reset properly), all the while ensuring that Julius was "refreshed with strong waters and other strengthening things." To be sure, Algermann may have exaggerated the details in order to turn this into a heroic story of bravery. If even the basic outlines are true, however, then one can only imagine the nerve it took for Julius to voluntarily undergo torture, essentially, at the hands of an executioner in order to improve his legs. "Although His Grace endured great anguish and pain," Algermann commented, "nevertheless he was so resolute and courageous that he suffered six turns [i.e., suffered the press to be tightened six times] that day; and thus the feet were restored to their previous state [i.e., before the childhood accident to which Algermann attributed the disability], and His Princely Grace could walk better and more skillfully than before."[60]

It is worth noting that this procedure would have been not only extremely painful, but also socially transgressive, for it violated the strict social rules that governed contact with executioners in sixteenth-century German culture. Generally speaking, the executioner was such a dishonorable social figure that even to be touched by him in a social or legal context was to risk becoming a social outcast oneself. The exception to this rule was executioners' medical practice, where they often enjoyed great renown.[61] In this context, executioners could touch bodies unproblematically, and they healed bodies of all social stations, including the nobility. While Julius's treatment at the hands of the *Meister* was, therefore, not entirely surprising, the particular form of his therapy violated the carefully constructed rules that made executioner medicine possible by blurring the boundary between polluting and acceptable interactions between honorable citizen and executioner. In this case, the medical treatment itself *was* a form of torture, virtually indistinguishable from the executioner's actions in the dishonoring context of legal torture. Perhaps this was partly why Julius's treatment took place out of the public eye and on the road to Antwerp. Fortunately, the socially and physically risky procedure paid off. Julius's mobility improved somewhat, although he retained a limp for the rest of his life. As a result, Algermann notes, he always wore boots (presumably to support his ankles?) and, although he pretended to be a good rider, continued to find it difficult to ride a horse and preferred to be carried in a litter.[62]

Shortly after this surgery, Julius's life as a student was interrupted when his father, Heinrich, suddenly summoned him back to Wolfenbüttel in 1553.

Julius's two elder brothers, Carolus Victor and Philippus Magnus, had both died on the same July day in the Battle of Sievershausen, leaving Julius, at age twenty-five, suddenly the sole heir to the duchy's throne. This deeply troubled the aging Heinrich, who no doubt grieved for his two oldest sons, but who also did not want the duchy to fall to Julius, whose Protestant leanings had long been a bone of contention in the family. Although Protestantism already had taken hold in cities like Braunschweig and Goslar, the defeat of the Schmalkaldic League meant that order and Catholicism nominally had been restored to the duchy. Heinrich was aware, however, that the future of Catholicism in his lands hinged on the succession. He had already seen two powerful Catholic territories in the north "fall" to Protestantism, in fact, when evangelical sons—Elector Joachim II of Brandenburg (1505–1571) and Heinrich II "the Pious" of Ernestine Saxony (d. 1541)—succeeded their Catholic fathers and promptly introduced the Reformation. Heinrich, therefore, began a campaign to find a new heir. In 1556 at the age of sixty-seven and fifteen years after his first wife died, Heinrich married Sofie, the daughter of King Sigismund of Poland, hoping for another son. When this marriage failed to produce an heir, Heinrich resolved to petition the pope to legitimate Heinrich's son with his longtime mistress, Eva von Trott, but the son himself objected. The souring relationship between Julius and Heinrich even drew the attention of the soon-to-be Holy Roman Emperor Ferdinand I (1503–1564), who sent a servant on a secret mission to Wolfenbüttel in order to speak with Julius. Julius asked the servant to beg Ferdinand to grant him refuge either at his court or at his son Maximilian's (future Emperor Maximilian II), so that Julius's "absence and honorable and good conduct will perhaps cause [Heinrich] to put behind him and reduce his paternal fury."[63]

Matters came to a head in 1558 when Duke Heinrich issued a mandate reminding his subjects that it was his duty to uphold Catholicism in his lands. Those who objected were welcome to leave the duchy, the mandate stated, but Protestants who stayed would have to face the consequences. Embracing the Lutheran confession, in other words, would be seen as a sign of political insubordination. Julius knew that this mandate was directed at him, at least partially. Soon thereafter, it became clear that Heinrich intended to incarcerate his son (no idle threat, since Heinrich had imprisoned his own brother Wilhelm for twelve years in order to secure his assent to the law of primogeniture in the Duchy of Wolfenbüttel, which Heinrich decreed in 1535). Julius fled to Brandenburg, where he sought refuge with his sister Katharina and her husband, Margrave Johann of Brandenburg-Küstrin (1513–

1571). There Julius met Hedwig, daughter of (Protestant) Elector Joachim II of Brandenburg and a Polish princess, Jadwiga (1513–1573). When the couple announced their engagement, Julius's father was furious. In the sixteenth century, parents, not children, decided on marriage matches, particularly among the nobility. Eventually, however, Heinrich relented and gave the couple his blessing. Julius and Hedwig married in Berlin on February 25, 1560, and then returned to Wolfenbüttel, where they took up residence in the Braunschweiger dukes' summer residence, Hessen Palace, about sixteen miles southeast of Wolfenbüttel.[64] Julius and Hedwig soon produced a daughter, Sofie Hedwig (1561–1631), and then a male heir, Heinrich Julius (1564–1613). Heinrich Julius's arrival, in particular, helped to ease tensions between father and son a bit, and even his name, a union of his grandfather's and father's, suggested some kind of rapprochement. Nevertheless, Heinrich and Julius's relationship remained frosty for the remainder of the old duke's life, and Julius continued to be treated as a pariah at court, rather than as the heir apparent.[65]

In June of 1568, Duke Heinrich died, and, despite all of his father's efforts to prevent it, Julius inherited the ducal throne. The new Duke Julius wasted no time in launching the Reformation, even burying his fiercely Catholic father with a mixed Catholic and evangelical ceremony. Less than two months later, on August 1, 1568, Julius issued a proclamation ending Catholic Mass in the duchy and embracing the Augsburg Confession, the 1530 statement of Lutheran faith, until further decrees. On October 8 of that same year he signaled his intention to take stock of the duchy's religious institutions and personnel by launching a general church visitation throughout the territory, whereby the clergy was examined, as were the conditions of the buildings in the parishes and monasteries. With this rapid series of decrees, Julius ended the official practice of Catholicism in Braunschweig-Lüneburg with little opposition. Reconstructing the territorial church anew and in a Lutheran mold, however, was a more challenging affair, requiring a suite of administrative, liturgical, theological, and social reforms. Julius recognized the importance and difficulty of this task, so he sought help from two prominent Protestant theologians: the Braunschweig city-superintendent (i.e., the politically appointed head of the Lutheran Church in the city) Martin Chemnitz (1522–1586) and the Tübingen professor Jakob Andreae (1528–1590). Chemnitz and Andreae came to Wolfenbüttel and together helped Julius craft his 1569 *Kirchenordnung*, a church ordinance that stipulated the new theology, liturgy, and ecclesiastical reorganization that would govern the duchy.

Chemnitz consulted on the theological and liturgical aspects of the *Kirchenordnung*, while Andreae helped work out the administrative and organizational details, following the Lüneburg and Württemberg models of church governance. In 1570, Julius appointed another theologian from Saxony, Nikolaus Selnecker (1530–1592), as general-superintendent (the Lutheran equivalent of a bishop, the highest official in the territorial church hierarchy, second only to Duke Julius himself), political councillor on church affairs (*Kirchenrat*), and court chaplain (*Hofprediger*) in Wolfenbüttel. Finally, in 1571, Julius founded a seminary in Gandersheim in order to train a new Lutheran clergy; this institution would evolve a few years later into the University of Helmstedt.[66]

When Anna, Heinrich, and Philipp arrived in Wolfenbüttel in 1571, therefore, Julius had just completed a set of major reforms that fundamentally altered the institutions and practice of Christianity in Braunschweig-Lüneburg. The duchy was now formally Lutheran, which finally brought ducal policy in line with long-standing popular support for the Reformation in this region. Julius had also established a new system of church governance to regulate Lutheranism in the territory. Guided by his advisers at court, as well as the more formal Synod and General Consistory, Julius stood at the apex of the ecclesiastical power structure in his lands, responsible for administering his territorial church and articulating the Lutheran confession to which his subjects were now to adhere. The precise form of that confession, however, had yet to be determined when Anna arrived at court in 1571, and this was simultaneously the most important and most potentially contentious decision Julius had yet faced as duke. Because Lutheranism had no centralized authority in the Holy Roman Empire—something akin to a pope, for instance—each ruler was responsible for issuing a set of documents known as a "body of doctrine" or *corpus doctrinae*, which would serve as the normative statement of confession in his lands. Julius had models to choose from. He could adopt Luther's disciple Philipp Melanchthon's *Corpus doctrinae Philippicum*, for instance, as had Elector August of Saxony, or he could opt for the *Corpus doctrinae Thuringiam*, the Gnesio Lutheran–influenced body of doctrine that Johann Wilhelm adopted for ducal Saxony in 1570 upon ascension to the throne following his brother Johann Friedrich II's defeat at Gotha and imprisonment.[67] All of these choices had implications, however, as each involved allying Julius with one of the various parties, including the Philippists and Gnesio Lutherans, that continued to vie for influence over Lutheranism in the 1570s. Ultimately, Julius would find his way through the thicket of Lutheran factionalism by crafting his own body of doctrine, the *Corpus doctrinae Julium*, but he would not settle on this until 1574, after several years of debate at court. Articulating a body of doctrine for Wolfenbüttel was no straightforward task, and this difficult set of decisions was at the top of Julius and his advisers' agenda for nearly the entire time Anna Zieglerin and her companions were at his court. Just like the urban and territorial infrastructure of the duchy, in other words, Lutheranism, too, was very much under construction in Braunschweig-Lüneburg in these exhilarating but precarious early years of Julius's reign.[68]

*   *   *

As Anna, Heinrich, and Philipp settled into Wolfenbüttel in the summer of 1571, then, Duke Julius had declared a bold new vision of his duchy by launching projects that would fundamentally transform its economic and religious landscape. He envisioned a prosperous and Protestant land, although much work remained to be done to secure that future. Anna, Heinrich, and Philipp clearly saw this as a moment of opportunity for themselves as well, but they knew enough about court life to know that they should proceed carefully. Sensing that their future in the duchy hinged on Philipp's success with the duke, Anna and Heinrich initially stayed in an inn just outside Wolfenbüttel in the prosperous city of Braunschweig. Philipp, meanwhile, completed his negotiations with Julius. He presented letters of credit from Jodokus Pellitius and from his erstwhile patron Duke Johann Friedrich II.[69] He and Duke Julius then formalized their new alchemical partnership with a contract, just as Philipp had previously done in Gotha. In exchange for an advance of 2,000 thaler, Philipp promised to produce one *loth* (about fourteen and a half grams) of alchemical tincture within a year; if he failed, he promised to return the money.[70] Once again, Philipp Sömmering was employed as a court alchemist, working with Heinrich as his assistant. With an entire year to make progress on his tincture, the future at last seemed to offer security and stability.

CHAPTER 3

Courting Julius and Hedwig

In September 1571, Anna Zieglerin, still in Braunschweig, wrote to Philipp Sömmering, now settled at court in Wolfenbüttel. "Today my dear husband came to me in Braunschweig," she began, "and reported how extensively involved you are with the Most Gracious Highborn Prince and Lord, Duke Julius of Braunschweig."[1] Philipp's first year or so at Julius's court was indeed remarkably successful, and his influence with the duke soon expanded to include not only alchemical matters, but also, thanks to his training as a pastor, religious policy. Philipp's good fortunes benefited Anna and Heinrich as well. By December of 1571, the couple moved from Braunschweig to a house in the newly developed Heinrichsstadt to the east of Wolfenbüttel Palace, supported by a one-time stipend of 500 thaler from Duke Julius (presumably for Heinrich's work as Philipp's assistant).[2] Finally, four years after her flight from Gotha, Anna's situation seemed promising.

Anna's optimism was tempered, however, by her knowledge that the shadow of Gotha still had not disappeared. In her letter to Philipp, she expressed some hesitation about the reception she appeared to be getting in Wolfenbüttel. "You know that my poor husband and I, his poor wife, went to [Duke Julius's] court as loyal servants," Anna wrote, "and were repaid with such stunning heartache, insult, shame, and peril to our bodies and lives." Anna attributed her woes to the songs of one particular "mockingbird" (*spotfogel*): Duke Julius's vice-chancellor, a legal scholar named Lukas Thangel (also Daniel or Tangelius, d. 1590).[3] Indeed, Thangel knew Philipp and Heinrich, and perhaps Anna as well, from ducal Saxony, where Thangel had served Duke Johann Friedrich II as his envoy to France on religious matters. Thangel also appears to have tried his hand at alchemy during those years, but, as in Sömmering's case, the angels disapproved of his proposal to the duke.[4] By 1565, Thangel had transferred his loyalty and his service to Johann Friedrich II's brother, Johann Wilhelm, severing ties with the increasingly embattled Johann Friedrich II (unlike Philipp and Heinrich, who remained loyal to the end).[5] Eventually, Thangel found new employment with Duke Julius, and when he learned that Philipp, Heinrich, and Anna were making their way to Wolfenbüttel in July of 1571, he sounded a first warning. He then followed up in writing, expressing his reservations directly to Duke Julius in August, and then made his displeasure about the group known more broadly at court.[6] "Because we have been vilified at His Grace's court by Doktor Daniel without any justification and have been denounced as rogues and loose people," Anna complained to Philipp, "you can see easily how I truly will have to bear that cross when I come to Wolfenbüttel."[7]

This mockingbird's song was an unwelcome reminder that Heinrich, Philipp, and Anna still had to contend with the stigma of their affiliation with Duke Johann Friedrich II's court. Of course, Anna had even less to do with Grumbach, Hänschen, and Gotha than her husband, Heinrich, and her collaborator Philipp, but clearly she felt that their ignominious past would be her cross to bear as well. Fortunately, Duke Julius did not act on Thangel's warnings, perhaps dismissing his vice-chancellor's animosity as the result of old Saxon rivalries that had no bearing on the Wolfenbüttel court. Regardless, Thangel's early attempt to oust Anna, Heinrich, and Philipp must have only reinforced what they already knew: their presence at Duke Julius's court was precarious, and the duke's favor was crucial to their continued presence there. Each in their own way, they would have to work to secure and maintain Julius's patronage and protection if they wanted to stay in Wolfenbüttel.

Philipp and Heinrich's tasks were clear. As alchemist/religious adviser and assistant, their roles were fairly well defined, as evidenced by the stipends they received for their work from Duke Julius.[8] The expectations that accompanied Anna's move to Wolfenbüttel, however, were murkier. She arrived not on her own, but as the wife of Heinrich Schombach, assistant to the Wolfenbüttel court alchemist Philipp Sömmering. This status shaped Anna's opportunities in important ways, giving her access to Duke Julius and Duchess Hedwig. At the same time, it also tied her fate to her male companions, making it difficult to carve out her own role at court.

The figure of the courtier in early modern Europe, seeking out position, influence, and support, is well known. As early modern courts emerged as centers not only of political and military might, but also of patronage, social advancement, and employment in growing bureaucracies, all kinds of people looked for ways to make connections to the ruling families of Europe. Nobles moved through these courts via administrative service and marriage. Meanwhile, those of lower social rank but nevertheless with expertise to offer looked to princes to support their scientific, technological, or artistic projects, hoping for the financial remuneration, as well as the intellectual and social status, that princely patronage could confer. The Holy Roman Empire, fragmented into over two hundred political and ecclesiastical territories, had more opportunities than other parts of Europe. Courtiers circulated throughout the political nodes of the empire, and those who knew each other from one court might well cross paths a decade later once again. This could be a problem when old rivalries resurfaced, thwarting a courtier's attempt to refashion himself in a new context. On the other hand, the interconnected world of early modern courts also created access and opportunities. Moreover, as individuals gained connections and influence, they could take on the role of "broker" or "expert mediator," helping make connections between patrons and those with expertise to offer. It is easy to understand Philipp Sömmering and Heinrich Schombach in this context. Out of an alchemical collaboration born at the ducal court in Gotha, they continued their partnership several years later in Wolfenbüttel, where they now worked on a different duke's alchemical project and in different roles. Whereas Heinrich served as a broker for Philipp in the 1560s, persuading Duke Johann Friedrich II to extend support for Philipp's alchemical projects, Philipp now returned the favor for Heinrich, among others, in Wolfenbüttel.[9]

Anna Zieglerin, too, sought a place at the Wolfenbüttel court, but she could not act as a courtier in quite the same ways as Philipp and Heinrich. Most obviously, gender limited her opportunities, for there were fewer positions open to women at early modern courts. Consorts, of course, had a great deal of power. Like their male partners, these ruling women played absolutely central roles at court as patrons, political brokers, and custodians of the physical and moral health of their lands. Not everyone could be a duchess or margravess, however. Otherwise, noblewomen served as *Hofdamen* or *Kammerfräulein* (ladies-in-waiting), positions they held either as a stage in life on the way to marriage or by virtue of their husband's position at court, while commoners worked alongside husbands employed as surgeons, for example, or perhaps as physicians or apothecaries. The domestic positions, especially, were in some ways parallel (if not always equal) to the more recognizable positions usually occupied by men at court, and the women who held them often wielded substantial power and influence. Women might also be more loosely affiliated with a court through their families or husbands, with their standing largely determined by these men, rather than by the women's own expertise or knowledge.[10]

Although Anna Zieglerin's connection to Philipp Sömmering certainly gave her an entrée at court, her husband Heinrich's rather lowly position as a laboratory assistant, or *Laborant,* did not promise much influence. Moreover, Anna clearly aspired to being more than Heinrich's wife. Just like Heinrich and Philipp, she, too, sought patronage, power, and influence in Wolfenbüttel. Although the figure of the courtier is typically imagined to be male, Anna Zieglerin's actions at Hedwig and Julius's court suggest that it is productive to view her, too, as a courtier, just like her male companions. Eventually, Zieglerin would make a bold bid for Duke Julius's support for her alchemical work, and this in itself reminds us that some women did seek patronage, and therefore should be counted among the innumerable projectors, authors, artists, and others pursuing support for their expertise in the courts of early modern Europe. In Anna's first years in Wolfenbüttel, however, her attempts to secure a place at court were more conventional and modest. Her early efforts to find a foothold, particularly in contrast to Philipp Sömmering's, throw into relief the way that gender shaped the strategies that all courtiers could deploy as they aspired to power at early modern courts. While Anna's male companions could establish their place at court by demonstrating their expertise and patronage networks in any number of ways, the terrain on which she could earn credibility and influence was far more limited. As we shall see, Anna's moral and sexual integrity, policed through powerful information networks and reputation, were primary, and extremely difficult for her to control or overcome.

Anna Zieglerin's First Steps

Philipp established himself in Wolfenbüttel through the typical strategies of the courtier, working first through brokers such as the salinist Johannes Rhenanus, providing letters of credit from Hamburg physician Jodokus Pellitius and Duke Johann Friedrich II of Saxony, and eventually signing a

FIGURE 6. Duke Julius und Duchess Hedwig von Braunschweig-Lüneburg. Elias Holwein, *Warhafftige Contrafactur Weyland [...] Herrn Julii [...] Frawen Hedewig [...]; Von Gottes Gnad [...] Frewd und Wohn. B.S.D.* (Wolfenbüttel, 1620). Woodcut. Herzog Anton Ulrich-Museum, EHolwein WB 3.4.

formal contract with Julius to establish the terms of his employ as an alchemist. Ordinarily, a woman like Anna, purportedly of noble birth, would have sought an entrée with the most prominent woman at court, Duchess Hedwig, by activating social and political networks that could bolster her credibility.[11] Zieglerin, however, does not appear to have done this. Perhaps her claims to noble social status were exaggerated or outright false, or perhaps whatever social capital she had grown up with in Saxony had been destroyed in Gotha, or in her difficult youth in Dresden. Whatever the reason, Anna clearly did not seek to marshal the resources that might have eased her path to Duchess Hedwig's favor, and Anna never did manage to win her over.

Not much is known about Duchess Hedwig, unfortunately, but the little that is known suggests that she and Anna may have had some shared interests. Hedwig and Julius had been married for eleven years when Anna, Heinrich, and Philipp arrived in the duchy, and the fruitful marriage had already produced four daughters and two sons; another five children would follow, and of Hedwig and Julius's eleven children, seven girls and four boys, nine would survive until adulthood (Figure 6). When Hedwig died in 1602, contemporaries described her as an exemplary Lutheran duchess and *Hausmutter*, emphasizing her charity, piety, and role in upholding the Lutheran morality of the court. As one funeral book put it, "Thus her royal Highness (even in her final weakness), like all her daughters and her ladies of the chamber, never sat idle but always had work in her hands, in particular, like Tabitha, sewing garments for the poor." Hedwig's charity and piety had other aspects as well, which might have made Anna Zieglerin especially interesting to her. "Thus, when the time was right," the funeral book continued, "her royal Highness and her ladies prepared apothecary's wares to dispense to the poor and for this purpose gathered violets and roses, carnations and redcurrants, and peeled quinces."[12] As this commemoration suggests, Hedwig was typical of many German noblewomen in that her charity and piety also included medical expertise.[13] She supervised the court apothecaries and would eventually spearhead the selection of herbs and flowers for a new garden in the 1580s, which came to be known as the "Garden before the Mill Gate."[14] In other circumstances, Hedwig and Anna may well have enjoyed each other's company, exchanging knowledge about the distillation and concoction of herbal and alchemical medicines and experimenting with different materials and techniques to produce them.

Hedwig and Anna never had these kinds of conversations, however. To the contrary, Hedwig was deeply suspicious of Anna, Heinrich, and Philipp from the outset, and almost immediately her suspicion would turn into a fierce campaign to expel the alchemists from court altogether. What exactly first sparked Hedwig's aversion to the alchemists is unclear. The duchess may have known about the "mockingbird" Thangel's warning, or she may have come to her own conclusions, knowing that Anna and her companions once had been affiliated with Duke Johann Friedrich II's court. The arrival of alchemists at Hedwig's court in the fall of 1571, however, may also have resonated with something much more personal. Earlier that year, in January of 1571, Hedwig's own father, Elector Joachim II of Brandenburg (1505–1571), had died unexpectedly. Blame for the elector's death quickly fell on the Jewish financier and official Lippold, or Lipman ben (Judel) Chluchim (before 1530–1573). Lippold had come to Berlin around 1550 to serve as the elector's treasurer (*Schatullenverwalter*) and the administrator (*Oberältester*) of the Jewish community in Brandenburg. He was responsible for ensuring that precious metals and gemstones did not leave Brandenburg and that the region's Jews delivered silver and other tributes in exchange for being allowed to remain in the territory. In 1556, Lippold became the master of the mint as well. Most likely because of his combined financial and metallurgical expertise, Lippold was also involved in managing the finances of Elector Joachim II's alchemical projects, and, although it is impossible to say from the surviving sources, could potentially have engaged in those alchemical projects as a practitioner as well. At the time of Hedwig's father's death, Lippold was one of his most trusted confidants.[15]

Historians have since acknowledged that the events that followed Elector Joachim II's sudden passing were nothing but blatant scapegoating, most likely in order to eliminate the enormous debts that the profligate elector left upon his death.[16] At the time, however, Joachim II's family, including Hedwig and her brother Johann Georg, who inherited Joachim's electoral title, were convinced that Lippold was involved not only in financial improprieties, but also in their father's death. The suspicion was that Lippold had poisoned the elector, since he had given Joachim II a glass of wine before bed, mere hours before he died. The legal process to prove this proceeded slowly over the next two years, unfolding in Berlin just as Anna, Heinrich, and Philipp were settling in to a life at court in Wolfenbüttel.[17] Lippold's final confession, and thus the full accounting for Elector Joachim II's death, would not come until 1573, almost two years after Anna arrived in Wolfenbüttel, but even in 1571 Lippold had fallen under suspicion, and this suggested to Hedwig a troubling pattern. Was Lippold another Grumbach? Did Hedwig's own father succumb to a treacherous adviser, just as Duke Johann Friedrich II had only a handful of years before? More troublingly, might this happen again in Wolfenbüttel? The accusations of poison, sorcery, and alchemy that swirled around both Grumbach and Lippold may well have added additional weight to Hedwig's own suspicions about the alchemists' arrival, underscoring the stakes for her family and her reign.

Anna perceived Hedwig's hostility from the outset, as she explained in a letter addressed directly to Duke Julius in December of 1571, just as she moved from Braunschweig to Wolfenbüttel: "Gracious Prince and Lord, Your Beloved Consort has hurled her blazing fury at us, indeed against God and all merit." Hedwig was not merely unfriendly; apparently, she actively tried to block the alchemists' arrival at court, turning first to Duke Julius, and then to other family members, mostly likely through personal conversations that are lost to us today. "And it is troubling," Anna continued, "that, since Her Grace was unable to move Your Grace to drop us from Your favor, she has now dispatched other people of rank to Your Grace"—namely, Hedwig's brother Elector Johann Georg of Brandenburg. "In all of my days," Anna concluded, "I have done no harm to the pious princess. Why, then, does she hurl her blazing fury at us?"[18] This was a question Anna would ask repeatedly over the next few years, as Hedwig's campaign only intensified.

Anna would not find a place in Wolfenbüttel as part of Hedwig's retinue, this much was clear. Instead, after her arrival, Anna found another way to demonstrate her value to the court, rather unusually circumventing the duchess in favor of the duke. Evidently Anna, too, like Tabitha and Hedwig, could sew, and when Julius asked her to make a pair of shirts and some other small things for him, she complied. She also procured clothing and "Spanish boots" for the two young princes at court, as well as shirts and, with Philipp's help, a hat for Julius in Goslar.[19] In these small but intimate ways, Anna made herself useful and carved out a place for herself at court apart from her marriage to Heinrich or friendship with Philipp. More importantly, she also established a connection directly to Duke Julius.

Philipp Sömmering, Pastor-Alchemist

By comparison, Philipp Sömmering launched himself at court quite successfully from the outset, drawing on a broad range of experience and expertise to meet a very different set of expectations. While Philipp arrived under a contractual obligation to perform alchemical work in Wolfenbüttel, Julius quickly found his alchemist to be a versatile and useful adviser in other areas as well. Just as the duke depended on Braunschweig City-Superintendent Martin Chemnitz for advice in church matters, and on the Flemish artist Hans Vredeman de Vries to direct various architecture and garden projects in the duchy,[20] Julius soon began to rely on the pastor-alchemist Sömmering on alchemical matters more generally as well as on ongoing efforts to establish the Reformation in Braunschweig-Lüneburg. In these intertwined roles, Philipp not only carried out his own alchemical projects, but also drew on his own networks to serve as a broker or "expert mediator"[21] to help Julius staff both his alchemical laboratories and his church administration by bringing in alchemists and theologians. Philipp also advised the duke's own alchemical work directly, and helped guide Julius toward consequential decisions about establishing Protestant doctrine in the duchy.

First and foremost, however, Philipp was in Wolfenbüttel as an alchemist, and so as soon as he arrived, he set to work, assisted by Heinrich Schombach, in his lodgings and an ad hoc laboratory in the Old Apothecary (Alte Apotheke), a building adjacent to the palace to the east. Unlike some of his fellow princes who custom-built laboratories of their own,[22] Julius improvised, possibly in part by repurposing some of the old distillation equipment and furnaces once used to make medicines, perhaps even by Duchess Hedwig, to make alchemical tinctures instead (or as well). Julius would continue his interest in entrepreneurial forms of alchemy (primarily alchemical techniques for improving the numerous mines in the Duchy of Braunschweig-Lüneburg) even after Philipp Sömmering was gone, but the duke's projects with Philipp seem to have been his first substantial foray into the art. To some extent, then, Philipp and Julius launched a ducal alchemical enterprise together, and Julius certainly depended on Philipp's alchemical experience and knowledge to organize it.[23]

Because alchemical processes often took months, if not years, and required careful attention to furnaces, instruments, and materials along the way to the final tincture or philosophers' stone, Philipp needed laboratory

assistants, or *Laboranten*. Thus, one of his first tasks was to staff the laboratory.[24] He drew on his own networks to identify trustworthy and skilled
assistants, functioning as an alchemical adviser to Julius, just as he had in
Duke Johann Friedrich II's laboratory a number of years before.[25] In Wolfenbüttel, Philipp drew on his expertise to identify and screen alchemists and
assistants once again. Because there was no obvious credential that potential
alchemists or assistants could present as proof of their skill, Philipp's judgment was all the more crucial, and Julius relied on him to bring the skilled
and knowledgeable practitioners to court while avoiding the incompetent or
fraudulent. Alchemical laboratories required all kinds of work. While it was
Philipp's job to devise and oversee the operations, he needed others to tend
fires, carry out operations, and generally keep the laboratory supplied and
functioning. Matz Rotermund, for example, did not offer much alchemical
experience, but he was trustworthy. Philipp first met Matz while they were
both in Schmalkalden several years before coming to Julius's court, and Philipp seems to have employed him off and on as a servant in the intervening
years. Matz could certainly assist with the most basic tasks in an alchemical
laboratory (delivering wood, for example), but his greatest asset appeared to
be discretion. When Philipp offered him employment in Wolfenbüttel, he
required Matz to take a special oath of secrecy—namely, "that he would serve
him [Philipp] and Illustrissimus [Julius] truly and carry out the things [i.e.,
do his work] and not let slip what he hears and says."[26]

   Some of Philipp's assistants did have prior alchemical or related experience, of course. Heinrich Schombach, for one, seems to have had some expo-
sure to courtly alchemical operations while he was employed at the court of
Duke Johann Friedrich II in Saxony. One of Philipp's other *Laboranten*,
Sylvester Schulvermann, also had some knowledge of metals, although he
had certainly pursued other lines of work in his life as well. Originally from
the northern German trading city of Lübeck, and the son of a goldsmith and
Wardein (an official inspector of ore and/or coins), Sylvester went north to
Sweden around 1563 to fight as a mercenary soldier for a period, before leav-
ing his family and deciding to seek his fortunes as a processor of gold ore
(*Goldscheider*) in the Holy Roman Empire. Around 1567 he arrived in the
Harz mining city of Goslar, presumably to begin this new career, but when
he stabbed a local *Bürger* he fled north again to Holstein, where he took
up his old life as a soldier. Schulvermann joined another mercenary band
(*Freibeuter*) and fought by turns for both the Danes and the Swedes in their

war against each other. In 1569, he seems to have found himself in prison in Prussia after he and two other men robbed a Lübeck merchant named Hans Kapell.[27] Sylvester managed to escape, however, with a (possibly forged) letter of employment from the Swedish crown. He spent a short period of time at the Cistercian Oliwa Abbey near Danzig—perhaps trying once again to turn over a new leaf, perhaps simply lying low—before hitting the road again to Königsberg, Talinn, Prussia, Saxony, Bohemia, and, eventually, Erfurt, where, in the winter of 1570, he met Philipp and followed him, Anna, and Heinrich to Eschwege. Sylvester's shady past made him a risky collaborator, to be sure, but Philipp must have thought that Sylvester's experience as a *Goldscheider* outweighed his past misdeeds, for the alchemist offered the former soldier work as a *Laborant* at Julius's court. For Sylvester, this proposal offered yet another opportunity to make his fortune, and he joined the group in Wolfenbüttel.[28]

Duke Julius appears to have had some role in staffing the alchemical laboratory as well, suggesting that he was not willing to turn everything over to Philipp. Discussing his assistants, Philipp said, "When the artisans arrive, Illustrissimus presents them to him [Philipp]," suggesting that the duke personally oversaw the selection of at least some of Philipp's alchemical assistants and collaborators. This was certainly the case with the Strasbourg goldsmith and engraver Franz Brun (fl. 1559–1596), for example, who was also working at Julius's court in the 1570s.[29] Primarily known today for his production of copperplate engravings of rustic and hunting scenes, military figures, and his *Christ, the Twelve Apostles and St Paul* (1562–1563) cycle, Brun described himself in 1574 as a "goldsmith and painter" who "had [trained] with a goldsmith when he was young and learned painting in Italy." Brun also noted that he had been engaged in alchemical transmutation long before he met Philipp Sömmering at Julius's court. In Strasbourg, he claimed, he attempted to make gold out of silver and copper for someone named Benedict Gulden and successfully created a tincture that tinged silver into gold (this, he said, he learned from a man in Cologne). The line between working with metals as an artist and working with them as an alchemist was indeed a fine one, so Brun's forays into alchemical practice are, on the face, unsurprising.[30] According to Brun, his collaboration with Sömmering began when he, Duke Julius, and Philipp were in Julius's gardens one day. When the talk turned to alchemy, Brun said, Philipp's ears pricked up, and later, Philipp summoned the goldsmith to discuss the issue further.[31] It seems that Duke Julius also encouraged Brun and Philipp to work together on alchemical projects. The precise nature of this collaboration, however, is a bit unclear. Philipp later listed Brun as one of his *Laboranten*, suggesting that he was more of a laboratory technician than someone with ideas of his own to contribute. Yet Philipp also distanced himself from Brun's alchemical work, reporting that

> he learned nothing from Franz Brun; he [i.e., Brun] had a completely different process. He wanted to tinge in six weeks. He didn't want to apply himself to the work [i.e., Philipp's work], but rather to his own art. Illustrissimus praised the *Laborant* [Brun] well, and said that he makes tincture. . . . The *Laborant* had another method with the process and the spirit of wine. He wanted to use a vial or a philosophical vessel for it. He called the vials a philosophical vessel or an instrument. He lost an entire year with those glasses [i.e., trying to get the necessary glassware]. His glasses were supposedly the absolute best.[32]

This description suggests that Brun was working on his own alchemical project, also supported by Duke Julius, one in which Philipp was interested but ultimately not involved (and perhaps even saw as too inefficient). In selecting alchemists to work alongside Philipp, Duke Julius clearly took a more active role than had Duke Johann Friedrich II in Saxony, who seems to have been a much less knowledgeable and more absent patron of his alchemical laboratory. Philipp had to adjust a bit to Julius's style, which he clearly viewed, in part, as micromanaging. Philipp complained, for instance, that "Illustrissimus himself got in his way and loaded him down with other hands."[33] Perhaps Philipp resented the duke's interference because the alchemist saw the laboratory as his own small sphere of patronage and influence; perhaps he simply wanted to select his own assistants. Either way, it is clear that Julius took an active role in his alchemical laboratory from the start. It was clearly a lively space, staffed with a cross section of people with varying claims to alchemical skill, all of whom had been vetted either by Philipp or by Duke Julius himself.

Julius's involvement with alchemy seems to have deepened after Philipp's arrival at court. As Philipp explained to Julius in a letter written about six months after arriving in Wolfenbüttel, moreover, some observers worried about the duke's increasingly serious interest in the art. "I have come under suspicion here and elsewhere," Philipp lamented, "for supposedly encouraging Your Grace to pursue chymical works, trials, and arts." Like many other rulers who participated in a culture of "court experimentalism" in this period, Duke Julius evidently had begun actually to *practice* alchemy, rather than just reading about it (as Duke Johann Friedrich II, for instance, seems to have done).[34] Indeed, Philipp noted, he had "seen, heard, and experienced that Your Grace has tried quite a number of things yourself, and that you have allowed your valets to calcine, roast, and conduct trials on your behalf." Duke Julius, in other words, was apparently conducting his own alchemical experiments with his servants, in addition to the alchemical projects that Philipp was overseeing in his laboratory. Philipp wanted to make it clear, however, that the duke had done this on his own volition. For the record, he stated, "I neither told, encouraged, nor swayed Your Grace to [do] such things." In fact, he reminded Julius, he had specifically warned the duke *not* to undertake trials with particular minerals and materials. Here Philipp revealed the true source of his fear—namely, that Julius would injure himself doing alchemy and that Philipp would be held responsible. Indicating his awareness that Julius was roasting "arsenic" (possibly ore bearing arsenic sulfide) in his chambers through two different methods, reverberation and calcination, Philipp strongly urged the duke to stop:

> Your Grace will not believe what horrific poisons lurk both in the transmitter of the mineral spirit [i.e., in the roasting ore] and in the raw minerals, therefore, God forbid Your Grace should attempt to practice over my heartfelt warning. . . . Your Grace should avoid [working with arsenic] not only on account of Your life and limb, on which the entire Christian church [in Braunschweig-Lüneburg] rests, but also think of Your young little lords and ladies [i.e., Julius's children]. . . . And in addition to all of this, Your Grace should also spare me, upon whom suspicion will fall, if Your Grace should be injured (may God protect you) through attempting such poisonous things.[35]

Here we see Philipp in action, both as alchemical expert, warning Julius not to undertake a potentially dangerous operation, and as savvy courtier, hedging against the possibility that he would be blamed if the duke's alchemical experiments jeopardized his health.

Philipp made himself equally useful to Julius on religious matters, making a virtue of his extensive experience in the fierce polemics in ducal Saxony in the late 1550s and 1560s that had pitted the more "Philippist" Lutherans (followers of Luther's collaborator Philipp Melanchthon) against the dissenting "Gnesio" or "Flaccian" Lutherans (followers of Matthias Flaccius Illyricus).[36] Eventually, the matter in ducal Saxony was settled when Johann Friedrich II's brother, Johann Wilhelm, took power in 1567 and issued a new statement of doctrine, the 1570 *Corpus doctrinae Thuringiam*,[37] firmly throwing in his lot with the Gnesio or Flaccian wing of the Lutheran movement. These controversies are now mostly of interest only to the most dedicated historians of early Lutheranism. In the 1560s, however, they were monumental, confronting ordinary pastors like Philipp Sömmering with extremely difficult choices, splintering the community of faith in ducal Saxony, and creating long-lasting enmities. Moreover, in an era when princes served as the heads of their territorial churches, doctrinal issues were always political as well. The struggles between these factions in ducal Saxony led to dramatic arrests, imprisonment, high-stakes debates, excommunication, and exile for the pastors and theologians who vied for Duke Johann Friedrich II's support for their position.

As a pastor in the 1550s and 1560s, Philipp Sömmering had been right in the middle of these controversies. In 1562, he faced particularly intense political pressure when Duke Johann Friedrich II ordered all pastors in his lands to sign a controversial statement of doctrine, known as the Striegel'sche or Viktorinische Declaration (after its author, Viktorin Striegel), hoping finally to put an end to doctrinal strife. Bowing to political pressure, Philipp signed the document, although then he later joined many of his fellow pastors and withdrew his signature. For Philipp, the storm around signing Striegel's 1562 Declaration became a personal turning point. The experience soured him on his career in the church, prompting him to consider leaving his work as a pastor altogether (interestingly, for what he perceived to be the safer option of alchemy). Philipp's brief support for Striegel's Declaration also pitted him against Gotha City-Superintendent Melchior Weidemann, whose lingering enmity over this issue may have led him to oust Sömmering from his position as the pastor of Schönau in 1567.[38]

When Philipp Sömmering arrived at Julius's court in 1571 nearly a decade later, he once again found himself in the middle of conversations about doctrine. At issue for Duke Julius during these years was forging a *corpus doctri-*

nae, or official body of doctrine, to serve as a normative statement of confession in the newly Protestant duchy. The central question was whether Julius should adopt one of the existing *corpora doctrinae*—for example, the more moderate articles of faith drafted by Philipp Melanchthon or the more conservative articles adopted by Duke Johann Wilhelm in ducal Saxony—or whether he should craft one of his own. In order to explore his options, in 1570 and 1572 Julius staged two disputations on the Eucharist, Christology, and free will, pitting his Philippist General-Superintendent Nikolaus Selnecker against more orthodox-leaning theologians. The result of these disputations was that Julius moved toward a form of Lutheranism that was firmly orthodox but nonetheless pragmatic, somewhere between the two poles of Philippism and the orthodox position Johann Wilhelm had taken in ducal Saxony. This middle position that Julius forged at home helped pave the way for his later leadership role in the successful efforts to find common ground among Lutherans and forge the unified statement of Lutheran faith, the 1580 *Book of Concord*.[39]

Philipp Sömmering's influence with Duke Julius reached its height in the midst of these debates in 1571 and 1572, and he no doubt drew on his own expertise to help guide Julius's decision process. Philipp's hard-won experience in ducal Saxony led him to advocate the pragmatic yet orthodox form of Lutheranism still most closely associated with Duke Johann Friedrich II's reign in ducal Saxony, and this meshed well with Julius's own inclinations.[40] More practically, Philipp also used his influence at court to help Julius fill new positions in local political and ecclesiastical institutions. When in 1572 Julius dismissed his vice-chancellor, Lukas Thangel (the "mockingbird" Anna had complained about), for example, Philipp brokered a position for Josias Markus (1527–1599), a professor of theology trained at the University of Jena in Saxony, as Thangel's replacement. Similarly, when Nikolaus Selnecker left Wolfenbüttel that same year, after apparently failing to convince Julius in the debates, Sömmering suggested that Julius replace Selnecker with Philipp's countryman and distant cousin, the orthodox Lutheran theologian Timotheus Kirchner (1533–1587). Philipp also persuaded his acquaintance Georg Kommer (or Kummerer), a Rostock professor and Mecklenburg ducal chancellor, to come to Wolfenbüttel in 1573 in order to serve on the Wolfenbüttel Consistory, which advised Julius on matters concerning his territorial church. Julius deployed Philipp's spiritual expertise, in other words, as deftly as he did his alchemical expertise, tapping into networks that extended beyond the Duchy of Braunschweig-Lüneburg to help attract knowledgeable advisers of all sorts.[41]

In a sense, then, Philipp Sömmering and Anna Zieglerin's early experiences as they tried to make their way at the Wolfenbüttel court appear to be quite different. Consulting with the duke on alchemical and church affairs, Philipp quickly moved into a visible, powerful, and influential position at court; in contrast, Anna seems to have been marginal, doing little more than some meager sewing and shopping for Duke Julius. Duchess Hedwig, however, saw things quite differently. Whether because of Thangel's warning, her suspicions that sorcery and alchemy may have led to her father's death, or perhaps something else lost to us, Hedwig did not simply ignore Anna. The duchess may also have viewed Zieglerin's sewing as a sign of her growing presence in the domestic side of the ducal household, and she knew better than to trivialize or dismiss this part of the court. To the contrary, Hedwig understood that Anna could exert just as much influence at court as Philipp did as alchemist or *Kirchenrat*, and she made her objections known.

Duchess Hedwig's Power

Duke Julius bore primary responsibility for bringing Philipp Sömmering, and with him Heinrich Schombach and Anna Zieglerin, to Wolfenbüttel. Duchess Hedwig had opinions of her own about Julius's decision, however, and evidently, she did not hold back on sharing them with her husband and the court. Having failed to convince Julius to ban the alchemists from court in the first place, Hedwig trained a watchful eye on them once they arrived. Anna was aware of this, and did her best to counter the duchess's campaign. About a year after coming to Wolfenbüttel, in September of 1572, Anna wrote to Duke Julius again to refute Hedwig's accusations, most of which were deadly serious. "God pity me for saying the truth," she began, "but in the past five years my troubles have not gone away."[42] First, Anna reminded the duke of the sewing and shopping she had done for him,[43] gestures that seem to have further provoked Hedwig. "I heard a credible report," Anna wrote, "that the Duchess, Your Grace's Beloved Consort, has turned this against me most nastily and heinously, [claiming] that I have done this [i.e., made the shirts] through magic and sorcery [*zauberey und schwarzkunsten*]. Also, [I heard] that she supposedly told Your Grace with the most nasty crying and carrying on that I am thoroughly evil."[44] Needless to say, Anna proclaimed her innocence, swearing to Julius "before the dear true countenance of God and before the entire world" that Hedwig's accusations of sorcery were false. It is not clear whether Anna and Julius intended these to be magic shirts designed to bring good luck, as Anna would later confess,[45] or whether they were entirely ordinary shirts. Either way, Hedwig saw foul play. "It is a mystery to me," Anna lamented, "why Her Grace has hated, vilified and groundlessly disparaged me, a poor woman, to Your Grace for an entire year without once letting up."[46]

Magic was not the only charge Hedwig leveled against Anna, however. Anna also reported a second rumor that she attributed to Hedwig: adultery. As Anna's letter to Julius continued, "Her Grace reported to Your Grace that people in town supposedly saw Herr Philipp and me sleeping in a bed together as if we were married, fornicating with each other contrary to God and all that is proper, which is absolutely and utterly untrue."[47] Both Anna and Philipp were married to other people, of course, and adultery was a serious legal matter. It was also a sin.[48] Accordingly, Duchess Hedwig, who was responsible for ensuring that the court served as moral exemplar for the duchy, saw this as reason to exclude Anna from the Lutheran community at court. Hedwig even refused to look at her, Anna told Julius, and banned her from the palace chapel.[49]

Hedwig's animosity stemmed from her own and others' observations about Anna's behavior in Wolfenbüttel, both at home and at court. Hedwig also received additional support from the other women in her family, who continued to exchange information about the alchemists.[50] Fueled by exchanges of letters and visits, this powerful women's network traded information and policed reputation and honor at Julius's young court, and, by extension, in the Duchy of Braunschweig-Lüneburg as a whole. Hedwig used this network to track down bits of intelligence about Anna and the others' past and reputation, as well as information about how other princes in the Holy Roman Empire perceived Julius's support for Anna, Heinrich, and Philipp.[51] In performing this important role at the Wolfenbüttel court, Duchess Hedwig joined numerous other German noblewomen in the second half of the sixteenth century in taking on more prominent roles at their husband's courts both as corulers of the princely state and as moral custodians of the princely household. Their correspondence networks, in particular, served a number of functions, including everything from marriage negotiations to the exchange of medical knowledge, and must be seen alongside the male communications networks—among ambassadors, diplomats, and spies—with which historians are more familiar.[52] Hedwig's exchange of information about Anna, Heinrich, and Philipp, in other words, was one of the many ways in which she participated in governance of the duchy. And on the issue of Julius's new court alchemists, the duchess did not see eye to eye with her husband.

In November of 1572, not long after Anna complained to Julius about Hedwig's slander, two of Julius's older widowed sisters, Margarete, Duchess of Münsterberg (1516–1580), and Katharina, Margravess of Brandenburg-Küstrin (1518–1574), came to Wolfenbüttel to visit. After touring the new construction at the palace, they sat down their younger brother Julius and asked him about the advisers he had assembled at court. Katharina, in particular, wanted to know whether it was true that one "Herr Philipp" was now in his service. Julius, of course, confirmed Katharina's information but assured her that Philipp bore letters of reference from his previous patron, Duke Johann Friedrich II of Saxony. Nevertheless, Katharina expressed concern about her brother's judgment and reputation. Echoing the phrase first attached to Philipp Sömmering in 1567 by Gotha City-Superintendent Melchior Weidemann, Katharina said that she had heard that Philipp was a "runaway pastor" who had left his post and his wife in order to take up with "that slut Zieglerin." Moreover, rumor had it that Philipp was gaining too much influence with the duke. He had so "ensnared and blinded" Julius, she claimed, that he persuaded the duke to distance himself from all of his old advisers, ensuring that, "eventually, no honest Junkers remained" at Julius's court.[53]

Katharina's intelligence, moreover, confirmed what Hedwig had been saying about Anna all along, although Katharina amplified Anna's infamy further. "It is known to all of the electors and princes in the Empire," Katharina claimed, "that she can do magic and all kinds of evil things, and that once she slept with a common soldier and became pregnant, and she murdered the child." With this last detail, Katharina transformed what Anna described as a rape into an illicit affair resulting in pregnancy and infanticide, presumably to cover it all up. What is more, "her husband, Cross-Eyed Heinz, was a fool, who proved himself to be untrue to his lord, Duke Johann Friedrich." Katharina, in short, chastised her younger brother for being "always poorly informed about these things," and she urged him to look into the matter again.[54] Most of all, however, she warned him that "he should expel these loose people from his court, lest they bring harm to all electors and princes"[55]

Julius's sisters left Wolfenbüttel after just a few days, but Katharina continued to wage her campaign from afar. It would be hard to overestimate Katharina's distress at learning that the trio continued to reside at the Wolfenbüttel court, for she feared that they could destroy Julius and Hedwig. Katharina might have felt doubly responsible for the Wolfenbüttel court, in fact, for not only was she Julius's older sister, but, thanks to the intricate patterns of intermarriage among German nobility, she was also Hedwig's aunt by marriage.[56] In a strikingly intimate letter to Hedwig in April 1573, Katharina remained adamant that Julius and Hedwig expel "those godless people" from their court, because they might do "damage, dishonor, or harm to Your young reign."[57] Katharina wrote again in June 1573, lamenting the news that "those unholy people, Cross-Eyed Heinz and Herr Philipp . . . with that loose hag [lose Vettel] Cross-Eyed Heinz's wife [i.e., Anna] . . . are around Your young regime almost daily and that they eat and drink with You." It is unclear whether the courtiers were as close to Duke Julius as his sister claimed, but it would have been a sign of their prominence at court if they were indeed dining regularly with the duke. Katharina was stunned, she said, that Julius and Hedwig surrounded themselves and their children with such evil people: "I want to say in all honesty, that I am extremely saddened and distressed, [and] I almost don't know what to do about my heartache. God in Heaven take pity that such godforsaken and dishonorable people should be around princely personages, their dear children, possessions, and all that God and nature have granted You, and [that they should] have Your confidence and are in Your service. It is well known throughout the land how they behaved at their previous Lord's [court]."[58] Katharina then repeated once again the charges that Philipp left his office as pastor and his poor wife and child in distress in order to take up with Anna, and she stressed the damage that Julius's support for them was doing to his reputation in the Holy Roman Empire. "Princes and lords and other good people" all knew that Julius was hosting such horrible people in Wolfenbüttel, and "that they now go about in velvet and silk. . . . Ignominious gossip will soon follow."[59]

Reputation, however precious, was not all that was at stake. "May Dear God's Grace grant that the vile Devil not manage to use these people as a tool for his work," Katharina implored Hedwig.

And indeed if Your Dearest does not intend to preserve Yourself and Your princely goods and reputation, then take to heart and bear in mind Your Dearest's precious young regime, as well as body, life, land, and people. . . . God knows I mean this truly and from my heart, I will not cease to ask God to enlighten Your Dearest [husband] and my Dear Brother's heart and spirit and turn it away from these cursed, sinful, godless people. Wicked company can cause things to happen which otherwise would never happen; remember that a lord is judged by his servants and advisers.[60]

As long as Anna, Philipp, and Heinrich resided in Wolfenbüttel, Katharina insisted, Julius and Hedwig risked not only their own reputations and the stability of the duchy, but also the safety of their children and perhaps even their own souls and lives.

Without question, Hedwig and Katharina saw Anna, Heinrich, and Philipp as threatening and perfidious figures. The sisters worried that Philipp, the "runaway pastor" who evidently had abandoned his holy office, was seducing Julius away from his trusted advisers. Anna Zieglerin, however, seemed to have different and even more frightening powers of seduction at her disposal. Rather than political persuasion, the sisters feared that Anna used magic and sexuality to destroy men's powers of reason; she embodied, in short, what Ulinka Rublack has called the "desirous woman," able to lure men into adulterous affairs and destroy the patriarchal house that was the bedrock of early modern society.[61] According to Julius's sister Katharina, Anna had already destroyed Philipp's marriage, not to mention her own marriage to Heinrich, when Philipp "abandoned his wife and post to take up with that slut Zieglerin."[62] Might Anna be able to seduce Duke Julius and destroy Hedwig's household as well? Equally frighteningly, Anna's purported powers of sex and magic revealed the fragility of princely rule, which could easily be derailed by the temptations of illicit sex, magic, and even, possibly, alchemy. In a patriarchal society where order depended on the powers of reason, discipline, and control ascribed to men, this vulnerability was terrifying indeed. If Wilhelm von Grumbach, as they saw it, had been able to manipulate his patron Johann Friedrich II with magic alone, Katharina and Hedwig could only imagine the destruction Anna could wreak by using her body as well.

Hedwig and Katharina had good reasons to be wary of Philipp and Heinrich, given their association with Duke Johann Friedrich II and the campaign against them in the wake of Gotha. In comparison, the sisters' claims about Anna Zieglerin lacked the same specificity, relying instead on more generic gendered assaults on her spiritual and sexual integrity. Whether or not any of Hedwig and Katharina's information was correct was in some ways beside the point. The women's status at court and relationship with Duke Julius lent their claims credibility, ensuring that Anna and the others would have to contend with them in some way. Moreover, the fact that Hedwig and Katharina felt compelled and justified not only to share their knowledge with Julius, but also to ask him to act on that information, demonstrates that they expected to be able to weigh in on decisions about who earned a place at court. These were powerful women indeed, and they made for formidable enemies.

"Unfounded Wrath"

As powerful as Hedwig and Katharina were, however, their condemnations were still not the final word on the alchemists at the Wolfenbüttel court. Duke Julius's views were equally, if not more, important. Sensing that Hedwig and Katharina's opposition was nearly insurmountable, Anna and Philipp repeatedly turned to the duke to defend themselves and plead for his support. There were significant differences in their strategies, however, as gender norms shaped how they could approach the duke and the grounds on which they could make their case. Philipp had a much more public role at court, which was both an advantage and a disadvantage; he drew a great deal more comment and scrutiny, but he also had multiple avenues by which to establish his value to the duke, as we have seen. Anna, on the other hand, attracted little attention beyond the duchess and the margravess. Nevertheless, Hedwig and Katharina's denunciations were relentless and focused, and they defined the terms on which Anna would have to appeal to Julius for support.

Like that of most women in this period, Anna Zieglerin's reputation, and thus ability to make her way at court, hinged on her ability to persuade the duke and duchess of her sexual and moral integrity, as the focus of Hedwig and Katharina's campaign against her made clear.[63] Anna, too, knew how important this was, and she astutely made her piety and purity a centerpiece of her strategy. Eventually Anna would start to shift the terms away from what she probably rightly saw as a losing battle, reframing her purity around her alchemical expertise instead. In 1571, however, Anna had not yet articulated her understanding of alchemy, and she took on the duchess on her own terms.

If Hedwig attempted to marginalize Anna at court through discourse and action, Anna did the same, turning directly to Julius to stop Hedwig's campaign. "I beg Your Grace for God's sake, as our dear sovereign and lord, dear lord father, graciously to suppress this disparagement once and for all," she implored the duke.[64] Anna also articulated a counter discourse that underscored her qualities as a pious and morally upright woman. First of all, Anna insisted, she herself never practiced magic. To the contrary, she stressed her belief that "magicians and sorcerers belong in the fire, both in the here and now and in all eternity," indicating her knowledge that sorcery blurred the categories of crime and sin. Anna also used the conflict with Hedwig as a way to emphasize her own Lutheran piety, particularly around the issue of communion. "I say now as I have before," she wrote to Julius, "that I have a terrific fear of God and a true . . . heart and mind."[65] She reminded the duke of her unsuccessful efforts to use communion to make peace with Hedwig. As David Sabean has noted, in the late sixteenth century Lutheran theologians and laypeople alike believed that "one could not go to communion with an agitated heart. In such a state one was unworthy [of God's grace] and liable to bring down judgment on oneself" and imperil one's soul. Sabean has drawn attention to the ways in which state officials often used communion as a way to coerce their subjects into "quiet behavior."[66] Anna, however, may have turned the tables by using communion to manipulate her patrons instead. She reminded Julius that during the previous Pentecost and at Julius's request, she had tried to reconcile with the duchess. Since Hedwig's own ability to take communion during Pentecost should have been contingent on resolving her conflict with Anna, Anna seems to have tried to use the occasion to compel Hedwig into a reconciliation. Sadly, Anna reported, her efforts came to nothing; in fact, Hedwig only renewed her opposition to Anna. Anna now placed herself in God's hands and in Julius's: "I hear that Her Grace enjoys reading the word of God. Perhaps eventually God's glorious spirit will illuminate Her Grace's heart, so that Her Grace will release me from her unfounded wrath."[67] In the meantime, however, Anna hoped in her great distress that Julius would continue to offer her his protection.

* * *

Anna Zieglerin's strategies as a courtier at the Wolfenbüttel court, like much of her life, were creative and in some ways unconventional. She did not (and perhaps could not) work through brokers and networks of noblewomen to secure the favor of the senior woman at court, Hedwig, or even, alternatively, to appeal to the duchess as an object of pity and charity. Hedwig and Katharina, meanwhile, did everything they could to block Anna, drawing on powerful information networks to inform a troubling image of Anna's spiritual and moral character. Their accusations centered on violations of female piety and honor, including magic, fornication before her first marriage and adultery thereafter, and infanticide. By contrast, Philipp's fortunes at court hinged on male information networks and notions of honor. He staked his reputation on his professional expertise in alchemy, mining, and church affairs, and drew on his own networks to attract alchemists, bureaucrats, and advisers into Julius's service. Hedwig and Katharina's charges against Anna were more difficult to refute, however, not least because they stemmed from two women with a great deal of credibility: Julius's wife and sister.

In return, Anna appealed directly to the duke, to whom she also had access, itself a striking fact, given that she was a woman of questionable social standing at court. She performed some small services, gaining trust and access, but she also turned to Julius for protection, not only in the wake of the Gotha Rebellion, but even from the campaign of Julius's own wife. In her letters to Julius, Anna put forward her own self-image, emphasizing her piety, firm rejection of magic, and desire to make peace with Hedwig. In Anna's first years in Wolfenbüttel, this was primarily a discursive battle, although Hedwig's move to exclude Anna from the court chapel suggests at least one way in which it was social as well. Moreover, charges of magic and adultery were far from trivial gossip, and the urgency of Hedwig's allegations would only be compounded by the rumors that continued to surface from Philipp and Heinrich's past in Gotha. Although little of this was related to Anna Zieglerin directly, her fate was inextricably bound to her male companions', making their histories hers as well.

Whether she realized it or not, in attempting to work her way into a place at court Anna Zieglerin trod on dangerous ground. Duchess Hedwig read Anna's halting attempts to play the courtier by offering her domestic skills to Duke Julius as sinister attempts to seduce and manipulate him. Perhaps the stereotypical male courtier's ability to dissimulate, flatter, and subtly influence his sovereign read differently coming from a woman, particularly a woman bold enough to try to counter a duchess's objections. As Anna Zieglerin would learn, she was on more steady ground in offering concrete expertise as an alchemist.

CHAPTER 4

The Lion's Blood

On April 1, 1573, in Wolfenbüttel Palace, Duke Julius's personal valet handed him a manuscript. A few weeks later, presumably having read the booklet and decided that its contents were worth preserving, Julius made a copy in his own hand.[1] The first lines of the small manuscript reveal why it caught Julius's attention. "In the name of God the father, the Son, and the Holy Spirit, Amen," it began, promising to reveal "that which I, Anna Maria Zieglerin, currently Heinrich Schombach's wife, observed with my own eyes of the noble, well-born Lord, Lord Carl, Count and Lord of Ottyng, and also carried out myself with my own hand, concerning the noble and precious art of alchemy."[2] What followed was a twenty-page collection of recipes detailing how to make and use a golden oil called the lion's blood, which had extraordinary medical, alchemical, and even agricultural powers (Figure 7).

By giving Duke Julius these recipes, Anna boldly took her fate at the ducal court into her own hands. She gave up the hopeless task of trying to appease Duchess Hedwig or merely hoping that Philipp's or Heinrich's success at court would translate into her own security. Instead, Anna approached Duke Julius directly and made her own claim to alchemical expertise. In fact, she had already laid the groundwork at court for this disclosure. In the December 1571 letter to Duke Julius in which she expressed her consternation about Hedwig, Anna appended an alchemical symbol below her signature. Although it seems out of context in this letter and she did not elaborate on its precise meaning, the symbol suggested, at a minimum, that she was familiar with the basic pictograms alchemists often used to represent metals, processes, and other common alchemical concepts.[3] Anna also had dropped some hints about the tantalizing figure of Count Carl of Oettingen, a Bavarian nobleman who also possessed a surprisingly illustrious alchemical and medical lineage. Although it

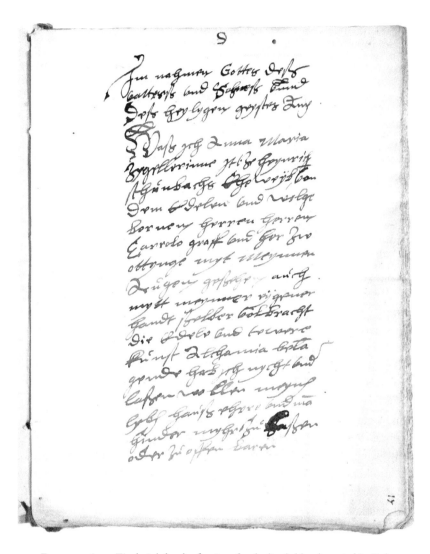

FIGURE 7. Anna Zieglerin's book of recipes for the lion's blood, copied in Duke Julius's own hand. Anna Maria Zieglerin, "die Edele und Tewere Kunst Alchamia belangende," 1. April, 1573, NLA WO, 1 Alt 9, Nr. 306, fol. 52r.

is difficult to reconstruct exactly what Anna told Duke Julius about this shadowy figure, clearly the duke's interest was piqued enough that in March 1573
he wrote Carl a letter inviting him to come to Wolfenbüttel and asked Anna
to send it to him. Unfortunately, the count did not immediately reply, forcing
Anna to explain to Julius that she was having difficulty locating him and providing new fodder for the mockingbirds at court who were skeptical about
Anna's claims.[4] In the meantime, Anna came up with something else: her book
of recipes, many of which, she said, she had learned from Count Carl. Perhaps
the recipes were meant to satisfy Julius while he awaited Carl's appearance at
court. If so, this strategy worked. By the fall of 1573, Anna had established a
laboratory of her own near Wolfenbüttel Palace, with at least one *Laborant*
to assist her operations, and she had successfully produced some preliminary
materials.[5] Duke Julius's subsequent correspondence with Anna no longer
focused on Hedwig's ire, but on Anna's alchemy.[6] In other words, Anna had
managed to transform herself from the wife of an alchemical assistant into
a courtly alchemist in her own right, even over Duchess Hedwig's fierce opposition.

It is easy to understand why Duke Julius responded so favorably to Anna
Zieglerin's gift. Far from vague alchemical promises, her recipes were detailed
and concrete, specifying precisely her materials, laboratory operations, and out-
comes. Moreover, Anna offered Julius not only her own alchemical knowledge
and expertise, but also access to others who were even more experienced. This,
of course, is exactly what Philipp Sömmering already had been doing for his
patron. While Philipp brought figures such as Timotheus Kirchner, Sylvester
Schulvermann, and Heinrich Schombach to the Wolfenbüttel court, however,
Anna now offered the duke access to the extraordinary figure of Count Carl.
Evidently Anna had observed Philipp's climb at Julius's court carefully, and she
would not merely emulate him, but even surpass him in Wolfenbüttel.

Anna Zieglerin's alchemy was more than just a clever ploy to shift the
balance of power in Wolfenbüttel, however, even if it did have that effect in
the short term. Anna's move from the margins to the center of the ducal court
brings her into focus more clearly, making visible her creative and original
engagement with the alchemical traditions in which she had long been
immersed. At the Dresden court of Anna and August of electoral Saxony, at
Duke Johann Friedrich II's ill-fated court in ducal Saxony, and even in her
own household, Anna was surrounded by alchemical texts, techniques, and
controversies. In this milieu, it should not surprise us that Anna Zieglerin took
an interest in alchemy and sought to learn more about it. It is unclear how

exactly she came to know as much as she did, but the recipes that Anna gave to Duke Julius in April of 1573 demonstrate that she took much more than a casual interest in the art. She engaged the rich and varied alchemical tradition knowledgeably and imaginatively, linking the alchemist's work in the laboratory to some of the most profound spiritual and corporeal questions of her world. It is worth slowing down a bit, therefore, to examine how Anna's recipes help us see her more clearly.

Alchemical ideas and practices in general appealed on many different levels in early modern Europe, offering practitioners intellectual insights into the workings of nature as well as entrepreneurial opportunities. The key to Anna Zieglerin's attraction to alchemy, however, lies in the name of the oil at the heart of her recipes: the lion's *blood*. Blood had multiple resonances in sixteenth-century Europe.[7] It was nourishing, essential to the generation of life, but also cleansing, and potentially poisonous. It had scholarly meanings in medical and alchemical literature, but also deep resonances in Christianity, as well as associations with magic in German folk culture. By calling her golden oil the lion's blood, Anna Zieglerin evoked all of these associations at once. Read carefully, Zieglerin's book of recipes not only offers some clues into her own claims to alchemical authority; it also makes visible a rich system of meaning that linked alchemy to the generation and redemption of matter, the knot at the heart of early modern Christianity.

The Powers of the Lion's Blood

What exactly was the lion's blood? Anna described the process for producing the lion's blood as follows. One began with alchemically prepared "tinged gold," itself produced by "projecting," as alchemists put it, a brown powder onto lead. Once the resulting gold had cooled, the alchemist filed it into small pieces, then heated it again for over nine hours in a small cucurbit with a distillation head (or alembic) affixed to the top. Eventually, a "flesh-colored mist" would appear in the alembic; this would continue to deepen its color until it was "brownish-red." This vapor, Zieglerin explained, was called the golden lion, and it would eventually "exude luminous, lovely blood." Zieglerin continued: "Collect this in the receiver vessel. One calls this extracting from the red lion his lovely, glorious, abundant blood. When the glass becomes white again, allow the fire to go out. The lion has emitted from itself as much blood as it could without injury and losing its red color."[8] In other words, with a slow, gentle heat, alchemical gold and lead would evaporate and then condense again into the precious oil that Anna called the lion's blood.

The lion's blood was, first and foremost, an alchemical substance meant to be productive. Not surprisingly, Anna dedicated a substantial portion of her booklet to the production of the *lapis philosophorum*, or philosophers' stone, which, in turn, could generate gold (among other things). She included two different recipes for making the *lapis*, the first of which involved repeating the process two more times, combining fresh lead with eighteen drops of "the lovely [lion's] blood" and heating it again until it produced an oil, then pouring it all into a mold until it cooled. Then, once again, the gold was cut into small pieces and heated again to "extract the blood again out of the lion."[9] To finish the stone using this method, Anna wrote, one must then take a small bird and put it in a little basket for six weeks, with nothing to eat but a few drops of the lion's blood each day in his water. After six weeks, the alchemist should roast this little bird until "it becomes like brown glass," then shatter it and pulverize it with a mortar and pestle. "Then you have the proper *lapis* for sure," Anna concluded, adding, "you can tinge greatly and copiously with this [powder]."[10] Although some alchemical authors distinguished between tinctures, which simply colored metals superficially, and the philosophers' stone, which transmuted base metals into silver or gold on a more fundamental level, Anna here used the terms *lapis* and "tinge" in the same passage, suggesting that she did not make a distinction between superficial "tingeing" and a more basic "transmutation."[11]

Anna's second recipe for the philosophers' stone was far more elaborate, but also more conventional. This recipe involved five steps. First, one created a small clear stone by taking mercury that had been cleansed with vinegar and salt, and "let the dew fall on the mercury on a beautiful night in May in a beautiful lovely meadow full of beautiful flowers."[12] Then one made a small brown stone using Hungarian gold that had been treated with antimony and quintessence. Equal portions of these two stones were then dissolved together in yet more quintessence and heated in a "philosophical fire" and decomposing dung for approximately six months. Gradually, Anna instructed, a "fruit" would begin to appear, lightening progressively from yellow to "flesh color," and eventually glowing, "shining through the glass like a carbuncle so that you think the glass is full of rubies."[13] At this point, the stone was nearly finished, save two more steps. The glass should be broken, exposing the stone to the air, which would soften it; then, it should sit in embers for another three days, whereupon "the stone is proper and complete."[14] Creating the philosophers' stone was still not the end of the "great work," however; two final steps remained. First, Anna explained, one should break the stone in two pieces, using one half to make more lion's blood and multiplying the other half, presumably for future use, by sealing it up in a flask with prepared mercury for several months. In order to actually use the stone to transmute metals, one had to soak it in the blood of a suckling lamb for nine weeks until it was "as soft as wax." Anna concluded with a curious final step: "Cast it into the shape of a lion, lindworm [a creature similar to a dragon but without wings], dragon's head or whatever similar thing pleases you."[15] These fanciful little molded capsules of the philosophers' stone, presumably, then finally could be "projected" onto base metals, transmuting them into alchemical gold, although Anna does not say this explicitly.

Although the philosophers' stone occupied center stage in Anna's booklet, the power of the lion's blood was by no means limited to metals. Anna also included recipes for using it to stimulate cherry trees, grapevines, and "whatever other fruit you want" to flower and produce fruit in winter rather than summer. If placed in a mixture of lion's blood and fresh water, then heated gently in manure, Anna explained, "the tree or branch will bloom and bear good, proper, natural, and, indeed, delicious tasting fruit." She added, "I have often tried this myself."[16] Similarly, she explained how to produce sapphires, diamonds, emeralds, and rubies by combining various plants and minerals with the lion's blood and a "quintessence." Finally, Anna extended the powers of the lion's blood to human life. Not only could it cure leprosy; it could promote fertility in women and nourish infants as well. "When you want to conceive a healthy child," she wrote, "then take nine drops of the above-mentioned oil [i.e., the lion's blood] three days in a row, evenings and mornings; give the same to your wife . . . ; while she is pregnant with the child, however, give her each and every day no more than three drops of the oil." When "with God's grace" the child is born, Anna continued, "then let the child taste no mother's milk, [and] give him nothing to drink nor to eat, other than three times a day, evenings, mornings, and at mid-day, each time no more than three drops in his mouth." Like Anna's other alchemical processes, this technique, too, promised to be even more effective than nature alone. "You can raise a child without any other food or drink for twelve entire years," she claimed; "also these children will be carried in the womb for less time than others because, by virtue of the great warmth [produced by the lion's blood], the fruits will properly form in the womb all the more quickly."[17]

Anna's collection of recipes, then, promised health and wealth, all derived through alchemical techniques and substances such as the lion's blood, antimony, quintessence, and the philosophers' stone. Where did she get these ideas, and to what extent were they her inventions? In fact, her recipes made complex claims about authorship. This is clear from the first line of the booklet, which introduced the figure of Anna's alchemical tutor, Count Carl, and positioned her simultaneously as Carl's student and as a practitioner in her own right. On one level, the booklet read as a statement of Anna's own experience, a record of past events that Anna had witnessed "with her own eyes" and carried out "with her own hand." To emphasize the historical specificity and veracity of particular processes, Anna sometimes used the past tense and added details of time and place. The first process she related, a simple process for tingeing lead into gold, illustrates this rhetorical strategy. "When the Count and Lord of Oettyngen came to the house of my mother Frau Clara Zyglerin, née von Schonborgk," she began, "about three days after his arrival he asked that he be given two pounds of lead." After heating the lead, Count Carl took out a small bag containing a brown powder (i.e., his alchemical tincture or philosophers' stone). Anna noted that she herself participated in this transmutation by projecting "two peas' worth" of the powder into the crucible, whereupon the lead began to coagulate, although not without a struggle. "It rose and fell and did not want to surrender," she recalled, until finally "the tincture conquered the lead, after which it swam to the top like oil; [it] was more beautiful than a small ruby." Once the substance cooled, she poured it into a mold, and "it was thirteen grades higher than the absolute best Arabian gold."[18] By reporting this transmutation as an event in which she was both witness and participant, Anna emphasized both her privileged access to Count Carl's secrets and her own experience with alchemical work.[19]

Anna did not present all of her recipes as historical record, however. She also drew on a more conventional recipe genre, either describing what "one" should do in general or addressing her reader directly, as if giving instructions. After discussing Count Carl's process for making tinged gold, for example, she went on to explain "how one can work with it further," including how to use the gold to make the lion's blood. Here Carl disappeared from the text; Anna was no longer describing what she and Carl had done at some point in the past; rather, she was passing on a recipe. "One should take the same tinged gold [and] file it very small so it is one hand thick and you can cut it up as small as you want; one puts this same filed gold into a small flask," and so on.[20] By describing the process as a set of steps that "one" should carry out, Anna detached the process from any particular operator, emphasizing instead the automatic efficacy of the recipe and implying that anyone could carry it out. The effect was the same when she switched to address her reader directly, using the direct pronoun "you" that was common in recipes.[21] In one instance she even explicitly noted that she had never tried the recipe she was about to relate, suggesting that she did not always see personal experience with a recipe as necessary to authenticate it.[22] This mixture of recipe and reportage offers some hints as to the various sources of Anna's knowledge. It may well be that the processes she claimed to have witnessed were in fact things she observed Count Carl or another alchemist do at some point, presumably either in Dresden or in Gotha. The processes Anna related in recipe format, on the other hand, she may have derived from books or even discussions with other alchemists, but had never seen or participated in herself.

Zieglerin divulged a great deal in her booklet, but she wisely did not give away all of her secrets. For example, she never explained how to make the quintessence that figured in several of her recipes. It was certainly some sort of distillate, but it could have been distilled from plant matter, wine, or even minerals. Moreover, some things apparently could not be conveyed in writing, but required demonstration in person. In order to complete her second recipe for the philosophers' stone, for instance, it was necessary to place the two small alchemically produced stones with some quintessence in a flask, seal it firmly "so that no harm comes to you," and put the flask in a philosophical fire. Zieglerin then added, "I will not write at the moment how it [i.e., the philosophical fire] must be prepared because I must show Your Grace in person."[23] Moreover, a careful reader of this booklet would have realized that even as Zieglerin disclosed the process for making the lion's blood, she managed to hold back one crucial piece of information. The first ingredient for lion's blood, it turned out, was the "tinged gold" that Count Carl made by projecting a brown powder onto lead, as described at the beginning of the manuscript. Zieglerin's booklet never revealed the contents of this powder, however, ensuring that Julius could not make the lion's blood (or, for that matter, carry out most of the tantalizing alchemical processes in Zieglerin's booklet) without her further instruction. By withholding details such as how to make the philosophical fire, quintessence, or tincture, Zieglerin asserted her own expertise at court, guaranteeing that Julius would still need her if he chose to carry out the alchemical processes she described. Zieglerin's booklet, therefore, simultaneously positioned her as a witness to successful past transmutations, transmitter of automatic recipes, and, occasionally, someone who could offer Julius personal instruction in the future.

This seemingly straightforward declaration of secrets, it turns out, made a complex set of claims about the authorship and transmission of knowledge and astutely positioned Anna Zieglerin as an alchemical expert at court.

Regardless of the precise sources of Anna's alchemy, both the content of the recipes and additional information she offered about Count Carl suggest where we might locate her more precisely in the rich and varied world of medieval and early modern alchemy. First, her emphasis on distillation and the quintessence positions her as heir to a tradition of distillation alchemy dating to the late Middle Ages. The notion that distillation could be used to extract nature's powers and refine substances into a tincture, elixir, or quintessence stemmed originally from the Persian author Jābir ibn Ḥayyān (known to Europeans as Geber, ca. 721–815) but was developed further by fourteenth-century Latin Franciscan authors such as Roger Bacon (1214–1292), Arnald of Villanova (ca. 1240–1311), Ramon Lull (ca. 1232–1316)—or rather the authors of numerous texts that appeared, pseudonymously, in their names after their deaths—and John of Rupescissa (d. ca. 1366). According to these authors, the technique of distillation could unlock the powers of nearly any substance. The quintessence of wine (alcohol) was particularly powerful, but distillations of herbs, animals, human blood, and, of course, the entire world of minerals and metals offered benefits as well, and authors all weighed in on both the materials one should begin with and the proper process for creating these elixirs and quintessences. Like Anna's lion's blood, the alchemical waters and oils extracted through distillation promised to be useful for all kinds of things. Used as a medicine, they were thought to perfect human bodies, but they could also be used to produce the philosophers' stone and thereby perfect metallic bodies as well. This tradition of distillation alchemy, particularly its medical applications, found its way into sixteenth-century Germany through such works as Hieronymus Brunschwig's *Kleines Distillierbuch* (1500), Philipp Ulsted's *Coelum philosophorum, Heimlichkeit der Naturen genannt* (1527), and Walther Hermann Ryff's *Das new groß Distillierbuch* (1545), as well as Latin and vernacular editions of late medieval alchemical texts.[24]

In their general outlines, Anna's recipes certainly call to mind this late medieval alchemical tradition. As had alchemists for centuries, for example, she associated colors with the various stages necessary to produce the lion's blood and the *lapis*, and her small clear and brown stones evoke the white and red stones, thought to produce silver and gold, respectively.[25] Anna also included the final steps of multiplying and softening the stone (often known as *multiplicatio*, or "multiplication," and *inceratio*, or "ceration"[26]) before it could be used in transmutation or projection. More precisely, Anna's inclusion of gemstones and plant fertilizer among the products of the lion's blood suggests that the pseudo-Lullian corpus might have been especially relevant for her, since the extension of alchemy to gemstones and plants was particularly characteristic of Lullian alchemy.[27] Since there is no sign that Anna could read or write Latin, most likely she was familiar with this late medieval tradition in a secondary way, which is to say, through conversations with other alchemists or through the growing corpus of vernacular German distillation and alchemical texts circulating by the mid-sixteenth century, rather than through the Latin originals.[28] In many ways, then, Anna's alchemy was fairly conventional, deeply steeped in the rich world of late medieval alchemy and its elixirs, quintessences, and stones. Yet she shrewdly recognized that this centuries-old tradition could be relevant in new ways to early modern princes such as Julius, who might well be interested in its applications to statecraft. Like many of his fellow princes, Julius was experimenting with a variety of new methods to exploit the natural resources in his territories. In this sense, Anna's proposal for the alchemical production of fruit or precious metals was perfectly consistent with these other activities, including the importation and cultivation of new kinds of fruit trees or the expansion of mining operations.[29]

It is notable that Anna never mentioned any of the most well-known medieval alchemical authors—Pseudo-Arnald of Villanova or Pseudo-Lull, for instance—in connection with her recipes, but rather associated the lion's blood with a much more modern figure: the controversial medical practitioner, author, and lay theologian Theophrastus Bombastus of Hohenheim, known as Paracelsus (1493/94–1541).[30] Anna linked herself to Paracelsus through Count Carl, whom she described not only as her alchemical tutor in the book of recipes, but also, in conversations with her fellow alchemists and with Duke Julius, as Paracelsus's son.[31] In emphasizing Count Carl's pedigree, she tapped into the revival of interest in Paracelsus in central Europe during the second half of the sixteenth century. Paracelsus himself had died decades earlier, in 1541, and, although he was a prolific writer on medical, philosophical, and theological subjects, only a few of his texts made it into print during his lifetime. Starting in the 1560s, however, both authentic and spurious Paracelsian texts began to circulate more widely in manuscript and in print, culminating in Johannes Huser's printed edition of Paracelsus's medical and philosophical works (1589–1591).[32] These texts, along with numerous commentaries and elaborations on Paracelsus's ideas, both reflected and sparked a flurry of interest in collecting, understanding, synthesizing, and elaborating on his notions of the origins of disease, composition of matter, and chemical medicine (or "iatrochemistry," the use of medicines made via chemical techniques or out of minerals).[33] Count Carl's remarkable lineage, therefore, gave him (and thus Anna) privileged access to Paracelsus's secrets precisely at the moment that central Europeans were keenly interested in recovering them. This surely worked to underscore Anna's status as alchemical expert at Julius's court.

Paracelsus is now appreciated for his innovative medical and philosophical ideas and practices, which introduced new theories of matter and the origins of disease as well as new remedies produced from minerals and via alchemical techniques. Above all, he is portrayed as a medical radical who broke with the reigning Galenic (humoral) medicine and Aristotelian matter theory. Modern scholars have also explored Paracelsus's complicated relationship to the alchemical tradition. On the one hand, he drew heavily on alchemical metaphors in his writings and understood the chemical separation or analysis of matter (*ars spagyrica*) to be a fundamental part of medicine. In this sense, Paracelsus continued the tradition of the elixir, with its emphasis on the extraction of quintessences from matter using distillation. On the other hand, Paracelsus rejected transmutation, departing from the medieval alchemical tradition, which tended to fuse alchemy's medical and transmutational aims. Paracelsus might seem to be a surprising choice as figurehead for Anna's alchemy, then, which emphasized the philosophers' stone and transmutation. And yet, as Anna's choice suggests, the image of Paracelsus that was circulating a few decades after his death only partially resembles the authentic Paracelsus that modern scholars have reconstructed. If modern scholars see him as a revolutionary figure who broke with his medieval heirs, Anna saw him instead as the culmination of the medieval alchemical tradition, including the transmutation of metals.[34]

Historical actors often perceive their worlds differently than historians do with hindsight. In this case, the tangled history of the publication of Paracelsus's works helps account for the discrepancy between Anna and her contemporaries' perception of Paracelsus and our own. Unlike modern scholars, Anna and her contemporaries did not have access to a large corpus of Paracelsian texts, but typically encountered him piece by piece instead, through scattered texts or commentaries. Perhaps more important, however, is the fact that many mid-sixteenth-century Germans derived their understanding of Paracelsus through texts that modern scholars now exclude from the Paracelsian corpus, especially a cluster of alchemical texts that circulated in Paracelsus's name during the 1560s and 1570s in both print and manuscript, including *Coelum philosophorum, sive Liber Vexationum* (ca. 1560)*, Das Thesaurus Thesaurorum Alchemistarum* (shortly after 1560), and *De tinctura physicorum* (ca. 1568).[35] Paracelsus's modern editor Karl Sudhoff identified these texts as spurious, but Anna Zieglerin and her contemporaries believed that they were authentic, and drew on them to construct a new mythology around the middle of the sixteenth century of Paracelsus as a "German Hermes Trismegistus," a celebrated magus with unparalleled insights into nature's secrets, above all the transmutation of metals.[36] This mythologizing was a process not only of replacing the primarily medical Paracelsus with an image of Paracelsus the "goldmaker," but also of recapturing and reintegrating into the Paracelsian medical corpus the tradition that the real Paracelsus had firmly rejected: transmutation. These spurious texts reconnected Paracelsus to the tradition of the alchemical elixir, in other words, allowing Anna Zieglerin and others to see him as someone with access to the secret not only of extraordinary medicines, but also of the philosophers' stone.

The pseudo-Paracelsian *De tinctura physicorum*, in particular, may have been one of the sources Anna drew on in conceptualizing her lion's blood. Composed by an anonymous author around 1568, *De tinctura physicorum* made quite a splash when it appeared in print under Paracelsus's name, appearing in at least six vernacular editions and one Latin edition between 1570–1572 alone, with additional editions to follow in the next few years.[37] Despite the Latin title, the text was originally in German, and an extant manuscript copy dated from 1568 (that is, from before it appeared in print) establishes that it circulated in manuscript even earlier. Anna Zieglerin may well have seen a copy of *De tinctura physicorum* just before she came to Wolfenbüttel, in other words, or at least heard about it secondhand. The short text describes a new (and superior) method of making a "tincture of the physicians," which, like Anna's lion's blood, had the power both to transmute metals and to restore health.[38] Moreover, one of two central ingredients for this tincture is something called the blood of the red lion. "Therefore I say that you [should] take nothing more than the rose-colored blood from the red lion and the white glue [*gluten*] from the eagle," writes the author. "After you have combined these two, then coagulate it according to the ancient process, and thus you will have the *tincturam physicorum*, which many have pursued, but only you few have found."[39] The author of *De tinctura physicorum* did not specify exactly what these substances were, of course, which left the text open to interpretation and speculation about the details. The point is not that *De tinctura physicorum* was the direct source for Anna's lion's blood (the recipes are not similar enough in content to make this claim), but rather that the idea of an alchemical substance called the "lion's blood," which could be used for both health and transmutation, was circulating in connection with Paracelsus in the mid-sixteenth century. Anna Zieglerin was neither the first nor the last to write about it.[40]

Anna's recipes, therefore, bore a complicated relationship to a broader alchemical tradition in sixteenth-century central Europe. By including quintessences, brown and clear stones, the lion's blood, gemstones, and fruit trees, Anna implicitly embedded her own recipes in the long tradition of the alchemical elixir. More scholarly alchemists might have used the opportunity of a recipe book to demonstrate their textual virtuosity and familiarity with the alchemical corpus, citing and commenting on competing theories about the philosophers' stone or methods for making it. Anna did not do this, in part because she probably had not read as methodically as some alchemical readers, but also because her primary goal seems to have been more immediate: to establish her credibility with the ever-pragmatic Duke Julius as an alchemical *practitioner*. In order to do this, she did need to mention all of the "keywords" one would expect to find in alchemical recipes, but, since neither Anna nor Duke Julius was primarily a scholar, she did not need to (nor, probably, could she) go on at length about her opinions about various authors. Moreover, by framing her recipes as "Paracelsian," she signaled that her alchemy was cutting edge and maybe even ahead of the curve, since through Count Carl she could access Paracelsian secrets that even the most avid collectors of Paracelsian manuscripts had not yet discovered. Even as Anna signaled that her recipes emerged out of a broader alchemical tradition, however, she still insisted on her own alchemical expertise. Not only did she herself possess experience with alchemical transmutation and the lion's blood, she claimed, but she (and only she) also could offer Duke Julius access to Count Carl and all of his secrets. She grounded her alchemy in bits of knowledge she had gleaned from other texts and practitioners, but, like any good alchemist, she digested that knowledge, added insights of her own, and shaped the whole into something that was distinctive. In this delicate balance between demonstrating familiarity with alchemical tradition and claiming authorship and innovation, in other words, Anna Zieglerin was a typical alchemist.

The red thread that ran through all of these recipes, however, was the lion's blood, and it is here that we can most clearly hear Anna Zieglerin's own alchemical voice. While substances referred to as blood—even the blood of the red lion—could be found in other alchemical texts, Anna's lion's blood seems to have been her own. More importantly, she made it the centerpiece of her recipes, and this was no accident. Blood was one of the most resonant concepts in late medieval and early modern culture, and this not only helped Anna conceptualize what the lion's blood could do and how it functioned, but also allowed her to maximize the meanings and thus extend the significance of her alchemical substance beyond her laboratory at Duke Julius's court.

Generation

It is nearly impossible to pin down all of the meanings of blood in early modern Europe. They proliferate so easily as to be almost uncontainable. First and foremost, however, blood symbolized life and fertility. Most obviously, when it was flowing properly through the body, blood was the fountain of life itself. According to the Bible, "The life of the flesh is in the blood" (Leviticus 17:11), and in classical medical theory, blood (along with black bile, yellow bile, and phlegm) was one of the four humors in the human body as well as part of the mixture of humors that actually coursed through one's veins.[41] It played a crucial role in the generation of new life in human reproduction as well. Uterine blood was widely thought to contribute the matter for the human fetus; thus a mother's blood literally turned into the fetus, granting it life and existence. Blood was also nourishing. Uterine blood fed the developing fetus while it was in its mother's womb and, after birth, turned into the breast milk that would continue to nourish the infant thereafter.[42] Alchemists incorporated the prolific power of blood into their own work, too. Several medieval authors, including Roger Bacon and Pseudo-Arnald of Villanova, for instance, proposed that the philosophers' stone or a healing elixir could be distilled from human blood, since blood contains both the four humors and the four qualities (hot, dry, cold, wet) of all matter. John of Rupescissa extended this notion, describing a quintessence of blood that could be generative, restoring the human body and preserving youth by actually recreating healthy flesh. Rupescissa, a Franciscan, was influenced in his thinking not only by medical and alchemical imagery, but also by the fecundity of blood in Christian iconography.[43] Christ's blood granted a different kind of (eternal) life, nourishing believers, as numerous iconographic depictions of Christ's wounds spilling and spurting blood attest, forming or birthing the very *ecclesia* itself out of his wounds, just as uterine blood formed the fetus.[44] Moreover, from the late Middle Ages into the early sixteenth century, Christ's blood took on a particular potency in the northeastern German lands, where the miraculous appearance of blood on communion wafers and other holy objects, as well as purported relics of Christ's blood, became the focus of numerous blood cults.[45] Holy blood was associated with fecundity here as well. Not only did host miracles grant prosperity and fertility of various sorts, but the language surrounding these blood cults was also suffused with generative imagery; they "throb with vegetal images of engendering, flowering, and sprouting."[46] Anna Zieglerin spent her life near or in the places where these cults had once thrived. In fact, since the thirteenth century the city of Braunschweig had possessed a blood relic that Henry the Lion (from whom Duke Julius was descended) had allegedly acquired on crusade to the Holy Land in the twelfth century.[47] Although Luther and other reformers challenged the status of relics, including blood relics, as objects of veneration, they retained surprising cultural power in this region even after the Reformation and into Anna's own day.[48]

In early modern medical, alchemical, and religious culture, then, blood was fecund, vivifying, abundant, fertile, prolific, fruitful, and generative. Not surprisingly, these associations fed into Anna Zieglerin's own alchemy as well. Her lion's blood could work across the natural world, generating animal, vegetable, and mineral life. Taken by both members of a barren couple, it could cure infertility, and if the woman continued to take it throughout her pregnancy, it could also nourish the fetus in utero. Fed to the infant once born, the lion's blood alone, without any additional food or drink, could nourish the child for an astounding twelve years. Dropped into water, the lion's blood could also work as a plant fertilizer, spurring a branch of a fruit tree to bloom and produce even in winter. And, of course, as the starting point for one of Anna's two recipes for the philosophers' stone (the one involving a bird), the lion's blood could ultimately proliferate gold, too.[49] Because Anna's small booklet was a collection of recipes, rather than a theoretical treatise on the generation of metals or production of the philosophers' stone, she did not articulate the logic behind the generative potential of the lion's blood. In order to understand how Anna might have understood the mechanism of the lion's blood, therefore, it is helpful to explore more fully some of the alchemical texts that were circulating widely in her day.

Zieglerin's terminology in her recipes is revealing, indicating fundamental connections among the generation of metals, minerals, plants, *and* humans. Like most using the vernacular German in the early modern period, she used the terms "child" or "fruit" to refer to what we would today call an embryo or fetus; but she also used the term "fruit" to refer to other products of alchemical generation, including her alchemical cherries, grapes, and even a preliminary stage of the philosophers' stone.[50] By describing all of these things as fruits, Zieglerin effaced the obvious and numerous differences among metals, plants, and humans, underscoring instead her view that they were all the result of similar generative processes. These processes unfolded on their own under normal conditions; but alchemy, Zieglerin claimed, could stimulate generation where nature resisted, as in the case of an infertile couple or cherry tree in winter. For Zieglerin and her readers, then, the metaphorical fruit served as shorthand for a much broader set of ideas about generation, offering her readers a key to unlock the universal power of her alchemy.

Zieglerin's use of the term "fruit" engaged a broad cluster of metaphors that went beyond just blood to link human reproduction, the generation of plants and metals, and the prolific powers of the philosophers' stone. Since antiquity, natural philosophers and physicians had drawn on vegetative metaphors to describe human generation. Just as seeds develop into trees when planted in the warmth and moisture of the earth, the analogy went, so, too, does the seed of the father (and, for some authors, the seed of the mother as well) coagulate the mother's uterine blood into a "fruit" or fetus, which then gestates in the warmth of the womb. Ancient and modern authors such as Aristotle, Galen, and Paracelsus articulated different views of this theory of conception, but the fundamental metaphorical analogies remained surprisingly constant, linking the womb to the earth in its power to nourish and generate life.[51] These vegetative and sexual metaphors were easily extended to the generation of metals as well. The Aristotelian notion that everything naturally strove toward perfection, for example, suggested that metals naturally "ripened" into silver or gold in the warmth of the earth. From antiquity through the early modern period, therefore, natural philosophers, physicians, and many others drew on what Kathleen Crowther-Heyck has called a "common reservoir of metaphors and images" that could be used to describe all natural generation in terms of earth, seeds, and fruits.[52]

This shared imagery also had its limits, however. As the ongoing fascination with alchemy in this period made clear, there was something uniquely challenging and intriguing about the world of minerals and metals. Unlike the generation of plants, animals, or humans, mineral generation was more difficult to observe, both because it normally happened deep underground and because of the long timeframes involved. It was even less obvious how one could manipulate the natural generation of metals or minerals to speed up the process, make it more productive, or produce a superior product. This expertise, of course, was precisely what alchemists had promised for centuries, generating both controversy and enthusiasm.[53] To help describe and conceptualize alchemical generation, then, they often turned to the vegetative and sexual metaphors with which their audiences would have been more familiar. Some drew on agriculture, arguing that one could isolate the "seeds" of gold or silver and combine them with *prima materia*, or fundamental, formless matter (usually an alchemical Mercury—not to be confused with common mercury), to create the philosophers' stone, which could, in turn, stimulate the seeds to multiply themselves, yielding a new crop of gold. The problem, of course, was how to acquire these seeds, a question to which alchemical authors offered numerous answers.[54] Although alchemy justly has a reputation for obscure language, recourse to this kind of analogy shows that alchemists sometimes sought to make their texts accessible, too.[55]

Because the alchemical tradition had long figured substances as male, female, and sometimes hermaphroditic, cultural ideologies about interactions between the sexes found their way into alchemical texts as well, further underscoring links between alchemical and human reproduction. Particular metals acquired gender in part because of their qualities. Following ancient Aristotelian and Hippocratic theory, for example, Sulfur represented the "hot" and "dry" principles in matter and was therefore male, whereas Mercury represented the cold and wet principles and was therefore female. Materials also derived their gender from associations with Greek mythology. Silver was typically female because of its association with the moon and the goddess Diana, while gold was male because of its connection with the god Apollo. Such imagery was ubiquitous in the alchemical corpus, both in word and, increasingly after the fourteenth century, in image.[56] Given this basic gendering of materials, metaphors of marriage, sexual intercourse, and procreation were an obvious way to describe the material interactions alchemists witnessed in the laboratory. So, for example, one medieval alchemical text that appeared in print in the sixteenth century explains, "For the conduct of this [alchemical] operation you must have pairing, production of offspring, birth and rearing. For union is followed by conception, which initiates pregnancy, whereupon birth follows. Now the performance of this composition is likened to the generation of man."[57] Passages such as this demonstrate how human sexuality offered a particularly rich set of images for depicting and explaining alchemical processes in the sixteenth century. Just as in human procreation, alchemical generation, too, required attraction between opposites (gold and silver), followed by consummation and fertilization (dissolution in a mercury bath), in order to produce offspring (the philosophers' stone).

One of the most striking and well-known examples of this kind of imagery in the German lands appeared in a text that Anna Zieglerin may have encountered at Duke Johann Friedrich II's court, since it was one of the two books on alchemy in his library: the *Rose Garden of the Philosophers*, or *Rosarium philosophorum*, which appeared as the second volume of a compendium of alchemical texts printed in 1550.[58] This was a complicated book because it incorporated two late medieval texts that had been circulating separately in manuscript for at least a century prior to their appearance in print together. The first, an illustrated poem (*Bildgedicht*), combined a sequence of images of Sol and Luna with German verses. The second component, a florilegium known as the *Rose Garden of the Philosophers* (*Rosarium philosophorum*) and most likely compiled in the fourteenth century by an anonymous author, wove together quotations in Latin from well-known ancient and medieval alchemical authorities into a virtual conversation about the philosophers' stone. The printer of the 1550 edition created a new enormously successful hybrid text by combining the German Sol-Luna illustrated poem with the Latin florilegium, publishing it under the title of the latter, *Rosarium philosophorum*.[59] Those who could read both the Latin and the German would have been rewarded with the compiler's interpretation and attempt to harmonize centuries of literature on alchemical transmutation.[60] Even for those who did not know Latin, however, the book was still partially accessible through the images and vernacular verse of the Sol and Luna series, and certainly some readers—including, perhaps, Anna Zieglerin—would have read only this part.

The twenty-one images in the series were also the most immediately compelling aspect of the book, because they depicted the "chemical wedding" with a series of visually arresting, sometimes explicit images that follow Sol (the king) and Luna (the queen) through a progression from marriage to coitus, conception, impregnation, and, eventually, proliferation as the white and red tinctures. In the midst of this, they die and are reborn multiple times. The chemical couple personified alchemical principles, demonstrating the way in which sexual difference could be used to represent any number of other dichotomous oppositions in nature: Sulfur/Mercury (the two principles that were thought to make up all metals), hot/cold, dry/moist, sun/moon, fixed/volatile, spirit/body, form/matter, active/passive. The series began, therefore, with the wedding between Sol (Sulfur) and Luna (Mercury), representing the union of these opposites, a crucial first step in the production of the philosophers' stone. In early modern Europe, of course, marriages were meant to be procreative relationships, and so it went in these alchemical marriages as well. As the accompanying Latin text of the *Rosarium* summarized, layering vegetative and sexual metaphors, "Sol is the man, Luna the woman, and Mercury the seed. However, in order for there to be generation and conception, the man must unite with the woman, and in addition, the seed is necessary, and so before fermentation there must be conception and impregnation, and when the material multiplies itself, one says that the infant is growing in the uterus of the mother."[61] After they marry, therefore, Sol and Luna are united by dissolution in a mineral bath before they consummate their marriage in a physical union (Figure 8) meant to bring together these two principles in perfect balance.

Despite the rather explicit imagery, human sexual reproduction is not the only metaphor at work in this series. The next woodcut is titled "Conception, or putrefaction" (Figure 9), reminding the reader that vegetative metaphors were an equally important resource for describing the production of the philosophers' stone. Death and putrefaction were central to this imagery. Just as seeds first had to decay in the earth before they could "regenerate" themselves as a new plant, alchemists surmised, so, too, must the alchemist's fundamental material (*prima materia*) first decay and "die" before it could be reborn as a more perfected substance. Alchemical authors frequently cited the Gospel of John as support for this notion: "Very truly, I tell you, unless a grain of wheat falls into the earth and dies, it remains just a single grain; but if it dies, it bears much fruit."[62] As the author of the possibly pseudonymous Paracelsian text *Nine Books on the Nature of Things* (1572), put it, "In general . . . one may say that in nature all things are born of the earth with the help of putrefaction."[63] This decay served two purposes for the author of *On the Nature of Things*. First, it transformed all matter into something else, turning poisonous and unhealthy things into precious medicines. Furthermore, the author explained, putrefaction was essential to generation: "Putrefaction, then, gives birth to great things, and we have a lovely example of this in the Holy Gospels, where Christ says: unless the kernel of wheat is thrown on the field to decay, it will not bear fruit a hundredfold. Thereby it is known that through putrefaction many things are multiplied so that they will bear noble fruit, because putrefaction is an inversion and the death of all things and a destruction of the first form and essence of all natural things, out of which the rebirth and new birth a thousand times better comes to us."[64]

After coitus, therefore, Sol and Luna proceed to the next step, "conception or putrefaction," in a tomb, signifying the death and decay that they must undergo in order to multiply themselves. The image is accompanied by the

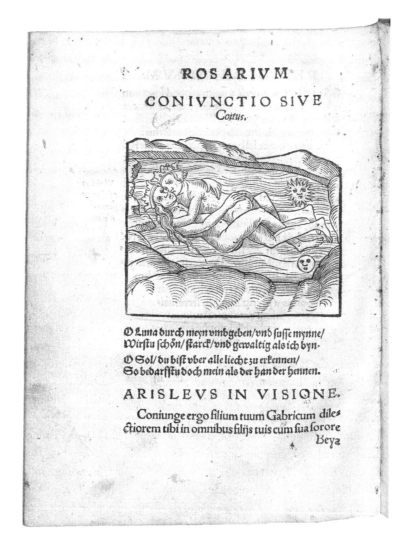

ROSARIVM

CONIVNCTIO SIVE
Coitus.

O Luna durch meyn vmbgeben/vnd ſuſſe mynne/
Wirſtu ſchön/ſtarck/vnd gewaltig als ich byn.

O Sol/ du biſt vber alle liecht zu erkennen/
So bedarffſtu doch mein als der han der hennen.

ARISLEVS IN VISIONE.

Coniunge ergo filium tuum Gabricum dile-
ctiorem tibi in omnibus filijs tuis cùm ſua ſorore
Beya

FIGURE 8. "Conjunction, or coitus." The German caption reads: "O Luna,
through my embrace and my sweetness, you will become beautiful, strong, and
as powerful as I am." "O Sol, you are to be recognized above all other lights,
thus you need me as the rooster needs the hen." *Rosarium philosophorum.
Secunda pars alchimiae de lapide philosophico vero modo praeparando, continens
exactam eius scientiae progressionem. Cum figuris* [. . .]. (Francoforti: ex officina
Cyriaci Iacobi, 1550), [Fiv]. ETH-Bibliothek Zürich, Rar 8230.

PHILOSOPHORVM.

CONCEPTIO SEV PVTRE
factio

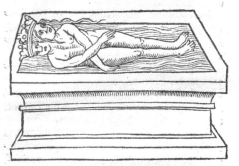

Hye ligen könig vnd königin dot/
Die sele scheydt sich mit grosser not.

ARISTOTELES REX ET
Philosophus.

Vnquam vidi aliquod animatum crescere
sine putrefactione, nisi autem fiat putri=
dum inuanum erit opus alchimicum.

FIGURE 9. "Conception, or putrefaction." The German caption reads: "Here
the king and queen lie dead, the soul is separated with great misery."
*Rosarium philosophorum. Secunda pars alchimiae de lapide philosophico vero
modo praeparando, continens exactam eius scientiae progressionem. Cum figuris
[...]* (Francoforti: ex officina Cyriaci Iacobi, 1550), [Gv]. ETH-Bibliothek
Zürich, Rar 8230.

Hie ist geboren die eddele Keyserin reich/
Die meister nennen sie jhrer dochter gleich.
Die vermeret sich/gebiert kinder ohn zal/
Sein vnd ötlich rein/vnnd ohn alles mahl.

Die

FIGURE 10. The white tincture. The caption, which is unusually lengthy and spills onto the following two pages, begins: "Here is born a noble and rich Empress, whom the masters liken to their daughter. She proliferates and bears innumerable children, who are immortal, pure and without blemish. The Queen hates death and poverty, she excels at making gold, silver, gemstones, [and] every kind of medicine, small and large. There is nothing like her on earth, and for this we thank God in Heaven." *Rosarium philosophorum. Secunda pars alchimiae de lapide philosophico vero modo praeparando, continens exactam eius scientiae progressionem. Cum figuris* [...] (Francoforti: ex officina Cyriaci Iacobi, 1550), [Mv]. ETH-Bibliothek Zürich, Rar 8230.

two-line verse "Here the king and queen lie dead / [and] the soul is separated with great misery." Eventually, the couple is reborn as a new substance altogether, depicted as a new winged, two-headed hermaphrodite. The prolific, vegetative power of this new material is indicated both by the plant next to it, heavily laden with a brood of infant "fruits," and by the lines of verse that accompanied the image: "Here is born a noble and rich Empress / Whom the masters liken to their daughter. She proliferates and bears innumerable children / Who are immortal, pure and without blemish" (Figure 10). The material then undergoes a further round of alchemical death and resurrection, finally resulting in the red tincture that can transmute metals.

The Sol-Luna series from the *Rosarium philosophorum* vividly depicted the blend of vegetative and sexual imagery that informed alchemical theory in the early modern period. These theories circulated more locally in Wolfenbüttel as well. Nearly five months after Anna presented her recipes to Julius, Philipp Sömmering decided to divulge alchemical secrets of his own. In a letter to Duke Julius dated August 25, 1573, Philipp articulated a theoretical understanding of alchemy that complemented Zieglerin's recipes and implicitly explicated their efficacy. As Philipp explained, at the Creation God imbued everything with a reproductive power. Plants store this power in their seeds, roots, or sap, while animals (including humans) store it in the "male seed," where it can be activated once the seed is cast into what Philipp called its "natural maternal vessel," where it is nourished until it is mature. "Therefore, it follows," Philipp went on, "that reproduction occurs the same way in all growing [things] and in all animals and people." The "ancient sages," he explained, wondered long ago whether this reproductive power might also be hidden in metals, enabling them, too, to reproduce themselves. After diligent research into the generation of metals, the ancients discovered that after the Creation, "all metals grew from a single root into the veins of the earth." The ancients had identified this as the "generative spirit" of metals. "In other words," Philipp concluded, "they clearly found that there is a prolific and reproductive power and spirit in metals, just as in other things." Furthermore, they reasoned that this "generative and reproductive power" must be "nobler" in metals than in other substances because metals remain unchanged by the fire and are not easily destroyed.[65]

Anna Zieglerin's recipes suggest that she drew on this same "metaphorical reservoir" of agricultural and sexual imagery and that it offered a framework for the efficacy of her lion's blood. Zieglerin's lion's blood was, essentially, the "blood of metals," the generative and reproductive power that, according to

Philipp, rested in all nature, but that was particularly potent in metals. Its power, after all, lay in its ability to stimulate metals, plants, and people to proliferate and produce "fruits," even when nature protested, as in the case of an infertile couple. It required only heat—which could come from decomposing dung for the fruit trees, for example, or from a barren woman's womb—to encourage the putrefaction that was the starting point for all generation. Unlike Philipp, however, who located his discussion of this *spiritus generativus* solely within an alchemical discourse, Anna chose a much more powerfully resonant term, blood, to communicate the potential of her alchemy. This choice, as we will see, was not an accident, nor was it merely a colorful metaphor. If we take her blood imagery seriously, we will see that it allowed Anna to claim much more than generative power for her alchemical lion's blood. For as much as blood signified life, fertility, fecundity, and nourishment in early modern European culture, it paradoxically connoted redemption and corruption as well.

Redemption

The redemptive, cleansing associations of blood lay at the heart of Christianity. In his sacrifice on the cross, Christ offered salvation to humanity. Although the crucifixion itself was not a bloody death, by the late Middle Ages devotional writers increasingly emphasized the role of Christ's blood in this central act of Christian soteriology.[66] Whether as relic, miraculous host, or part of the Eucharist, his blood washed, cleansed, restored, and cured bodily and spiritual afflictions alike in the Middle Ages.[67] Christ's real presence in the Eucharist, of course, became one of the most contentious issues in the Reformation, generating numerous interpretations of the relationship of both bread and wine to the body and blood of Christ.[68] The symbolism of the crucifixion was deeply embedded in early modern culture beyond the Eucharist or Lord's Supper as well, however, and this had implications for the redemptive and cleansing properties of blood in secular contexts. Early modern execution rituals, for example, evoked Christ's own death and thus sacralized the executed criminal's body, rendering many of its parts (and not just the blood) as powerful medicines. This may in part be why numerous sources document epileptics waiting at the scaffold to drink criminals' blood after beheadings, although it is important to note that the practice of epileptics drinking human blood is documented in classical sources such as Pliny the Elder's *Natural History* as well.[69]

Medieval Latin alchemists also integrated the redemptive powers of blood into their work. The notion that Christ's life could be mapped onto the stages of the philosophers' stone, in particular, rendered blood extremely potent.[70] Just as Christ had to be crucified in order to redeem humanity, the logic went, so, too, must the alchemist's materials undergo similar torments, even death and resurrection, before they could in turn "redeem" base metals as the philosophers' stone. The fourteenth-century alchemist John of Rupescissa offers only one example from this rich exegetical tradition: in his striking interpretation of Arnald of Villanova, Rupescissa likens the third stage in the production of the philosophers' stone, a distillation, to the crucifixion. Where alchemical mercury was digested and its vapors ascended to the top of the alembic, Rupescissa saw Christ's ascension on the cross: "Mercury is placed in the bottom of the vessel for dissolution, because what ascends from there is pure and spiritual, and converted into powdery air and exalted in the cross of the head of the alembic just like Christ, as master Arnald says."[71] Likewise, the alchemical vessel that enclosed the final stage of the red stone resembled "Christ inside the sepulcher." Once the flames had brought out the internal redness of the stone, the alchemist removed it from its vessel so that it would "ascend from the sepulcher of the Most Excellent King, shining and glorious, resuscitated from the dead and wearing a red diadem."[72] Rupescissa's emphasis on red in this passage—the red stone, Christ's red diadem—suggested that the analogy between Christ and the philosophers' stone extended to Christ's blood and the powers it was commonly understood to bear as well. As Leah DeVun describes this logic, "The creation of the stone through alchemy reenacts the narrative of the passion; hence, the stone may reenact the curative and salvific powers of Christ."[73] Just as Christ could cleanse the soul, therefore, so, too, could the philosophers' stone cleanse metals, absolving them of their impurities and transforming them into the noble metals, silver and gold. As the pseudo-Lullian *Codicillus* put it, "And just as Jesus Christ from the tribe of David assumed human nature in order to redeem and free humanity, imprisoned in sin because of Adam's disobedience, so, too, in our art will that which is tainted by wrong be absolved, washed, and released of every blemish by something that is opposed to it."[74] And it was the crucifixion, which by the Middle Ages increasingly was understood to have been an increasingly bloody event, that granted both Christ and the philosophers' stone their redemptive powers.[75]

Anna's lion's blood, too, had salvific power, which drew its meaning from this multivalent analogy between Christ and the stone. Most obvious, of course, was the fact that the lion functioned widely as a symbol of Christ in the Latin West (for example, the "lion of the tribe of Judah" in Revelations 5:5), so that the lion's blood would have resonated directly as the blood of Christ. Anna's exegesis was not nearly as sophisticated nor as explicit as John of Rupescissa's before her. Still, she suggested the connection when she framed the lion/tinged gold's emission of liquid/blood as a kind of sacrifice or gift: "The lion has emitted from itself as much blood as it could without injury and losing its red color."[76] The redemptive power of blood came through even more clearly in Anna's discussion of the little bird that ingested the lion's blood before giving its life to become the philosophers' stone. "If you also want to draw a heavenly analogy," she began, signaling the Christian context for this little bird's sacrifice, "then take a little bird from the forest and put him in a little basket." At the end of the passage, after describing how the bird should be roasted on a wooden spit and ground into a powder, Anna alluded to this "heavenly analogy" again: "On the timber of the holy cross the son of God tinged all of us poor sinners with his most holy rose-colored blood and took part [in earthly affairs], just as this small bird will take part in the earthly tincture."[77] Here, like Rupescissa before her, Zieglerin made explicit the redemptive powers of her alchemical operations: nourished by the lion's blood, the little bird died on a wooden spit and was resurrected as the philosophers' stone, reenacting Christ's sacrifice on the wooden cross. By being placed in a basket, the bird was grounded to earth, just as Christ was divinity incarnate and thus, in a sense, spirit grounded as well. And it was the bird's ordeal in the fire, like Christ's crucifixion, that bestowed its redemptive power as the philosophers' stone.

Unlike many of the peacocks, phoenixes, or Birds of Hermes that fly through alchemical texts, Anna's "little forest bird" is not an exotic specimen but a humble bird of the forest. One can imagine it sitting on a branch in the woods between Wolfenbüttel and Braunschweig, awaiting its spectacular transformation in Anna's laboratory. And yet, alchemists often wrote in code names (or, in German, *Decknamen*), using animals, in particular, to refer to their laboratory materials. Birds were common *Decknamen*, sometimes referring to the stages of the philosophers' stone because of their colors (thus, for example, the black or *nigredo* stage could be depicted with a raven) and other times appearing as vapors or other volatile materials that ascended to the top of the alchemical vessel, like birds taking flight. Anna's description of "the little forest bird," therefore, might not have been a real bird after all, but rather a volatile spirit. In this reading, the gradual addition of water and lion's blood in an alchemical vessel (basket) over a period of six weeks would stabilize the spirit into a solid. This process, in other words, would incarnate the spirit into an earthly body, just like Christ's incarnation, so that it could be roasted and ground up into a powder (the crucifixion) and thereby acquire its power to redeem metals. Although this passage was short, and Anna did not go on to elaborate on this "heavenly analogy" nearly as artfully as John of Rupescissa had before her, the pious Duke Julius would have immediately recognized the significance of her metaphors. Above all, the parallel suggested that Anna could reenact the central events of Christian soteriology by manipulating natural objects in the laboratory.

Anna's "heavenly analogy," therefore, allowed her to emphasize the power of the lion's blood to cleanse impurities and thereby redeem, just as Christ's blood redeemed humanity from sin.[78] This may also explain her mention of leprosy, which was the only medical application of the lion's blood Anna included in the booklet other than recipes for using it to conceive a child. "If you want to cleanse a leper," she wrote, "then every day for nine entire days give him three drops of the oil that you extracted from the tinged gold [i.e., the lion's blood]; on the ninth day, tap one of his veins, and the leprosy will run out of him, or rather out of the incision, like grains of sand."[79] One could read this reference to leprosy on multiple levels. Alchemical authors sometimes referred to impure base metals as "lepers" that needed to be purified, invoking the biblical story of Naaman the leper (2 Kings 5), who dips seven times in the river Jordan and is thereby healed, just as alchemists argued that the "leprous" metals could be cleansed by mercurial waters.[80] Anna may well have intended her lion's blood to cure actual human lepers, however, by releasing the disease from their bodies—an alchemical reinterpretation of the long-standing belief that blood could cure not only epilepsy, but also leprosy.[81] Finally, of course, Anna's cures for leprosy would have gestured at the alchemist's holiness, just as Christ's holiness was proven by his ability to cure lepers in the Gospels.[82] Leprosy's double sense as a disease of the body and soul raises the possibility that Anna may have conceived of her lion's blood as something able to redeem and "purify" not only bodies, whether of metals or lepers, but also their souls, just as Christ's blood redeemed humanity. We need not choose from among these interpretations; alchemical texts shimmer, moving back and forth among many meanings at once, an attribute that made them especially compelling to both authors and readers.[83]

Philipp Sömmering's letter to Duke Julius articulated a theoretical justification that may well have explained the cleansing powers of the lion's blood. After the ancient sages had identified the "reproductive power" in metals,

Philipp explained, they determined that "in metals the reproductive power is hidden and firmly locked up in their hard, fused bodies," making it difficult to access. The ancients, therefore, "decided that the metallic reproductive spirit could not achieve growth or reproduction unless the corpses are dissolved." Unlike plants, he explained, which store their reproductive power in readily accessible seeds or roots, the reproductive power in metals could be extracted only by destroying them. The metallic body must be dissolved gently, however, so its spirit can be separated from the body and then the body can be purified "internally and externally" before returning the spirit and thus creating the body of a more refined metal. "That," Philipp concluded, "is the entire secret of the natural noble art of alchemy."[84]

With this passage, Anna's fellow alchemist Philipp articulated an ancient and important thread in alchemical theory: the idea that metals have a spirit (and sometimes also a soul) as well as a body, and that the alchemist's task is fundamentally the separation and recombination of matter and spirit.[85] The *Rosarium philosophorum*, once again, offers a particularly rich locus for exploring this core idea, portraying the evaporation and condensation of alchemical vapors in the distillation apparatus as the rise and fall of spirits and cleansing dew. As we have seen, the first few stages of the chemical wedding drew on vegetative and sexual imagery to depict Sol and Luna's preparation for conception up to the point of putrefaction, at which point the soul was separated from the hermaphrodite, who lay "dead" in a coffin. The following three images in the series depicted essentially the same process that Philipp described in his letter to Duke Julius. Following the dissolution of Sol and Luna in a mineral bath, then their conception and putrefaction, we see the soul actually depart the body in the form of a small child. The caption explains, "Here the four elements separate themselves, the soul quickly departs the body" (Figure 11). While the soul is absent, the body is cleansed, indicating, as the caption explains, "Ablution, or *Mundificatio*: Here the dew falls down from heaven, and washes off the black body in the grave" (Figure 12). The body cleansed, the soul then returns to reanimate the corpse: "Here the soul sweeps down, and quickens [*erquickt*] the purified corpse again" (Figure 13). The ravens at the foot of the tomb represent the *nigredo* phase of the process, signifying that the philosophers' stone has reached a new stage. The soul works as a kind of catalyst on the dead body, stimulating it to rise again but in a more perfected and more prolific form. After the (unseen) alchemist carries out the entire process again, the stone, finally perfected, emerges from the alchemical vessel. The final image recalls the trope of Christ as philosophers' stone, depicting the *lapis* as Christ rising from the sepulcher, just as John of Rupescissa had discussed centuries earlier (Figure 14). The caption reads: "After my numerous afflictions and great martyrdom, I have risen, luminous and flawless."

In material terms, what this sequence describes is the action inside the alchemist's distillation vessel, which refines matter into ever more purified forms, eventually resulting in the perfected philosophers' stone. This kind of language stating the need to redeem matter in order to unlock its generative powers closely paralleled medical texts on human reproduction from the same period, suggesting once again that the process of alchemical perfection or resurrection was intimately tied to reproduction. The preparation of the philosophers' stone depicted in the *Rosarium philosophorum* required two stages. First, the material body had to be destroyed through mineral waters before conception and putrefaction could take place; then, the body could be purified and resurrected again. At this point it was capable of further perfection, but it could not complete its perfection without the return of the soul/spirit. As the Latin text of the *Rosarium philosophorum* put it, "Ortulanus and Arnald say that then the *anima* pours into the body and the crowned king is born. Moreover, the book *Turba philosophorum* contains the following lesson: dissolve the body and let it drink in the spirit."[86] Medical texts from the same period drew on similar imagery in describing the birth of children.

Consider, for example, a passage from Walther Hermann Ryff's 1541 book on anatomy, *True Description or Anatomy of the Most Excellent, Highest and Noblest of All Creatures, Fashioned by God the Master and Creator of All Things on Earth, That is, Man*. After describing the generation of the "fruit" from the mixing of the father's and mother's seed in the warmth of the womb, Ryff explains that "around the forty-fifth day the soul is poured into it, then it is no longer called a fruit, but a child."[87] Neither alchemical perfection nor human reproduction could be complete without this final step, the "pouring in" of the soul. This discussion of the "soul," whether of metals or infants, highlights what Kathleen Crowther-Heyck has called "the tremendous symbolic and spiritual significance ascribed to procreation in early modern Germany."[88] She has argued that "both religious and medical writers linked procreation to the narrative of the creation, fall, and redemption of mankind."[89] The action of creation in the womb reenacted the original creation of Adam and the world in Genesis, while the pain of childbirth was "imagined as redemptive," analogous to Christ's own suffering for humanity.[90]

By hinting at the "heavenly analogies" in her alchemy, Anna did something similar for alchemy, situating the alchemical redemption of metals within a

ROSARIVM
ANIMÆ EXTRACTIO VEL
imprægnatio.

Hye teylen sich die vier element/
Aus dem leyb scheydt sich die sele behendt.

De

FIGURE 11. "Extraction of the soul, or *impraegnatio*." The German caption
reads: "Here the four elements separate themselves, the soul quickly departs
the body." *Rosarium philosophorum. Secunda pars alchimiae de lapide
philosophico vero modo praeparando, continens exactam eius scientiae
progressionem. Cum figuris* [...] (Francoforti: ex officina Cyriaci Iacobi, 1550),
Hvii. ETH-Bibliothek Zürich, Rar 8230.

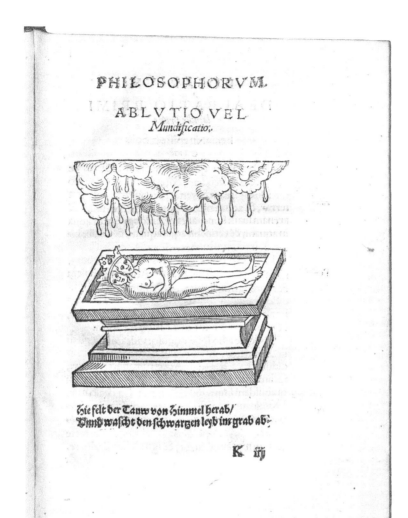

FIGURE 12. "Ablution, or *Mundificatio*." The German caption reads: "Here the dew falls down from heaven, and washes off the black body in the grave." *Rosarium philosophorum. Secunda pars alchimiae de lapide philosophico vero modo praeparando, continens exactam eius scientiae progressionem. Cum figuris* [...] (Francoforti: ex officina Cyriaci Iacobi, 1550), Kiiij. ETH-Bibliothek Zürich, Rar 8230.

PHILOSOPHORVM.

ANIMÆ IVBILATIO SEV
Ortus seu Sublimatio.

Die schwinge sich die sele hernidder/
Vnd erquickt den gereinigten leychnam wider.

L iij

FIGURE 13. "Jubilation of the soul, or *Rising or Sublimatio*." The German caption reads: "Here the soul sweeps down, and quickens the purified corpse again." *Rosarium philosophorum. Secunda pars alchimiae de lapide philosophico vero modo praeparando, continens exactam eius scientiae progressionem. Cum figuris* [...] (Francoforti: ex officina Cyriaci Iacobi, 1550), Liii. ETH-Bibliothek Zürich, Rar 8230.

PHILOSOPHORVM·

Nach meinem viel vnnd manches leiden vnnd marter (groß/
Bin ich erstanden/ clarificiert/ vnd aller mackel bloß·
& à lœßs

FIGURE 14 The philosophers' stone as Christ rising from the sepulcher. The German caption reads: "After my numerous afflictions and great martyrdom, I have risen, luminous and flawless." *Rosarium philosophorum. Secunda pars alchimiae de lapide philosophico vero modo praeparando, continens exactam eius scientiae progressionem. Cum figuris* [...] (Francoforti: ex officina Cyriaci Iacobi, 1550), [avi]. ETH-Bibliothek Zürich, Rar 8230.

larger soteriological framework and refiguring her alchemy as God's work. Anna's recipes did not merely propose generation, or even a more prolific kind of generation. They proposed to redeem matter, to improve it, and make it superior to the fruits, children, and precious metals and stones that nature could offer on its own. Just as her trees could do what nature could not (i.e., produce fruit in winter), therefore, her gemstones outdid nature as well. "These [gem]stones," Anna boasted, "exceed all other stones, [even] all other rarified stones, both in their power and virtue, as well as in [their uses for] medicine."[91] Precisely because of the redemptive power of her lion's blood, Anna's alchemical generation was superior not only in quantity, but in quality as well.[92]

Poisonous Menses

The lion's blood was a powerful substance. It offered practical utility and versatility in the laboratory, promising not only health but also fruit and gold. Yet these material benefits alone do not exhaust the messages that the lion's blood would have communicated at Duke Julius's court. The recipes in Zieglerin's booklet gained additional power from the way in which they played on the rich analogical culture of sixteenth-century Europe, resonating on many levels at once. By drawing on ideas and images associated with alchemy, agriculture, and generation, the lion's blood could simultaneously evoke the philosophers' stone, the generative force contained in seeds, and the cleansing powers of Christ's blood, thereby appealing at once to Julius's economic, intellectual, and religious interests. The lion's blood also accumulated meaning, however, from what it was not. For Anna Zieglerin, what mattered most was that the lion's blood encapsulated all of the nourishing, prolific, cleansing powers of blood without any of its dangers. Blood was not always a desirable substance in sixteenth-century Germany, of course, particularly when it left the body.[93] Gore, blood shed from the body, was filthy, a symbol of death and decay, and early modern Germans created a whole set of taboos around those who worked with it by trade (skinners, for example, or executioners).[94] Blood could be corrosive, too. Goat's blood, for example, appeared in craft recipes as a means of softening hard gemstones such as diamonds.[95] The lion's blood was not this kind of blood.

The lion's blood had a more complex relationship to another kind of blood, however: menstrual blood, about which there was a deep ambivalence in medieval and early modern culture. On the one hand, medical authors

rooted in the Aristotelian and Galenic traditions disagreed about the existence
of a female seed in generation, but they all agreed that the mother's uterine or
menstrual blood played a positive role in generation by nourishing the "fruit"
during gestation. Furthermore, menstruation itself was thought to be a normal,
healthy occurrence that purged the female body of excess.[96] On the other hand,
an equally ancient and venerable tradition, most notoriously encapsulated and
disseminated in the late thirteenth-century Latin treatise *On the Secrets of
Women,* viewed menstrual blood as a poisonous substance, potentially even one
that could cause leprosy if it contaminated the organs during pregnancy.[97] In
the sixteenth century, the ambivalence toward menstrual blood persisted in
both lay and learned medicine. In *Book about the Birth of Rational Things
Through Reason* (ca. 1520), Paracelsus granted women an important role in
generation, attributing to their wombs not only female seed, but also the crucial
role of pulling in the male seed and arranging it properly to create a "fruit"
(although ultimately he downplayed the role of the body in human reproduction, emphasizing God's intervention above all).[98] Menstrual blood itself, however, was more problematic in the Paracelsian tradition. The [pseudo?-]
Paracelsian text *Nine Books on the Nature of Things,* for example, explained the
monstrous potential of menstrual blood in the course of a discussion of a
"monster above all other monsters," the basilisk. As described by Roman author
Pliny the Elder, the basilisk was terrifying. "It kills bushes not only by its touch
but also by its breath, scorches up grass and bursts rocks," he explained. "Its
effect on other animals is disastrous: it is believed that once one was killed with
a spear by a man on horseback and the infection rising through the spear killed
not only the rider but also the horse."[99] *Nine Books on the Nature of Things*
linked the basilisk to menses, explaining that this creature carried an extremely
noxious poison in its eyes, "not entirely unlike a woman in her time of month,
who also has a hidden power in her eyes." With a mere glance, in fact, a
menstrual woman could cloud mirrors, render wounds untreatable, "and with
her breath and exhalation she can also poison, corrupt, and render powerless
many things." Her mere proximity could ruin wine, drain brandy of its strength
and gold and coral of their color, and even dull precious stones. Not only did
the basilisk's powers *resemble* those of menstruating women; "the basilisk grows
and is born from the greatest impurity of women, out of their *menstrua* and
out of the sperm of the blood."[100] In his commentary on this passage, William
Newman has suggested that "for the author of the *Nine Books on the Nature of
Things,* the basilisk is the epitome of the female itself," grotesquely embodying
all of the destructive power of (female) menstrual blood.[101]

As with blood's generative and redemptive associations, its poisonous powers, too, appeared in more explicitly alchemical texts. In addition to the alchemical processes that began with actual blood, one can find numerous references to substances with *Decknamen* involving blood as well. These substances came to be known as blood as a way of conceptualizing not only their physical properties (red, liquid, viscous, oily, ruby-like, etc.), but also their corrosive powers. Of the many "bloods" in alchemical texts, one of the most famous, perhaps, is the "blood of the green lion." Often (although not always) referred to as "oil of vitriol" as well, the blood of the green lion is usually a type of mineral acid and frequently a key ingredient in the production of the philosophers' stone. Like other *menstrua* found in alchemical texts, it is a solvent capable of dissolving metals. These solvents played an important role in the creation of the philosophers' stone, since they dissolved the alchemist's starting materials into the formless *prima materia* and thus (to use Philipp Sömmering's terms) prepared them for their redemption.[102] One must be careful about assuming that such terms stay constant across alchemical texts, for they often change meaning in the hands of different authors, and it is best to take each alchemical author's "blood" on its own terms.[103] Still, there is remarkable consistency around *menstrua* and other bloods in alchemical texts, all of them picking up on the corrosive (and misogynist), rather than the generative, associations with blood, especially menstrual blood.[104]

Menses, in fact, were also a topic of interest at Duke Julius's court. Mere proximity to a menstruating woman, the alchemists all reported, could corrode both living things and inanimate objects. When Philipp Sömmering was asked about menses, for example, he offered a story to illustrate their destructive power. According to the transcript of his trial, "He gave an example of a rose. When a man and a menstrual woman each pluck a rose, and each puts it in front of or near a window, the man's rose will wither slowly, and the woman's will be putrid on the third day" because of the corrosive power of her menses.[105] Even Anna may have shared this view of menstrual blood as toxic. Anna confirmed Philipp's view of the corroding powers of menses, explaining that "when someone ingests that [i.e., menstrual blood], he becomes . . . ill."[106]

When Anna and Philipp attributed noxious powers to menstrual blood, they drew on an age-old tradition that continued to be alive and well in sixteenth-century central Europe. But they also spelled out some of the problematic implications of this view of menses for practicing alchemists. After relating the example of the flower, Philipp addressed the court's question about menses with a second story that suggested a connection to alchemy: "Once, a female dog was brought into Theophrastus's [i.e., Paracelsus's] *Laboratorium*," Philipp continued, "and all of the glassware cracked open."[107] Although Philipp did not say so explicitly, clearly he had a menstrual dog in mind, since he related the story about Paracelsus's laboratory in answer to a question about menses. Anna also apparently thought menstrual blood could jeopardize alchemical work. Philipp reported that Anna "said to him that he should not go to his wife because of the *menstrua*."[108] During one of her own interrogation sessions, Anna confirmed this, stating that she "told Philipp that he must abstain from women, otherwise he will not be able to accomplish anything in alchemy."[109] Apparently Anna not only believed that menstruation could interfere with alchemy, but also understood the lion's blood as the inverse of menstrual blood; if the lion's blood was prolific and purifying, menstrual blood caused decay, illness, and corruption, including even the potential to spoil the alchemical reproduction promised by her own lion's blood.

The Homunculus

This alchemical aversion to menses was not unique to the Wolfenbüttel court. It was also central to the Paracelsian desire to create artificial human life in the form of the homunculus, a kind of pure, artificial human that embodied alchemists' loftiest aims. This homunculus suggests a point of comparison for the lion's blood, and perhaps even a tradition with which Anna was directly engaged as she worked out the implications of her alchemy. It embodied not only the ultimate generative and perfective potential of alchemy, but also a way that human reproduction could triumph over the poisonous potential of menstrual blood. Discussions around the homunculus, therefore, also make visible the extent to which the fault line between generative or redemptive blood and poisonous blood was also to some extent a gendered boundary.

As William Newman has shown, "Tinkering with natural human generation was a widespread topic of discussion in the Middle Ages and the early modern period."[110] In the Middle Ages, a handful of authors writing in Islamic, Jewish, and Latin Christian traditions considered the theme of the homunculus, but the topic really took off in the sixteenth century in the hands of Paracelsus and his followers.[111] The Paracelsian homunculus appeared in two texts: the possibly pseudo-Paracelsian text *Nine Books on the Nature of Things*, which appeared in print in 1572, the year before Anna Zieglerin recorded her recipes and gave them to Duke Julius, and *On the Homunculus* (ca. 1529–1532), an authentically Paracelsian tract. Together, these two texts explored the production of the homunculus and offered an ambivalent assessment of its moral meaning. The author of *Nine Books on the Nature of Things* addresses the homunculus following a discussion of how to create an artificial bird through putrefaction. "It is also known," he continues, "that people can be born without natural fathers and mothers. In other words, they are not born in a natural way from a female body, but rather by art and by the skill of an experienced spagyric [physician] they grow and are born, as will shortly be reported."[112] The text then goes on to describe how to make a homunculus as follows:

> Its process is thus: that the sperm of a man be putrefied by itself in a sealed cucurbit for forty days with the highest degree of putrefaction in a horse's womb [i.e., hot, decaying dung], or at least so long that it comes to life and moves itself, and stirs, which is easily observed. After this time, it will look somewhat like a man, but transparent, without a body. If, after this, it be fed wisely with the arcanum of human blood, and be nourished for up to forty weeks, and be kept in the even heat of the horse's womb, a living human child grows therefrom, with all its members like another child, which is born of a woman, but smaller.[113]

Here the generative potential of alchemy, especially the blood-based alchemical elixirs, reached their ultimate potential: the creation of human life. And yet, the alchemists can reach this potential only by eliminating the female principle that is otherwise present in alchemical processes (most obviously, of course, in the chemical wedding). For the author of *Nine Books on the Nature of Things*, the homunculus was remarkable not least because it was "born outside the female body and [without] a natural mother." In fact, as Newman points out, the author of *Nine Books on the Nature of Things* presents the homunculus as a sort of "masculine twin" to the basilisk, deriving its purity from an *absence* of menses, just as the basilisk owed its venomous qualities to its origins in menstrual blood.[114]

Perhaps Anna knew of this book or the tradition upon which it drew, and the Paracelsian ambition to make a homunculus fed into her understanding of the reproductive powers of the lion's blood. Like the author of *Nine Books on the Nature of Things*, Anna, too, shared the ambitious desire to extend alchemy's powers into the creation of human life. Yet she may also have shared the concerns that Paracelsus articulated in *On the Homunculus*, where, as William Newman has pointed out, the homunculus became "a potent image of sin."[115] In fact, in this text Paracelsus described a much more monstrous homunculus, claiming that it would lack a soul because it originated in human seed expelled outside of the proper context of marital reproductive sex. Perhaps responding to the frightening possibility of a monstrous homunculus without a soul, or perhaps simply rejecting a distinctly masculine fantasy of creating human life *without* a woman's body, Anna put her own twist on the homunculus tradition. She put the female womb, fertilized by the lion's blood, back at the center of alchemical reproduction, restoring the woman's body to alchemical procreation. This approach had many advantages, not least of which that it ensured that, unlike the homunculus, alchemical infants made in the womb with the lion's blood would have a soul. Anna was able to draw on the rich systems of meaning that linked the natural and supernatural, creation and procreation, in order to frame her alchemical process as holy work. But locating the reproductive work of the lion's blood in the female body also came with one serious risk, and it was clearly something that bothered Anna: the presence of menstrual blood could pollute the womb, possibly preventing the growth of alchemical "fruits" or, worse, producing a basilisk instead. Fortunately, Anna Zieglerin's own womb was in no danger of such monstrous production. As everyone in the Wolfenbüttel court knew, her body was unusually pure because she did not menstruate.

CHAPTER 5

A New Virgin Mary

The recipes Anna Zieglerin gave to Duke Julius articulated a series of connections between alchemical reproduction and redemption. These links were suggestive indeed, making laboratory operations an act of piety. The full import of the lion's blood, however, only became clear from an additional set of stories that Anna and others circulated at court. Unlike the recipes, these stories were not presented formally, nor were they written down. Through conversations and comments, however, Anna circulated an account of her own origins that highlighted the extraordinary qualities of her body and positioned her as uniquely able to realize the full potential of the lion's blood.

According to her husband, Heinrich, Anna said that while her mother, Clara, was pregnant, Clara's brother was stabbed to death. Overtaken with grief, Clara ran out of the house and fell through ice into some water. She did not survive, but fortunately Anna did. As Heinrich reported, "One Doctor Peter cut open her mother and took her out of the womb and anointed and greased her [body] with a hand cream." Emerging prematurely from her dead mother's body, Anna's entry into the world was dramatic enough, but a couple of additional details made her birth even more astonishing. According to Heinrich, this "Doctor Peter" immediately wrapped Anna up in a lubricated skin, intending that "when it was smeared with ointment, she would become plumper." Heinrich said that this skin "was supposedly taken from a woman's body; someone named Maria also supposedly lay in the skin for six weeks. The skin was supposed to have come from Poland. And it was a Turkish liquor or tincture that Dr. Peter had; supposedly it was a tincture." Philipp Sömmering reported slightly different details about Anna's birth:

"The Elector [Heinrich "the Pious" of Saxony] had a specially balsamed skin, in which she lay and grew for twelve weeks. Duke Georg, the old Elector's father, had left this skin and an eagle stone behind as a treasure. Frau Anna said it was a little skin, was lying in a mosque, and was from a woman's body."[1] Whatever the precise origins and nature of the skin and the tincture, together they were meant to serve as a kind of ersatz womb while Anna completed her gestation outside of her mother's body. Perhaps the skin was some kind of preserved placenta or even a caul (amniotic tissue); the tincture clearly evokes the lion's blood's generative powers.[2] While the fragile infant Anna lay in this early modern incubator, moreover, she was fed not with breast milk from a wet nurse, as one might have expected, but rather, Heinrich reported, "she was nourished entirely with a tincture and was specially purified thereby. . . . She had imperfect legs, very frail. She was nourished as if she had grown in her mother's womb, therefore, and was supposedly specially purified thereby. She was very weak in her legs." It is no wonder that Anna's story of her own birth made an impression on Heinrich. Removed from her dead mother's uterus, smeared with ointment, wrapped up in a skin from another woman's body, and fed on tincture, Anna entered the world in an extraordinary fashion indeed.[3]

Anna's story explained several features of her own adult body. First, as Heinrich explained, she had weak legs as a result of her premature birth. If this was true, then Anna shared this quality with her patron, Duke Julius (although his legs were injured in a childhood accident, while Anna claimed that hers were weak from birth). Indeed, this story may have offered a poignant point of connection between the two. Her appearance might have been unusual in other ways as well. According to Philipp, Anna said that her alchemical mentor, Count Carl, had found her interesting enough to draw up her physiognomy, but one can only wonder what personality traits he (or others) interpreted from her physical appearance.[4] More importantly, however, Anna, Heinrich, and Philipp all noted that her early nourishment with a tincture, rather than breast milk, had "specially purified" her body. As the story circulated at Julius and Hedwig's court, Anna's body became a focal point for alchemical and spiritual expectation in Wolfenbüttel. Anna framed her body as far more than a site of knowledge and expertise, in other words; it was an unusually generative space, both literally and figuratively, for exploring both the potential of the alchemist's art in the unfolding events of sacred time and the relationship between matter and spirit.[5]

Purity

What exactly did it mean that Anna's body was "pure"? Anna, Heinrich, Philipp, and others were asked repeatedly about her "purity" (*Reinigkeit*). Their answers reveal that they understood Anna's purity to have several dimensions, all of which clustered around her exceptional ability to escape the burdens of her carnal body. On a physical level, Anna's purification via tincture at birth explained what seems to have been a well-known fact about her adult body: she did not menstruate. As Philipp explained, "Frau Anna said that she lay in her mother's body for twenty-two weeks [i.e., nearly six months] and after that was brought up entirely with a tincture and had no *menstrua*."[6] Heinrich, who, as Anna's husband, presumably had at least some firsthand experience with her body, confirmed that she did not menstruate. He explained that "he had not sensed her flux, except that when she was angry the flux turned brick red. . . . The people who do the laundry did not report it to him," he added, "neither had the old women, and he had never experienced it in his entire life. She gave him the feeling that he was not allowed to ask her many questions about it."[7] When Anna was asked about the issue, she said only that "she didn't have her womanly time every month, especially when she was frightened."[8]

Missed or absent menstrual cycles (amenorrhea) could mean any number of things in early modern Europe, depending on one's views on the purpose and nature of menstrual blood itself. Many learned physicians would have understood Anna's missing "flux" either as a sign of pregnancy, or as a pathological retention of excess matter, regardless of whether they understood menstrual blood to be nourishing or poisonous.[9] Anna, however, interpreted her amenorrhea not as a failure to evacuate menstrual blood from her body, but rather as evidence that her body did not accumulate menstrual blood in the first place. More importantly, she saw this as a good thing, since in her view, menstrual blood was in fact a polluting, dangerous substance that she was glad to do without. She and others posited her amenorrhea, in fact, as the central sign of her purity.

Because Anna and Philipp, at least, seem to have believed that menstruating women had the power to disrupt alchemical projects, Anna's purity was crucial to her participation in the alchemical projects in Wolfenbüttel. According to Philipp, however, Anna's purity extended beyond her own body: "If one were with her, then those who were infected by other menstrual women could be cleansed by her."[10] Mere proximity to Anna might be enough, but one could also tap into the source of her purity more directly, for her cleansing powers seemed to originate in her blood and could be trans- ferred to others if they were willing to ingest it. Philipp, in fact, explained that he "had frequently drunk her blood because she was supposed to have been raised with tincture."[11] According to Anna, this was Philipp's idea, not hers: "She did not teach Philipp anything," she said; "he, Philipp, drank her blood on his own, no doubt . . . [but] she let from her vein the blood that Philipp drank."[12] Moreover, Philipp was not the only one to drink Anna's blood. "The Elector of Saxony also drank [my blood]," she claimed, "but [he did so] unknowingly."[13] Heinrich claimed not to know that the elector of Saxony had drunk it, but he, too, reported that "he knew well that Frau Anna's blood was supposed to be very good. The Count [i.e., her alchemical tutor, Count Carl] drank it when he could get it." Heinrich added that "he did not know of anyone who let her blood other than the Count and the King of Denmark, [who did it] out of great love. He took it in a goblet in red wine."[14] It was as if Philipp and others believed that Anna Zieglerin had something akin to the lion's blood—or even the cleansing blood of Christ— running through her veins.[15]

Anna's purity, therefore, was critical, not only for her own participation in the alchemical projects at the Wolfenbüttel court, but also for those of the male alchemists around her. Her ability to cleanse others would have ensured that even her male colleagues were fit for alchemical work, unspoiled by the noxious influences of their menstrual wives or other sources of pollution. The implications of Anna's uncorrupted body for alchemy, however, paled in comparison to the larger claims it allowed her to make about her own role as a mediator between heaven and earth. Hers was a peculiar body indeed, since it appeared to be free of the earthly corruptions and carnality that weighed down other women. In fact, Heinrich learned from both Philipp and Anna "that she did not have the flux and [that she had] many other virtues, and that she was like the angels."[16] Philipp Sömmering confirmed this in his own comments on Anna's "purity." He said that he "praised Frau Anna a great deal concerning her purity, and compared her to the Mother of God, just as she herself did. . . . He . . . praised Anna greatly, and that he perceived her to be equal to the Virgin Mary."[17]

Whether Anna Zieglerin resembled angels or the Virgin Mary, the com- parison implied a body that was able to transcend some or all of its sinful, carnal nature. Although there was no clear consensus about the specific char- acteristics of angels in the sixteenth century, they were generally thought to be incorporeal beings without bodies at all.[18] The Virgin Mary, meanwhile, did have a normal human body, but one that was still somewhat different from the bodies of other women: she was human, but pure, even after she gave birth to Christ. While a Scholastic consensus in the Middle Ages had established that Mary was pure and untainted by original sin, the precise nature of Mary's purity remained a subject of debate.[19] Theologians and popular devotional practices offered competing visions of when and how Mary came to be free of original sin, for example, in their debates around the Immaculate Conception. Was Mary pure at her conception, or purified later, during her gestation in her mother's womb, or even at the Annunciation?[20]

Moreover, debates about Mary's purity were always as much about her body as they were about her soul, as Mary's body, like Anna's, was a space in which matter and spirit were integrally bound to one another. Thus the question of whether Mary menstruated generated a surprising amount of attention on the eve of the Reformation. Adherents of popular Marian devotional practices reasoned that her spiritual purity required her to be exempt from menstruation as well. This popular view that Mary was "without spot" in both a spiritual and a physical sense, however, was difficult to reconcile with medical understandings of generation, which (whether Aristotelian or Galenic) required menstrual blood. Moreover, there was agreement that Mary nursed Jesus as an infant, which offered further evidence of her menstruation, since medieval medical theory understood menstruation and lactation to be excretions of the same fundamental matter. Even more troubling were the "unthinkable consequences" of a nonmenstruating Mary in terms of Christology.[21] Quite simply put, the logic of Christology required that Mary, the Mother of God, be human in order to transmit her humanity to Christ. Not only was she theologically required to menstruate like all other women in order to prove her humanity; she was also required to transmit this humanity to Christ in utero, and this required menstrual blood. The paradox of Mary's virginity and motherhood, therefore, made the issue of her menstrual blood both deeply problematic and absolutely crucial.[22] In the end, late medieval theologians insisted on Mary's carnal humanity even as they insisted on her unique purity. As Charles Wood has put it, "Logic might tell them that a Virgin with an Immaculate Conception ought not to menstruate, but Christology, observation, and common sense told them she must have. So she menstruated."[23] Such Christological concerns were less likely to trouble most medieval lay Christians, however, who were more willing than the church doctors to extend Mary's exemption from sin to a broader exemption from other bodily markers of that sin, such as menstruation.

When Anna and her peers compared her to the Virgin Mary, they invoked this long tradition of late medieval popular Marian piety. The Blessed Virgin offered a powerful example of a young, nonmenstruating woman exempt from the ordinary bodily functions that mired other women in sin, carnal fertility, and menstruation. Moreover, the comparison to Mary allowed Anna to refigure her amenorrhea as spiritual purity, and at the same time, to extend alchemy's redemptive powers from matter to spirit. After all, Anna attributed her purity to her treatment at birth with alchemical tinctures.

It is important to remember, however, that Anna was a Lutheran living over half a century after Luther and the early reformers first prompted a radical reconfiguration of Christianity in Western Europe. If Anna was a new Virgin Mary, in other words, she was not only an *alchemical* Virgin Mary, but also a *Protestant* Virgin Mary. In central Europe, the Reformation did not alter the traditional image of Mary as much as we might think. The early reformers (as well as their audience of potential converts), of course, grew up in the climate of deep Marian devotion characteristic of the years around 1500, and this might have influenced their willingness to retain some aspects of traditional Mariology.[24] Luther, for example, defended the notion that Mary remained a virgin even after she gave birth to Christ, as well as the notion that she was without sin, and at times even seemed to support the possibility of her own immaculate conception.[25] Likewise, some cities, such as Nuremberg, showed a surprising willingness to maintain continuities in medieval visual and liturgical culture around Mary even after implementing the Reformation.[26]

These general continuities are noteworthy, and may have been particularly strong in Wolfenbüttel, where the Reformation was accepted officially only a few years before Anna Zieglerin arrived.[27] At the same time, there were important differences between the medieval Mary and the Lutheran Mary. Protestant reformers' attitude to Mary was influenced by their attack on the cult of the saints generally as unwarranted distractions from the worship of God. Given the strength of the Marian cult on the eve of the Reformation, Luther and other early reformers found Mary particularly problematic, worrying above all about the dangerous tendency to grant her an especially powerful ability to intercede with Christ because she was his mother. Luther, therefore, demoted Mary from her medieval role as an intercessor with privileged access to God to a *Fürbitterin* (one who prays for us) whose prayers could do no more than those of any other Christian. Moreover, the reformers also rejected Marian festivals that were not based on events in the biblical canon—namely, feasts commemorating Mary's conception, birth, and assumption—and they deemphasized Mary's virginity, transforming her into a model of the obedient Protestant wife and mother who demonstrated faith, humility, and piety before God.[28] Mary remained virginal, but Protestants, who rejected clerical celibacy, emphasized how unique she was in this regard, and did not hold up her virginity as a model for pious Christians. For Protestants, Mary became a symbol of a new model of piety that embraced the family and encouraged women to demonstrate their faith as wives and mothers rather than as nuns.[29] In a nutshell, as one scholar has put it, Lutherans transformed Mary "from a divine intercessor into a humble *Hausmutter*."[30]

Anna was raised with this new Protestant understanding of Mary, and this may have informed her emphasis on Mary as mother. She may also have had another very different point of reference for understanding the Virgin Mary, however: Paracelsus, to whom Anna linked herself through the figure of her alchemical mentor and partner, Count Carl.[31] Paracelsus was as much a lay theologian as he was a natural philosopher, physician, or (in his pseudonymous guise) alchemist. His writing and ideas about nature were deeply influenced by his theology and vice versa.[32] Nowhere is this clearer than in his texts about generation, which easily moved back and forth between considerations of the role of the womb in conception in general and the physical contribution of the Virgin Mary to Jesus's body. Paracelsus's religious ideas were idiosyncratic, fitting easily into neither Catholic nor reformed theologies.[33]

Paracelsus grappled with Mary's nature in several different texts over his life, and it is important to pay attention to the evolution of these ideas.[34] Nevertheless, several constants emerge. Like Catholics and reformers alike, Paracelsus accepted the basic dogma that Mary was the mother of Jesus, and that she remained a virgin even as she bore Christ. He emphasized her purity, from the Immaculate Conception to her death.[35] Paracelsus departed from Catholic and reformed theologians alike, however, when he emphasized Mary's divinity, elevating her into what might be thought of as a "quadernity." Paracelsus suggested that Mary was either the wife of God the Father or even his female half. She was, in this view, a "goddess" who was to be venerated independent of her brief earthly role as Christ's mother. Moreover, Paracelsus maintained Mary's preexistence; as he put it in *On the Invocation of the Blessed Virgin Mary* (ca. 1527), "She was the first creature, before heaven and earth was constituted."[36] In other words, Mary existed in heaven long before and long after her brief stint on earth, where she became flesh in ways almost analogous to Christ's incarnation. Her conception was indeed immaculate, in the sense that her parents, Anne and Joachim, had nothing to do with it; rather, her conception was miraculous and a product of her own will to launch the incarnation. In fact, in his treatise *On the Birth of Mary and Christ*, Paracelsus emphasized these parallels between Mary and Christ to the point of viewing her as co-redeemer.[37] This was not the Mary late medieval Catholic theologians tried to tie to earth in order to secure their Christology, and was even farther from the reformed Mary, who was dependent on God's grace for her role in the incarnation. Rather, Paracelsus's Mary was pure, divine, eternal, and sovereign, a powerful equal to God and Christ both.

When Anna compared herself to the Virgin Mary, therefore, she had a complicated constellation of ideas about the holy virgin from which to draw. Anna did not accept uncritically the reformers' vision of the Lutheran Mary, but used alchemy to articulate her own vision of the Mother of God. Above all, Anna did not identify entirely with the new Lutheran image of Mary as *Hausmutter*. Married at least once (and maybe twice) but still childless, living in a household with not only her husband Heinrich but also their companion Sylvester Schulvermann, and fiercely independent in charting her own life course, Anna did not model herself on the humble, obedient, Lutheran Mary.[38] Instead, her ideas more closely resemble the Paracelsian understanding of Mary as active, powerful, and autonomous (at least to a point; as we shall see, she did require Count Carl's assistance in order to fulfill her fated role). If Anna rejected Mary's demotion to humble housewife, however, she did identify with and find power in Mary's motherhood, as well as the attribute of the Virgin Mary that had most successfully withstood the Reformation, that is, her spiritual and physical spotlessness.

Anna's purity, moreover, had a purpose. She interpreted her own lack of menstruation not only as the result of her alchemical purification at birth but also as a sign that she had a special destiny. Like the Virgin Mary before her, Anna would play a central role in sacred history by fulfilling an exceptional form of motherhood. And like Mary, whose husband Joseph did not father her child, Anna would not conceive a child with her own husband, Heinrich. Rather, in Anna's alchemical Holy Family, the father of her children would be Count Carl, the shadowy presence who had accompanied her claim to be a holy alchemist from the outset.[39]

As it happened, Carl, like Anna, also entered the world with a bit of fanfare. According to Philipp Sömmering, "When the Countess [von Oettingen], the count's supposed mother, was impregnated by Theophrastus [i.e., Paracelsus], he went to the emperor's court and said that the countess would give birth to a child who would surpass Theophrastus, Hermes, and all of the philosophers." Even nature attested to Carl's future philosophical prowess. "And when the count was born," Philipp continued, "there was a white eagle that fell out of the heavens. And then when she [the countess] rode out after it, she saw the eagle lying on the ground and then it soared back up toward the heavens."[40] As Anna explained, Paracelsus prophesied that Count Carl would join Anna and together they would put the fertile lion's blood to use. As Philipp explained, "When the Count grew up and received his father Theophrastus's [i.e., Paracelsus's] books, he found there that his father had sought the highest art, and that a maiden would be born who was free from the monthly flux, and with her . . . the Count was to have children who would live until the final days."[41]

One might wonder how this arrangement was to work. First of all, Anna was already married to Heinrich. Philipp later reported that Carl had addressed this obstacle, making a special arrangement with Heinrich to facilitate Anna's union with Carl. "The count wanted to have Frau Anna," according to Philipp, "and in exchange he offered her husband Heinrich his own sister plus [two?] tons of gold. And Heinrich was willing, and said that if he got the money, he would gladly give her to the count."[42] Since Carl never appeared, the marriage arrangements remained hypothetical, but Anna seems to have put quite a bit of thought into the nature of the extraordinary children that she and Carl would produce when and if they got the chance. Echoing Anna's own birth, Anna and Carl's babies would be born prematurely and fed with a tincture. According to Heinrich, Anna "professed that the Count could create a child with her every month, and if a child lay in the womb for six weeks or less, it was supposed to be finished [*vorkommlich*] and would be raised further by the Count with the tincture."[43] Anna herself was a bit more modest than Heinrich. According to her, Count Carl said that the children would be born in just over six *months*, rather than six *weeks*. Still, she confirmed that this feat could be accomplished alchemically, just as her own birth had been, for the babies would be nourished after their premature birth with a tincture.[44] Like Anna's, these babies' unusual births would result in exceptionally pure bodies, nearly untouched by the postlapsarian plights of disease, corruption, and death. Anna maintained, for example, that "the children would never become ill, would outlive their normal lifespan, and would die when God so desires."[45] Heinrich, meanwhile, confirmed that "the children shall grow up faster than others, [and] the females shall not be burdened with the flux . . . and if Adam's fall had not happened, they could well have lived forever, but now they will become as old as Methuselah and other patriarchs."[46]

Who were these alchemical children, unusually pure and as long-lived as the Old Testament patriarchs? They offer yet another glimpse of Anna's holy alchemy, which posited the alchemical tincture as a means to restore the world in order to hasten its end. First of all, the lion's blood could restore the fertility that had been diminishing ever since the Fall. As Luther explained, "If the human race had continued to remain in innocence, the fertility of the women would have been far greater. . . . I nevertheless have no doubt that if there were no sin, women would have given birth to a much more numerous offspring. Now those who are most fertile give birth at most to one child in a single year, and that shameful and heinous lust has been added to it. All this reminds of the enormity of sin."[47] For Luther, lower rates of fertility were just one of many examples of the degeneration of the world since the Fall. Count Carl's fecund alchemical tincture, however, could restore the unbridled fertility of the Garden of Eden and thus truly fulfill God's command to Adam and Eve to "be fruitful and multiply." Moreover, this alchemical conception promised to produce not only more bodies, but also *different kinds of bodies*, just as it had altered Anna's own body and thus prepared it to host this spectacular reproduction. Anna and Carl's alchemical babies, too, would be nourished with the tincture from birth, which would not only confer on them the great longevity of the patriarchs, but also ensure that Anna's daughters would be as spotless as she was. The tincture not only stimulated fertility and promoted longevity, therefore—twin claims that alchemists had made for centuries about the elixir and the philosophers' stone—but also purified bodies at birth and exempted them from the worldly suffering brought about by original sin.

Anna, in short, envisioned her alchemy as a means to bring about what had happened only once before through a miracle, when the Virgin Mary was released from sin through God's grace and purified so that she might bear the Savior. As Martin Luther explained it in a 1522 Christmas sermon, the holy Virgin

> gave birth without sin, without shame, without pain, and without injury, just as she also conceived without sin. The curse of Eve

doesn't apply to her, which says that "in pain shall you bring forth children," but otherwise it happened to her exactly as it does with any other woman giving birth. For grace did not promise anything, and did not hinder nature or the works of nature, but improved and helped them. In the same way she fed him naturally with milk from her breasts; without a doubt she did not give him any stranger's milk or feed him with any other body part than the breast. [Her breasts] were filled with milk by God without any uncleanness or injury.[48]

In Luther's view, Mary's bodily experience of reproduction was the same as any other woman's, to the extent that she conceived, delivered, and breastfed the infant Christ. Likewise, Anna Zieglerin envisioned herself conceiving and bearing real children (although she would feed them with tincture rather than breast milk).[49] God's grace rendered Mary's reproduction spotless, removing the pain, shame, and corruption that had attended motherhood for all women since the Fall; alchemy did essentially the same thing for Anna. An alchemical tincture had already purified her body, after all, rendering it capable of a superior form of reproduction, and now it promised not only to stimulate prelapsarian levels of fertility and antediluvian levels of longevity, but also to purify the bodies of her own sons and daughters.

Anna framed this spectacular, alchemical reproduction in eschatological terms, linking it to the end of earthly time. Just as Mary's son, Christ, had offered humanity salvation once before, so, too, would Anna's children play a crucial role in restoring the world once again, paving the way for the end of the world. "The Count said he would take a wife, so that he could bring about a new world thereby," Anna proclaimed, adding that "this would take place with the tincture, this he said often."[50] Heinrich simply pronounced, "From such children would arise a new world."[51] Paracelsus had recorded this prophecy in a book, they said, which Count Carl had received from his mother and then passed on to Anna. In a telling phrase that highlights the fusion of practical and spiritual alchemy that characterized Anna's work, she said that she "had an entire *Kunstbuch* that Theophrastus created and wrote. . . . She doesn't know how many hundreds of pieces [i.e., recipes] were in it. One calls it the cabala."[52] Indeed, she said, Paracelsus's handbook not only contained his prophecy, but was also "about alchemy and astrology," and contained instructions for sewing clothing, curing deadly illness in sheep and horses, "and other things from which a horse might be bewitched."[53] This mixture of practical techniques, medical recipes, alchemy, and astrology was typical fare for *Kunstbücher* in this period; indeed, it was not unlike Anna's own collection of recipes for the lion's blood.[54] But Anna also called this book a "cabala," which connoted something more—a book that might decipher the divine secrets hidden in the book of nature. Indeed, Philipp, who had seen the book, called it a "book of the angels," and he recalled reading in it "how the angels in paradise prayed to God, and that in God's council it was decided that God wanted to let mankind fall, and then create the world again through the Count and Frau Anna and help mankind."[55] This made the Paracelsian manuscript far more than just a collection of household recipes—it contained heavenly secrets as well, including Paracelsus's stunning prophecy that Anna and Count Carl would use the lion's blood to fulfill God's plan and usher in nothing less than the final act in the drama of sacred history. Just as the Virgin Mary's body had once served as the vessel for the incarnation of Christ, with the help of the lion's blood this new Virgin Mary, Anna Zieglerin, would turn her body into an alchemical vessel to produce innumerable spotless children who would restore the world and bring about the Last Days.

Revealing the End Times

Just as Anna's alchemy allowed her to reinterpret contemporary Lutheran ideas about the Virgin Mary, so, too, did the art offer her a way to address one of the most pressing political and religious issues of her day: the coming end times. A convergence of disruptive and noteworthy events in the sixteenth century—the fracturing of Christendom and religious warfare, famine, social unrest, new diseases such as syphilis, war with the Ottoman Empire, astral events, and so on—all contributed to a pervasive sense of living at the end of time, as well as to a proliferation of texts connecting current events to prophecies about the Last Days.[56] In the German lands, numerous astrological prognostications, or *practica*, linking medieval prophecies to contemporary astrological events rolled off the presses and found eager readerships among artisanal urban readers and emperors alike.[57] Events on earth, particularly the social, religious, and political upheaval that accompanied the Reformation, were equally likely to inspire prophetic fervor and the expectation of an earthly paradise before the end times. This eschatological anticipation was particularly intense among Luther's followers. Luther himself expected the Day of Judgment to arrive any day, and he developed a detailed biblical exegesis interpreting the travails of his own age as fulfillment of biblical prophecies. After his death, Luther's followers collected, printed, and elaborated on the reformer's prophetic statements, further intensifying eschatological speculation and expectation. "All agreed that Christ would return in the flesh to judge mankind, that the Second Advent was not far off, that the Reformation was the final flowering of the Gospel, and that the current decadence of the worldly regiment was the result of Satan's last and greatest effort to win souls," according to Robin Barnes. "Members of every [Lutheran] theological faction, along with pastors and laymen who were members of no faction at all, shared in the popularization and intensification of Luther's expectations.[58]

The imminent end was not simply a matter of concern for pastors and their congregations; scholars directed a great deal of their energy to the matter as well. Luther himself was relatively conservative about the sources of prophecy, focusing largely on scripture and rejecting direct prophetic revelation, but after his death in 1546, this caution dissipated. Prophecies proliferated, as Luther's followers began to find evidence of God's plan everywhere. Lutherans launched a variety of what Barnes has called "prophetic research" projects involving not only the examination of scripture or collections of ancient prophecies promising prophetic insights into the present, but also the observation of nature, analysis of political events, and deployment of mathematical tools to calculate biblical chronology or astrology to read the stars. By Anna Zieglerin's lifetime, it was widely assumed that God had provided clues to his unfolding plan in two parallel books of revelation, scripture and the "book of nature" (that is, the natural world), and that skilled interpreters could detect those clues. The aim was to look for signs that might indicate when and how the Last Days would arrive (or whether they already had).[59] Natural wonders, disasters, strange celestial events, spiritual corruption, social upheaval, and famine all pointed to the fact that the tribulations of the end times had already begun. As Barnes put it, "The imminence of cosmic upheaval and the Last Judgment became, for the generations after Luther's death, a principle or framework by which all human experience could be organized. . . . [It was] an all-embracing assumption."[60]

But what did it mean to live at the end of time? Given their certainty that the world was reaching its final moments, sixteenth-century Europeans naturally wanted to know what to expect. Here they had recourse not only to scripture, particularly the Revelation of John, but also to a long exegetical tradition stretching back to the earliest church fathers, and including a flood of more recent interpretations. Not surprisingly, there were perhaps as many interpretations of the events of the end times as there were speculations about when they would begin. While all agreed that epic battles with Antichrist would precede Christ's return to earth before the Day of Judgment, there was a variety of opinions about how to interpret the passage in Revelation 20:1–6 about a thousand-year kingdom of Christ. Some, following Saint Augustine of Hippo, rejected the idea that this "millennium" referred to an earthly kingdom to come, interpreting it instead as a reference to the "life of the Church in the present."[61] Luther, trained as an Augustinian monk, shared Augustine's pessimism about the state of the world, seeing evidence everywhere of spiritual, natural, and social corruption. He greeted this with a sense of joy, knowing that the world would soon dissolve; there was little point in trying to reform it in its few remaining moments. A parallel eschatological tradition, however, was more optimistic, foreseeing a period of happiness, peace, spiritual insight, and renewal, either before or after the appearance of Antichrist.[62] The most important medieval figure in reigniting discussions about an earthly millennium was the twelfth-century abbot Joachim of Fiore, who predicted that after Antichrist's annihilation would arise an era of the Holy Spirit, a period of increased spiritual understanding and peace on earth before the end of time. Significantly, Joachim's ideas influenced Arnald of Villanova (ca. 1235–1311) and John of Rupescissa (ca. 1310–ca. 1365), both of whom not only authored (or were falsely associated with) alchemical texts, but also contributed to eschatological speculation about the nature of the end times. Notably, both Villanova and Rupescissa not only argued for an earthly millennium in their eschatologies; they also made the period of peace and insight after Antichrist's demise longer than had any of their predecessors. Villanova extended it to forty-five years, but Rupescissa argued for a literal millennium, a full thousand years, a scenario that no one had endorsed since the early church fathers.[63] When Anna Zieglerin claimed that her alchemy could produce a divinely ordained "new world," she may have had something akin to an earthly millennium in mind.

Rupescissa's view was extreme in its time, as his persecution at the hands of the church made clear. In the early years of the Reformation, however, chiliasm gained renewed currency among radical reformers and more establishment figures alike, who shared the belief that the new age had already dawned and would be accompanied by social justice as well as increased gifts of prophecy and biblical interpretation among both learned and lay.[64] Paracelsus, too, shared with his contemporaries not only the deeply held conviction that he was living at the end of time, but also this more optimistic view of the end times as a period of both tribulation and promise, at least for the elect. Characteristically, he put his own twist on this tradition. Whereas previous eschatological texts emphasized the profound spiritual understanding that would accompany the earthly millennium, Paracelsus argued that it would include a renewal of knowledge (particularly in medicine and the arts) in a final burst of reform before the return of the Messiah. Paracelsus's emphasis on the reform of knowledge led him to a novel interpretation of the Old Testament prophet Elias, who had long been a central figure in understandings of the end times. According to Christian and Jewish tradition, Elias (or Helias or Elijah) would return, his very arrival signaling the coming end. Whereas traditionally Elias was conceived of as a bearer of perfect spiritual knowledge, Paracelsus's "Elias artista" would restore knowledge of *nature and the arts*, including the secret of alchemical transmutation. Thus, by integrating the arts, even alchemical transmutation, into the knowledge Elias would reveal, Paracelsus framed the *very appearance* of someone with knowledge of the philosophers' stone, among other arts, as the fulfillment of the biblical prophecy linking revelation of nature's secrets to the Last Days.[65] For Anna Zieglerin and her contemporaries, therefore, Luther's pessimism about the fate of the world in its last moments was somewhat tempered by the optimism of Paracelsus, among others, about a final burst of intellectual and practical reform before the (imminent) last day.

In this eschatological atmosphere, numerous individuals claimed to have special insight into God's plan, either because they were particularly skilled at deciphering the clues he provided in nature and/or scripture, because they had a more direct access to the divine, or both. Paracelsus, for instance, foresaw a special role for himself, both at the forefront of reform in medical and natural philosophical knowledge, and as one of the few whom God had endowed with the ability to decipher the signs of his plan. His prognostications were some of the very few of his writings that appeared in print in his own lifetime, cementing his reputation as not only a physician and alchemist, but also a prophet.[66] The Elizabethan mathematician John Dee (1527–1608/9), meanwhile, turned to angels to help him decipher the natural world and participate in its restitution, while the French diplomat, scholar, and prophet Guillaume Postel (1510–1581) claimed both prophetic gifts and a mission to promote religious concord, bolstered in part by the tutelage of a Venetian holy woman, the "Virgin of Venice."[67] The most powerful culturally available model for understanding this kind of extraordinary spiritual and intellectual insight was the biblical prophet Elias, and, indeed, many made comparisons between Elias and Postel or Dee. Postel, for instance, believed that his project for spiritual and intellectual reform heralded a new era of "restitution," and, accordingly, he thought of himself as a "new Elias."[68] Likewise, Dee's ongoing conversations with angels led him to believe that he, too, was heir to the biblical prophets Elias, Esdras, and Enoch.[69] Some even identified Luther as Elias, although he did not encourage the comparison.[70]

Dee and Postel were both erudite denizens of the international Republic of Letters who pursued sophisticated (if risky) prophetic and scholarly projects with the resources of privileged men of learning and social status. Their learning, status, and masculinity made comparisons with Elias conceivable in a way that it was not for others. Still, other kinds of prophetic gifts proliferated among average men and women, often in connection with angelic revelation.[71] Scripture, of course, was filled with examples of angels acting in the world, and although theologians differed about matters such as whether individuals had guardian angels, for example, or whether it was acceptable to represent angels visually, there was general agreement in the early modern period that angels could serve as guides and messengers.[72] Some individuals, most famously John Dee, sought angels out, going to great lengths to find a way to communicate with the celestial beings. In most cases, however, angels simply appeared, usually dressed in white robes, to lay prophets who were to spread their message.[73] Such angels usually delivered a message of repentance, most often for the sin of pride, and admonished the prophet to whom they appeared to share the message with the local pastor, who was to preach it to his congregation. Most prophets were unlettered: more like Hänschen in Gotha than John Dee. They could be men or women, rural or urban, and could come from a variety of social backgrounds, although noble prophets were rare. Typically they were not radicals, in that usually they did not seek to undermine the established church, although angelic revelations did give common people a way to speak out about the sinful ways of their contemporaries. In the Lutheran lands of northern Europe, where saints could no longer appear to spread divine messages, hundreds of "new prophets," as they were called in contemporary parlance, received angelic revelations in the sixteenth and seventeenth centuries, and their stories circulated widely in printed woodcuts and in sermons. Indeed, according to Jürgen Beyer, who has studied these Lutheran prophets most closely, most people would have heard about contemporary angelic revelations in early modern Lutheran lands. They were simply part of the fabric of Lutheran life.[74]

When Anna Zieglerin introduced Paracelsus's prophecy to Duke Julius's court in Wolfenbüttel and claimed for herself a central role in the events to come, therefore, she was participating in one of the widespread preoccupations of her day: speculations about the timing, nature, and progression of the end times. Nor was she alone in finding a connection between alchemy and prophecy in particular. One point of intersection between these two traditions in early modern culture more broadly was allegorical. If the philosophers' stone could be likened to Christ (and vice versa), then, by implication, the entire alchemical opus—the process of making and using the philosophers' stone—was, in some sense, soteriological; working with the bodies and souls of metals, perfecting nature and matter, the alchemist touches a fundamental truth of faith, its promise of salvation. As Chiara Crisciani has argued, moreover, "Alchemy could be defined as a concrete and operative prophecy. It is an announcement of truth and salvation because it attains them in operations and facts, in the verified transformations which the alchemist produces through his doctrine, *experientia* and *industria manuum* [hands-on work]." By reenacting sacred history in the laboratory, in other words, the Christian alchemist also announced Christianity's truths, serving as a kind of prophet. "The prophets' words come true both in the real history of Christ and in the material process of the elixir, the saviour of matter," Crisciani writes. "Therefore, the events of the Son of God explain and orient the material phases of the *opus*."[75] Even Martin Luther appreciated the way in which alchemical work could, in a sense, ratify prophecies about the fate of the world, although he focused not on the analogy between Christ and the philosophers' stone, but rather on parallels between the alchemical separation and purification of matter and the Day of Judgment. "The science of alchymy I like well, and, indeed, 'tis the philosophy of the ancients," he reportedly told his companions. While Luther welcomed "the profits it brings in melting metals, in decocting, preparing, extracting, and distilling herbs, [and] roots," he was drawn to it in the context of his faith as well. "I like it also for the sake of the allegory and secret signification, which is exceedingly fine, touching the resurrection of the dead at the last day," he observed. "For, as in a furnace the fire extracts and separates from a substance the other portions, and carries upward the spirit, the life, the sap, the strength, while the unclean matter, the dregs, remain at the bottom, like a dead and worthless carcass; even so God, at the day of judgment, will separate all things through fire, the righteous from the ungodly. The Christians and righteous shall ascend upward into heaven, and there live everlastingly, but the wicked and the ungodly, as the dross and filth, shall remain in hell, and there be damned."[76] Perhaps not surprisingly, given how large the Day of Judgment loomed in Luther's consciousness, he found in alchemy not just an allegory for redemption generally, but specifically an allegory of humanity's final reckoning.

Medieval alchemists who not only wrote about alchemy but also practiced it certainly appreciated these parallels as well, but typically they also found a more practical role for their art in the end times. Roger Bacon, (Pseudo-)Arnald of Villanova, (Pseudo-)Ramón Lull, and John of Rupescissa all combined the study of nature (and of alchemy in particular) with their understanding of the Last Days.[77] Rupescissa offers the most powerful point of continuity with Anna's own apocalyptic alchemy. He argued that alchemical technologies would prove to be crucially useful tools in the epic battles against Antichrist as the world came to an end. On the one hand, transmutation could provide a supply of money to the true Christians once their tormenters had seized church goods. On the other, the healing properties of the elixir promised to fortify and heal a select group of "evangelical men" who would play a leading role in the final confrontation with Antichrist. As Leah DeVun has argued, Rupescissa made alchemy central to his innovative claim for human agency in the end times; by studying nature, specifically alchemy, these evangelical men could take fate into their own hands, using alchemical technologies to defeat Antichrist and hasten the millennium.[78]

The parallels between Anna's apocalyptic alchemy and Rupescissa's suggest that she may have been familiar in some way with earlier scholars' explorations of the eschatological possibilities of alchemy. Even if the lines of influence were diffuse, Anna's ideas place her firmly within this long tradition, stretching from Rupescissa to Newton, which examines intersections between alchemy and the end times.[79] Anna's inspiration may also have come from something much more immediate—namely, the heady combination of alchemy, prophecy, and politics she witnessed firsthand in the 1560s while she was associated with Duke Johann Friedrich II's court in Gotha, where Hänschen, the peasant boy who conversed with angels, demonstrated forcefully both the promise and the perils of prophecy. Although it is not clear whether Anna was actually in Gotha itself during the 1567 siege that Hänschen's angels brought about, she must have been distressed by the degenerating social and political situation inside the city walls, and fears about her own fate (both earthly and eternal) certainly would have been foremost in her mind.

As awful as Anna's experience in Saxony that winter must have been, Hänschen's failure clearly did not shake her belief that the end of time was at hand. For Anna as well as nearly everyone else in the sixteenth century, prophetic error could easily be explained away, leaving the fundamental truths of Christian eschatology firmly intact.[80] Rather than a new skepticism about prophecy or the immanent eschaton, the lessons Anna took away from her affiliation with Duke Johann Friedrich II's court during these extraordinary last years of his reign seem to have been more nuanced. First and foremost, she learned about the power of prophecy at court, particularly the ways in which prophecy could command bold action. Hänschen was merely an unlettered peasant boy, after all, but his prophetic utterances (once they were certified as credible) gained him influence that few others could hope to have. This may have emboldened Anna several years later when she shared the Paracelsian prophecy about her alchemical reproduction with Count Carl in Wolfenbüttel. Moreover, in Gotha Anna also witnessed the radical move of taking action and participating in the end times, rather than merely "looking forward and waiting," as the historian Robin Barnes put it, for them to come.[81] This, too, might have encouraged her to believe that she might have a part to play in the final moments of the world.

And yet, Anna's eschatology was far more cautious (and less violent) than Duke Johann Friedrich II's campaign to promote the gospel and reform the world in anticipation of the coming end. As a prince of the empire guided by angels, Duke Johann Friedrich II resorted to holy war, cataclysmic violence, and a military strategy that defied political logic. Anna did not promote the idea of a final, climactic battle; even Antichrist is strikingly absent from her vision of the end, at least as that vision was documented at Julius's court. Anna seems to have embraced the more optimistic eschatology articulated by figures from Joachim of Fiore and Rupescissa to Paracelsus, dedicating her energies to the restitution of the world in its final moments and the revelation of knowledge about the workings of nature. This may have been part of Anna's appeal, from Duke Julius's perspective. Anna offered him a way to use his position as a prince to uncompromisingly serve Luther's cause—to restore the gospel and a fallen world in its last moments—*without* risking the political destruction that had befallen Johann Friedrich II in Gotha. As Julius was seeking to establish himself as a leader in the delicate political and religious affairs of a fractured Holy Roman Empire, this kind of alchemical intervention would have been immensely appealing. What is distinctive about Anna's view, moreover, was her emphasis on the crucial role of *generation* in her apocalyptic alchemy. The lion's blood was not just an opportunity for a general meditation on the truths of Christianity, or even a more practical medicine or source of wealth in the final battles with Antichrist, but rather a way to regenerate the world.

Anna Zieglerin's holy alchemy, in the end, reveals a profound interplay between her alchemy and her belief, as each informed the other and even inspired creative interpretations of long-standing traditions. Anna assembled her understanding of the lion's blood and its workings out of several ideas that were culturally available to her, including strong associations between a lack of menstruation, spiritual purity, and the Virgin Mary; alchemy's powers to stimulate fertility, promote longevity, and purify matter; and a long tradition that posited alchemy as a way to contemplate sacred history and even intervene in events to come. Where Anna was innovative in her own time, and instructive for us, is in the way she assembled and reinterpreted long-standing ideas and debates about alchemy, the Virgin Mary, and the end of time to craft a narrative that was both extraordinarily creative and bridged the deeply personal with the epic scale of sacred time. Thus, while alchemists before her had envisioned the philosophers' stone as Christ, achieving its redemptive power only through a creative process that resembled the passion, Anna saw the Virgin Mary, likening the alchemist's exceptional generative powers—working at once within and beyond nature—to Mary's own motherhood. This connection allowed her to posit a new role for the alchemist in sacred time, as someone who could generate not only medicines, precious metals, and gemstones, but also life itself, restoring prelapsarian levels of fertility on the brink of the eschaton.

Equally striking, however, is the way in which Anna situated her own body at the center of her alchemical work, making it the primary site where her belief and alchemical work came together and found meaning. While the lion's blood could cure leprous or infertile bodies and produce the philosophers' stone and other earthly delights for anyone, its sacred purpose would be revealed only through its action on and in Anna's body, itself made extraordinary through an alchemical intervention at birth. The thread that ran through Anna's entire story about herself and the lion's blood was her claim that alchemy could break down the boundary between matter and spirit, between the natural and the supernatural, such that manipulating matter became a way of manipulating the sacred as well. This was the radical claim of Anna's holy alchemy, and the reminder to us that in early modern Europe, alchemy offered not only a way of engaging the reigning scholarly debates of the day, but also a means to work out questions at the heart of Christianity.

CHAPTER 6

Unraveling

In the summer and fall of 1573, Anna Zieglerin and Philipp Sömmering were both busily working on their own alchemical projects and communicating regularly about the art with their patron, Duke Julius.[1] Duchess Hedwig and her sister-in-law/aunt, Margravess Katharina of Brandenburg-Küstrin, had already expressed disdain for the alchemists in their private correspondence, but Duke Julius himself did not appear to share the concerns of the women in his family. He supported the alchemists' work alongside his many other ducal projects. As fall turned to winter, however, new voices joined Hedwig and Katharina's campaign against Anna, Philipp, and Heinrich, and even Julius began to waver. In the spring of 1574, perhaps detecting that the tide finally had turned against them at Julius and Hedwig's court, Anna and Heinrich left Wolfenbüttel and headed up into the Harz Mountains to the mining city of Goslar. Anna continued to pursue her alchemical work. As Philipp later reported, "Frau Anna worked in a little pantry in Goslar, and claimed to have made a solution of mercury. It is in the pantry in a little flask."[2] Philipp soon joined them. He, too, hoped to complete his alchemical work in Goslar, and estimated that it would take him six weeks to finish his tincture.[3]

Anna would later report that she experienced terrifying omens, apparitions, and dreams on the road to Wolfenbüttel nearly three years before. As Anna left Wolfenbüttel now, perhaps she reflected on these disturbances once again. She had interpreted the visions as signs that they would perish in Wolfenbüttel, so she may have felt some relief at leaving the ducal city alive after all.[4] Or perhaps the sense of foreboding followed her all the way to Goslar, since even there, where the Wolfenbüttel dukes held a great deal of influence, she was still not entirely beyond Julius's reach. Indeed, it was not long before Anna's visions were borne out, her prophetic powers proving themselves in ways she certainly would have preferred they had not.

What triggered Duke Julius's change of heart and encouraged him not only to drop his support for the alchemists, but ultimately to use the full force of the law to obliterate them? No single event seems to have set their downfall in motion, nor did Duke Julius or Duchess Hedwig accuse the alchemists of failing to complete their work. Moreover, the alchemists' transformation from valuable, if controversial, courtiers into criminals was never inevitable. For the most part, alchemy was taken seriously in the courts of early modern Europe because sovereigns prized and understood its potential to bolster their economic, intellectual, and political agendas. At the same time, patrons were not naïve. They recognized that some alchemists were not as skilled as they claimed to be, and that some might even be downright deceptive, merely posing as alchemists in order to take advantage of potential patrons. In response, patrons like Julius developed a variety of ways to protect themselves from fraud and incompetence, while still continuing to support alchemy that they deemed to be legitimate.[5]

Fraud charges eventually emerged in Wolfenbüttel, but they were not the main cause of Anna Zieglerin and her fellow alchemists' downfall. Rather, the turn against them in Wolfenbüttel brings into focus a different dimension of courtly alchemy in sixteenth-century central Europe: its associations with sorcery, poison, and political sabotage. For a number of Duke Julius's fellow sovereigns in the 1560s and 1570s, alchemical expertise came to be linked not only with the production of metals and medicines, or even, more negatively, with fraud, but also with the ability to make substances that could manipulate the affections and bodies of dukes and duchesses and thus destabilize princely rule in the Holy Roman Empire. In Wolfenbüttel, this possibility was raised most vividly by the confessions of the sorcerers and poisoners Grumbach and Lippold, both of whom also seemed to be linked to alchemy, as well as by another incident involving accusations of poisoning and witchcraft at the court of Julius's cousin, Duke Erich II of Braunschweig-Lüneburg-Calenburg. These scandals caused Julius to see his alchemists in a new light, and he began to consider the possibility that they, too, might bring disorder, even death, to his court. Duchess Hedwig and her brother, Elector Johann Georg of Brandenburg, expressed genuine anxiety about this prospect, and they continued to pressure Julius to take it seriously. When Anna and her companions arrived in 1571, Julius was not at all convinced that alchemy had anything to do with sorcery or poison, and as long as he supported his alchemists, they seemed to be able to withstand the full force of the rumors that swirled around them from their years in Gotha. By the spring of 1574, however, the climate around courtly alchemy had shifted, and the duke begin to waver about his support for his own alchemists. Eventually, he decided to launch a legal investigation into the truth about Anna Zieglerin, Philipp Sömmering, and their companions.

Kettwig's Tale of Fraud

In November of 1573, one of Julius's military and financial advisers, Jobst Kettwig, wrote to the duke to convey a simple and powerful message: Anna Zieglerin and Philipp Sömmering were frauds.[6] They had misrepresented their alchemical expertise and they had forged numerous instruments of identification, including titles, seals, and even names. In accusing Philipp and Anna of both alchemical and identity fraud, Kettwig deployed the powerful image of the alchemical *Betrüger*, or fraud, which had gained cultural currency in the sixteenth century, shaping the trajectories of numerous alchemists and their patrons.[7] Kettwig assured the ducal court that his claim was more than a literary trope, however; he could back up these assertions with proof, for he had traveled around the Holy Roman Empire (at great personal cost and danger, he underscored) investigating the assertions that Anna and the others had made at the Wolfenbüttel court. In his travels, he said, he had located a number of witnesses who would challenge Anna and Philipp's own narratives about themselves.[8] He explained that he felt compelled to reveal all of this to Julius "not out of hate or envy," but rather to protect the duke because of his "committed obligation, the oaths and duties, that I have vowed and sworn to my Gracious Prince and Lord."[9]

The particular circumstances under which Jobst Kettwig brought this information to his patron's attention suggest a more complicated set of motives than he let on, for he wrote this letter to Julius from a Braunschweig jail. Kettwig, a former mercenary captain from Holstein, had entered Duke Julius's service in 1572 through his connection with Anna, Heinrich, and Philipp's companion Sylvester Schulvermann, with whom Kettwig had worked as a mercenary soldier before coming to Wolfenbüttel.[10] Julius evidently found Kettwig useful, for from the summer of 1572 to the spring of 1573, the duke sent the former soldier on several trade/financial missions around the Holy Roman Empire—for example, to help recruit skilled miners and prospectors to come to the duchy to help develop the mining industry.[11] In November of 1573, however, a warrant issued by the Danish crown and seeking Kettwig's arrest on charges of banditry and disturbing the peace appeared in Wolfenbüttel. When Julius learned from Philipp that Kettwig was in Braunschweig, the duke requested that the Braunschweig city council place him in custody.[12]

It was at this moment, sitting in a Braunschweig jail while Duke Julius and the Braunschweig city council argued about jurisdiction and extradition, that Kettwig accused Anna and Philipp of fraud. The timing was no accident. Kettwig himself linked his denunciation of the alchemists to his own arrest, claiming that he had discovered the truth about "the pastor and the whore" and was on the verge of reporting it to Julius's ducal officials, prompting Philipp Sömmering to denounce him preemptively. All he wanted now was the opportunity to bring what he had learned before the "eyes and ears" of Duke Julius himself, in writing or in person, and to refute the "false claims or complaints" that Philipp Sömmering had made. He simply wanted "his innocence and the godly truth" to come to light.[13] In short, Kettwig hoped to save his own skin by discrediting the "rogue, whore, and knave,"[14] that is, Philipp, Anna, and Heinrich, whom he blamed for filling Duke Julius's ears with the "false reports and lies" that led to his arrest. (Never mind, of course, that the entire affair began with a warrant from Denmark.)

Kettwig's invective flowed freely. Anna and Philipp were "godless, diabolic, lying people," who had committed "damnable, criminal, egregious, outrageous, [and] diabolic vice, harlotry, deceptive lies and fraud."[15] Moreover, as Kettwig explained, in his recent travels throughout the southern German, Swiss, and Austrian lands, he investigated Anna, Philipp, and Heinrich's stories about their pasts, their alchemical expertise, and Count Carl. First, Kettwig said, he traveled to the Swiss lands "of the deceased philosopher Theophrastus Paracelsus" in order to speak with the physician's acquaintances and family members, as well as to a monastery just a few miles outside of Zurich. He then roamed through Bavaria, Innsbruck, and Augsburg to Swabia and the lands of the Counts of Oettingen to the Freiherr von Mentzingen,[16] and, finally, to interview the brother and friends of Herr David Baumgartener, the Augsburg merchant and alchemist who had been executed in Gotha in 1567 for his involvement in the Grumbach Feud.[17] It was not easy to identify individuals in early modern Europe,[18] but against Philipp and Anna's own word about who they really were, Jobst Kettwig sought to marshal the personal testimony of others who had known them (or whom they had claimed to know).[19]

What Kettwig had learned on this trip, he revealed, convinced him that Philipp Sömmering and Anna Zieglerin had committed a twofold fraud, and he offered details that must have given Duke Julius pause. They had falsified seals, titles, and names, even the existence of Count Carl himself, and they had claimed abilities *in natura rerum* that "from the origins of the world never were and never would be."[20] The actual Counts of Oettingen were astonished to learn about Anna's tales of Count Carl, Kettwig reported. They lamented that their good reputation, name, title, signature, and seal "were misused for such false *Betrug*," and felt that they should be compensated accordingly.[21] Moreover, the person that Philipp and Anna had used these instruments to conjure up—Paracelsus's son Count Carl von Oettingen—was not, in fact Count of Oettingen, Kettwig argued, but rather Carol, Freiherr von Mentzingen, who was associated with the former Holy Roman Emperor and King of Spain Charles V.[22] Kettwig's conclusion? It was imperative for Duke Julius to understand that "never before have lies and fraud more outrageous and deceitful been heard and fabricated than those that Anna Maria and the pastor Philip Schreiber [*sic*] have brought forward and recited against My Gracious Prince and Lord."[23] The social fraud, Kettwig claimed, went hand in hand with the intellectual fraud of claiming alchemical expertise. The only way they could produce alchemical gold, Kettwig asserted, was through sleight of hand: "If the pastor and the whore ever make gold with their alchemy, I shall find it under the stairs, indeed, under the soles of their shoes."[24] Rather, he claimed, upon their arrival in Wolfenbüttel, the trio held council and resolved to sustain the charade for as long as they could get away with it before moving along to another place. They were swindlers, pure and simple.

If Kettwig was hoping that denouncing Anna and Philipp somehow would lead to his freedom, then he was mistaken. His letter to Duke Julius did not result in his immediate release from the Braunschweig prison, although he was soon moved from the Lauenturm (or Löwenturm) on the Kohlmarkt in the heart of Braunschweig to a more comfortable confinement in the residence of the bailiff Hans Hoyer, where Kettwig's family and others, including the Braunschweig city councillors, could visit. Apparently this move was precipitated by the fact that the harsh conditions in the Lauenturm had led him to fall repeatedly into "apoplexy"—a sudden loss of consciousness. Whatever the reason, the new quarters offered Kettwig an opportunity, and he did not hesitate. Only a few weeks after denouncing Anna and Philipp, on December 13, 1573, Kettwig escaped his new confinement.[25] Duke

Julius was furious with the Braunschweig city councillors for allowing this to happen (and perhaps even for colluding in Kettwig's escape, he suggested), and announced that if they did not deliver Kettwig within six weeks and three days, the period of time customary in Saxon law [the so-called *sächsischer Frist*], the city would forfeit all legal privileges.[26] Meanwhile, the duke set a manhunt in motion, circulating wanted posters and letters to princes and others seeking help in capturing Kettwig.[27] It took another nine months for the search for Kettwig to yield fruit.[28]

In the end, therefore, Kettwig's November 1573 letter to Duke Julius did him little good. In another sense, however, it did exactly what it was supposed to do: provide a powerful alternative to the narratives that Anna and the others told about themselves. Whereas Anna saw herself as an alchemist and a new Virgin Mary, Kettwig claimed that she was nothing but a "whore," a liar, and a fraud who had invented not only her identity as an alchemist, but also Count Carl von Oettingen (not to mention the letters and seals to make him credible). Count Carl was not who Anna said he was.[29] Moreover, this "Carl" had a history of leading sovereigns into debt. According to Kettwig, he had been associated with none other than Holy Roman Emperor Charles V, who had spent so much money on Carl's processes for making gold and gemstones that the emperor fell into debt and had to pay off creditors in land.[30]

There is something decidedly formulaic about Kettwig's claims. By the 1570s, the twinned social and intellectual fraud of the alchemist was emerging as a trope in central Europe, making it culturally available as an easy way to discredit anyone claiming alchemical expertise. Certainly, Anna and Philipp were not alone in facing these kinds of charges.[31] Likewise, Kettwig's repeated accusation that Anna was a "whore" was a generic-enough insult for early modern women to weaken its punch.[32] Finally, one wonders if Kettwig really did take the journey to investigate Anna and Philipp's story, as he offered no corroborating evidence (although he promised to provide witnesses). Were there really any of Paracelsus's "relatives" left to talk to in the Swiss lands, more than thirty years after his death? Who was this Freiherr von Ment-zingen, and if the Count von Oettingen was in fact outraged that his name and title were being abused in Wolfenbüttel, why did he not contact Duke Julius directly? Kettwig put forward a number of different pieces of information, and it is difficult to sort out what truth lay behind his charges. Never-theless, the claims that Kettwig made in his November 1573 letter raised troubling questions.

Anna and Philipp did not let Kettwig's assault go unanswered, naturally, and both wrote directly to Duke Julius to defend themselves. "I must confess that we handled things badly," Anna wrote, "but if anything happened deliberately or intentionally [as Kettwig claimed], then my dear Lord Jesus Christ's passion and death on behalf of myself and of all of us is for naught."[33] If anyone was a perpetrator of fraud, she said, it was "Kettwig and his retinue," not Anna and hers. "Therefore, I ask, for God's sake, that Your Princely Grace will rule me out of such suspicion, and by the grace of God not send me away." Invoking a sentiment that she would repeat during her subsequent trial, Anna said that she preferred her fate to rest with God. She would face divine justice if indeed she had committed any sort of fraud, for "God knows all hearts, and judges them according to their deserts and his divine knowledge."[34]

Shortly before this letter to Julius, Anna also wrote to her onetime patron, the imprisoned Johann Friedrich II of Saxony. Mostly this letter dealt with a practical matter, the assets she had lost during the siege of Gotha nearly a decade prior, suggesting that she may have been trying to gather resources in case she lost Duke Julius's favor. She reminded Johann Friedrich II that she and Heinrich had transferred all of their assets to the ducal household at the Grimmenstein fortress before the siege of Gotha, whence they had presumably been confiscated after Johann Friedrich II's defeat. She now asked the erstwhile duke to help retrieve their assets so that they no longer had to depend on Duke Julius's generosity in Wolfenbüttel.[35] Beyond this practical matter, Anna also described her suffering in the years since Gotha in vivid terms, lamenting that she had endured "such protracted misery, enacted in such a way that I have often wished that I had not been born."[36] She saw Julius's decision to offer her, Heinrich, and Philipp refuge as an act of God and praised the duke for taking them in, offering his protection (*Schutz und Schirm*), and even helping them financially. Without this intervention, she wrote, "Against God and all fairness, we would have been sacrificed on the altar, just as they [i.e., her enemies] have ranted and raged daily since."[37] If these two letters are an indication of Anna's mind-set as things were unraveling at the Wolfenbüttel court, she was increasingly willing simply to throw herself on the mercy of God and her patrons. She saw the work of the devil in her suffering, but she also saw the hand of God: "Our God cares for us, and keeps paternal, godly watch over us," she told Julius; "and [God] enjoins the Devil to suspend his damnation and [stop] afflicting and plaguing us."[38] Anna fiercely maintained her innocence and rebuked her enemies, to be sure, but she also understood her suffering as part of her worldly lot.

Philipp Sömmering's letters to Julius during these months were much more effusive, as well as a bit more complex in their strategy. Most immediately, Philipp addressed Kettwig's claims by simultaneously stating his own innocence and attacking Kettwig's credibility. He also returned to the Grumbach Feud as the origins of his trouble, rehearsing his persecution at the hands of his old nemesis, Gotha City-Superintendent Weidemann, and he deployed his training as a pastor by quoting scripture about the persecution of the righteous. Unlike Anna, Philipp was not content to let God decide his fate, but rather he tried to use Julius's credibility to bolster his own; he reminded Julius how useful he had been at the Wolfenbüttel court and asked that the duke either vouch for Philipp's character in print or release him from service so that he could start over (again) elsewhere. Finally, and perhaps most interestingly, Philipp addressed head-on the persistent claim that his involvement with court alchemy somehow linked him to the practice of sorcery. His strategy worked on many levels, addressing imperial politics and gossip, his own history with Julius, and the precarious status of the courtly alchemist.

Duke Julius appears to have let Philipp "see and read" a copy of Kettwig's denunciation.[39] In response, Philipp quickly countered Kettwig with his own letter to Duke Julius in which he expressed his astonishment at "how shockingly, cruelly, and outrageously that rogue [Kettwig], who deserves to be broken on the wheel seven times over, assailed me."[40] After rather dramatically claiming that a thousand witnesses could attest to his innocence, Philipp drew on his training as a pastor and cited scripture to frame his "persecution" as the suffering of the righteous.[41] Even as Philipp tried to position his foes as enemies of Christ, the alchemist also returned Kettwig's invective with a string of his own, fulminating that Kettwig deserved to be broken on the wheel, quartered, and hanged on the four roads into town as a "godless, unchristian [man]; repeat perjurer; dishonorable, desperate, arch-rogue, traitor, wretch, land- and sea-bandit" (among other things).[42] Having painted his detractors as criminals and sinners of the worst sort, Sömmering, like Anna and also Kettwig, concluded with a heartfelt declaration of his own innocence: "I bear witness before God and the world . . . that in all my days I have never lied to nor deceived His Princely Grace about anything."[43]

After Kettwig's flight from prison, in January of 1574, Philipp Sömmering wrote to Julius again. Naturally, given the delicate situation, Philipp began his letter by addressing the particularities of his own position in Wolfenbüttel. What makes this letter especially illuminating, however, is the way that Philipp moved beyond his own story, speaking to alchemy's reputation more generally. Philipp put his finger on the crucial problem for the alchemists and the duke alike: in some of the princely courts of central Europe alchemy had developed a dangerous reputation. Moreover, in identifying a problem that would affect everyone at Julius's court, Philipp argued for a crucial role for princely power in breaking the false link between alchemy and sorcery.

Philipp rooted his particular problems in the Grumbach Feud, and so he began by retelling his entire story, from the 1567 siege of Gotha to the precarious situation he now faced in Wolfenbüttel. Once again, Philipp repeated his tale of unrelenting persecution at the hands of the Gotha City-Superintendent Melchior Weidemann, Philipp's old nemesis from the fractious Lutheran world of ducal Saxony, who had hounded him out of his homeland after the Grumbach Feud came to an end.[44] Philipp traced his suffering to Weidemann's false charge that not only had he abandoned his post as pastor in Schonau, but also, and much worse, he had been one of Grumbach's associates, and thus was guilty of some involvement in poison, sorcery, and political sabotage.[45] Julius had already investigated and rejected these charges,[46] Philipp reminded his patron; moreover, he added, during his time in Wolfenbüttel Philipp had served truly and honorably as Julius's adviser in political, church, and mining affairs. As a religious adviser, Philipp took credit for helping Julius navigate the "dangerous labyrinth" of Lutheran factionalism, protecting his churches and schools from the "poison" of both the radical Flaccians and the Sacramentarians. In mineral matters, Philipp recalled, he prevented Julius from smelting arsenic in his chambers and warned the duke that he could endanger his life. Philipp also helped the duke avoid impostor alchemists, just as he had done for Duke Johann Friedrich II before.[47] Finally, Philipp was proud of his contributions to Julius's mining industry, to which the duke and his mining officials would attest. In short, he insisted, he had been extraordinarily useful to Julius, and finally felt optimistic about his future.

And yet, Philipp lamented, no matter what he did, he could not escape the old rumors that he was a magician and a sorcerer (*zeuberer und schwartzkünstler*). "Just when I had hope again," Philipp lamented, "it was crushed by those desperate rogues and knaves Kettwig and Schulvermann, who destroyed honor, property, body, life, even salvation." Once again, thanks to

Kettwig, old rumors that Philipp was a sorcerer, even that he had renounced Christ, resurfaced, endangering not only Philipp's reputation, but Julius's as well. Philipp feared that his enemies would not rest until he was destroyed; he did not even dare to eat at court for fear of poison, he added. And so Philipp, finally, proposed a solution. First, he implored Julius to let him, together with Heinrich and Anna, put their troubles behind them once again and leave Wolfenbüttel, perhaps retreating to another nearby city (Braunsch-weig, Helmstedt, even Goslar). Philipp also offered to leave Julius's service altogether, perhaps to join the court of another prince, if the duke preferred to distance himself. But Philipp also begged Julius to put his own honor and credibility on the line, "to give witness before God and the world that you have never noticed us do anything unchristian or to do with sorcery," echoing his earlier request that the duke issue a printed statement in support of Philipp.[48]

With this last suggestion, Philipp pointed explicitly to patrons' power to bolster reputations and offer security.[49] "I am a Christian," Philipp contin-ued, "and since my youth, I have zealously pursued honor, virtue, piety, and the liberal arts. When I came of age, I gave myself over to the study of theology just as keenly. I also served the ministry, however unworthily, for 15 years. . . . I was unimpeachable in my office, teachings, and life and my reputation in my town was such that they stood by me nearly as much as the Superintendent of Gotha [Weidemann] pursued me." He went on to recount his turn to alchemy:

> Then I had a particular desire [to understand] the lovely, glorious generation of metals and minerals, and so came into the pleasure garden of the philosophers, where, thank God, I tasted a little of the manna of heaven and the balsam of the green earth. Whence I was not ashamed to profess to be a philosopher and did not balk at disputing with any faculty or advanced school in all of Germany, with the most learned scholars, about nature and her wonderful, unfathomable works, by which one can come to know the workshop of the Creator of Nature. And because of this profession, I am now taken to be a magician or a sorcerer.[50]

With this lovely image of wandering in the philosophers' garden, Philipp turned to address the accusation that his practice of alchemy had led him into sorcery. This was entirely unfounded, he believed, and he made his point with a surprising comparison to one of his contemporaries, the physician and alchemist Leonhard Thurneisser zum Thurn (1531–1596), whose patron was Duchess Hedwig's own brother, Johann Georg of Brandenburg.

"The Elector of Brandenburg has one Lenhart Dorneisen [i.e., Leonhard Thurneisser]," Philipp explained, "who since his youth has not studied anything, but rather is supposed to have been a miner and a smelter and yet knows every language, writes books, [and knows how to] distill urine, by which art he knows how to distill quite a few of the most felicitous medicaments." With this brief description, Philipp encapsulated Thurneisser's practice quite nicely. During the 1570s, Thurneisser worked and lived in the Graues Kloster in Berlin, where, as court physician to Johann Georg and Sabine von Brandenburg, he produced books, medicaments, and diagnoses for an aristocratic clientele. His numerous publications and practice of alchemical uroscopy—drawing on Paracelsian theories and techniques, he distilled patients' urine to diagnose their illnesses—have assured his place as an important figure in the history of medicine, alchemy, and book production in early modern Europe. This reputation was already clear to Philipp Sömmering, who noted that the source of Thurneisser's alchemical and medical practices never fell under suspicion the way Philipp's did.[51] "Should one say," Philipp continued, " 'Where did he learn this art? The Devil must have taught him! He must be a sorcerer and friend of the Devil!' I predict that anyone who said this soon would be silenced. Ah, then why do people want to take me for a sorcerer, just because I claim to be able to do things even less curious than he?"[52]

Philipp's identification with Thurneisser was aspirational, certainly; not only was Philipp's patronage far less secure, but he had neither the thriving medical practice, the growing list of publications, nor the acclaim of Leonhard Thurneisser. Nevertheless, Philipp was not wrong to see in Thurneisser a similar type: a wanderer who managed to parlay experience with mining and alchemy into a promising career at court, thanks to the support of a powerful patron. "If the Elector offered protection to one like this [i.e., Thurneisser] and took him into his service," Philipp asked Julius, "then does Your Grace not have the power to do the same?"[53] In comparing his own relationship with Duke Julius to Thurneisser's relationship with Elector Johann Georg of Brandenburg, Philipp thus articulated neatly the power of patronage. He positioned Julius as a prince whose status could confer intellectual and spiritual credibility, a maker of alchemists and Lutherans alike, and who could clear Philipp Sömmering's reputation once and for all.

FIGURE 15. The execution of Brandenburg Mint Master Lippold ben (Judel) Chluchim, January 28, 1573. Detail, L.T.Z.T. [Leonhardt Thurneysser zum Thurn], *Warhafftige Abconterfeyung oder gestalt/ des angesichts Leupolt Jüden/ sampt fürbildung der Execution/ welche an ihme/ seiner wolverdienten grausamen und unbmenschlichen thaten halben* ([N.p], [ca. 1573]). Bayerische Staatsbibliothek München, Einbl. II,14.

Thurneisser's vast output of published books dealt primarily with his medical, alchemical, and astrological understandings of nature. Among these scholarly publications, however, was a single print that Philipp Sömmering probably had not yet seen, but that was more relevant than anything to his efforts to think through the relationship between alchemy and sorcery: a 1573 broadsheet commemorating the grisly execution of the Brandenburg court Jew and master of the mint Lippold ben (Judel) Chluchim (1530–1573) (Figure 15).[54] The investigation into Lippold's "crimes," which had begun in 1571 when Duchess Hedwig's father, Elector Joachim II of Brandenburg, died unexpectedly, finally ended two years later. Under torture, Lippold confessed to sorcery, poisoning the elector, and possession of a book that contained

these nefarious recipes. On January 28, 1573, Lippold was executed with spectacular brutality, just as Grumbach had been nearly six years before.[55] Lippold
was broken on the wheel and, like Grumbach, was disemboweled and quartered, his body parts then dispersed and hung on the edges of town. Moreover, Lippold's execution, too, was commemorated in print, ensuring that
the message it was intended to send would endure beyond the moment of
his death.[56] Thurneisser's depiction of Lippold's execution was elaborate. The
top half of the broadsheet is crowded with imagery of biblical and secular
justice. Lippold's execution takes center stage, the scaffold surrounded with
innumerable soldiers holding pikes. A portrait of Lippold, head bowed and
with the book allegedly containing his recipes chained to his neck, appears
just below. The text denounces him as a magician, poisoner, sorcerer, mur-
derer, thief, fiend, and arch traitor, as well as for fornicating with Christian
women. The broadsheet framed his crimes primarily in anti-Semitic terms,
as an assault against Christianity. The lesson of the "grisly and inhuman acts
. . . which he committed against innocent Christian blood" was driven home
on the text-heavy bottom half of the broadsheet by "a few stories of why no
Christian should trust a Jew," beginning in the year 575.[57]

Leonhard Thurneisser most likely was not the author of the incendiary
text of this broadsheet; in all likelihood, he printed it because he ran the only
printing press in Brandenburg and thus would have been the logical person for
his patron, Elector Johann Georg of Brandenburg, to ask to document the
spectacular and highly politicized execution. Nevertheless, the broadsheet
offered Thurneisser an important opportunity to distance himself from Lippold, whose metallurgical and alchemical expertise was implicated in his downfall and was also dangerously close to Thurneisser's. In his final confession,
extracted through torture, Lippold admitted that his book of instructions for
all kinds of harmful magic and necromancy also contained recipes for alchemy.
Moreover, the poison that he supposedly used on the old elector contained
mercury sublimate, a distinctly alchemical substance that he was able to obtain,
he confessed, "because he was a minter."[58] Thurneisser, who, as a Paracelsian,
used minerals and alchemical techniques in his own practices, must have taken
note of the way that Lippold's confessions placed alchemy in the wrong sort of
company, proximate to sorcery and poison, rather than medicine, mining, and
precious metals. By printing the anti-Semitic broadsheet that so graphically
depicted Lippold's execution, and thus participating in his condemnation,
therefore, Thurneisser dissociated himself, and perhaps alchemy more broadly,
from what Lippold had come to represent.

In his letter to Duke Julius responding to Kettwig's accusations, Philipp Sömmering attempted to make the same kind of distinction, firmly rejecting the connection that others had made between his alchemical work and sorcery, rather than merely appealing to the mercy of God and patrons, as Anna did. It is not clear how Julius responded to Philipp's and Anna's pleas, or more generally to the rancorous dispute among his courtiers. The duke must have taken note of the new (and detailed) accusations that Kettwig let loose before he escaped; on the other hand, Kettwig's flight must have cast some suspicion on his motives. Regardless, Julius did not have much time to reflect before he received yet another condemnation of Anna and Philipp, this time from family: Duchess Hedwig's brother, Elector Johann Georg of Brandenburg.[59] Writing only a year after Lippold's execution, the Brandenburg elector urged Julius to make precisely the sort of connections between alchemy, sorcery, and poison that alchemists like Thurneisser and Sömmering rejected. Duchess Hedwig and Julius's sister Katharina, Margravess of Brandenburg-Küstrin, of course, had already laid the groundwork. For years they had sounded the alarm and urged Julius to sever ties with Anna, Heinrich, and Philipp.[60] Julius had long dismissed these warnings, but when Elector Johann Georg wrote to him that February Julius seems to have listened more closely. Perhaps he was more open to words of caution coming from an esteemed brother-in law rather than from his wife or sister, or perhaps the events surrounding Kettwig's arrest and escape had made him more receptive.

"It has come to our attention," Johann Georg began, "that Your Dearest has surrounded yourself with several evil, infamous people who are suspected of fraudulent dealings. Those in the other places where they have been have nothing good to say about their conduct." The elector acknowledged that his advice may have been presumptuous, but noted that the danger was great enough that it warranted the intrusion into another prince's affairs: "Although we know well that it is not our business with whom you should surround yourself, still we consider it our due to warn our good kin, Your Dearest, about harm and damage. And it would be truly painful to us if Your Dearest, a godly and worthy prince, were led by such evil people to damage your body or soul, or if Your Dearest were aggrieved by them."[61] The elector alluded to Lippold, wishing that his own father, Joachim II, had received such a warning. Unfortunately, the old duke "was also deafened and won over by evil people like this through their magical arts, as later came to light."[62] Johann Georg's message was clear: Julius's very unwillingness to listen to the warnings of his wife and sister could be proof that he was *already* under the spell, quite literally, of his wicked courtiers. "Therefore we ask Your Dearest again and once more," he concluded, "with the best of intentions and stalwart persistence, to watch out for these people's magical and deceitful dealings and ability to dazzle you."[63] In short, Julius's Brandenburg family argued that, just like Grumbach in Gotha and, more recently, Lippold in Berlin, Anna and Philipp, too, were using a potent mix of alchemy and sorcery to rob Julius of good judgment, if not his very life and soul.

Despite this powerful plea from his brother-in-law, Julius still did not drop his support for the alchemists, at least not yet. The duke's deliberative approach is striking, in fact; he allowed Philipp to see the elector's letter and asked for a response, which Philipp drafted on March 5, 1574.[64] Once again, Philipp underscored his loyalty and service to Julius's family, land, and people.[65] He stated emphatically (and he reminded Julius that he had been saying this for years in response to Julius's sister Katharina's accusations) that he was not a sorcerer, nor was he ever in Grumbach's retinue; these claims were nothing but rumors traced easily to his enemies in Gotha.[66] Nearly two years earlier, he reminded Julius, he had begged his slanderers to substantiate their claims, to "show or prove that I have done something unchristian or sorcerous," but no one had ever been able to prove any of this.[67] Philipp begged Julius once again to use his princely power to settle the matter by offering a public "counter report" affirming Philipp's innocence of any involvement with sorcery or magic.[68]

By this point, Philipp's defense against the accusation of sorcery was well rehearsed. He was right, to a point, to focus on the persistent shadow of Gotha on his fortunes in Wolfenbüttel. The Brandenburg elector's comparison to Lippold added a new context and urgency to these old rumors, however, and Philipp does not seem to have fully grasped the implications or dangers of this new development. "The example of the Jew is indeed awful," Philipp concurred, "and from the misdeeds of such accursed rogues all Christians should learn to be wary of the enemies of Jesus Christ." As a pastor, Philipp regarded himself as a good Christian, and he underscored this throughout his correspondence with Duke Julius. The main worry he expressed in his letter to Julius was that the Lippold case would make the duke more suspicious of outsiders: "Your Princely Grace may remember this example, and not trust all strangers so quickly and so much wherever they came from."[69] This was of immediate concern, since, in the mental geography of early modern Europe, Philipp, a Thuringian, would have been regarded as an outsider in Wolfenbüttel. And so he turned to scripture. Picking up on a reference to "Joseph in Egypt" earlier in his letter, Philipp cited Exodus 23:9 ("You shall not oppress a stranger, for you were strangers in the land of Egypt") and Deuteronomy 27:19 ("Cursed be anyone who deprives the stranger, the orphan, and the widow of justice").[70] Still, Philipp seems to have missed the point. Johann Georg warned Julius about Philipp not because the Thuringian alchemist, like the Jew Lippold, was an outsider, but because both were trusted advisers with unparalleled access to their patrons and also were (confessed or potential) sorcerers and poisoners. Philipp concluded his March 1574 letter to Duke Julius by reiterating his request to be released from service: "Therefore I humbly hope . . . that if Your Princely Grace will not give me permission by Easter, then, should God will it, [I hope that] I will receive my merciful leave from Your Princely Grace by the following Trinity Sunday."[71]

The Braunschweigers Speak

Sometime in March,[72] Duke Julius went to Berlin to visit Elector Johann Georg. We do not know what they discussed, but one can easily imagine that the elector followed up on his February letter to Julius with an in-person plea to take seriously the dangers of keeping Anna and Philipp at court. Julius, meanwhile, would have had Philipp and Anna's letters fresh in his mind as he spoke with his brother-in-law. As Duke Julius mulled over the private advice he received from his family members, the court of public opinion was having its own say on the matter. At some point, news of the duke's involvement with the alchemists, as well as Jobst Kettwig's escape, had seeped out of the privileged halls of Wolfenbüttel Palace and the Braunschweig city council and into the streets. Shortly after Kettwig's flight, as 1573 turned to 1574, a satirical ballad (*Spottlied*, or "pasquil") began to circulate in Braunschweig. It was clearly inspired by Kettwig's escape, but it lampooned Duke Julius, the alchemists, and his councillors more generally as well.[73] The opening verses promised an "altogether unusual tale" of a reputation sullied. Until now, the ballad proclaimed, Duke Julius was widely acclaimed. "We believed he was sent by God; he had good advisors and good people, who were true to him," but "now everything has been turned upside down, his acclaim built on sand."[74]

According to the unknown author(s), the chief cause of Julius's tarnished reputation was Philipp Sömmering, along with his companions Heinrich,

Anna, and others. Philipp arrived as an impoverished pastor who promised to make great lords "powerful and rich" with "a certain art." Despite the duchess's caution, Julius embraced the entourage: "Heinz, Philipp, Anna, joker, rogue, whore Contobrina, to whom he bound himself with a sworn oath. They took the sacrament on it, they took a confessor into their crowd, the prince staked land, people, honor, assets, body, life, and blood on them."[75] The ballad's mocking tone continues throughout, portraying all three mostly as frauds and criminals and Julius as a tragic fool. The author puts the following words in Philipp's mouth, for instance: "I am a fine *Archemist*, a Theophrastian Philosopher, a fine, enterprising adventurer, acts of roguery often pay off. The true art of alchemy is stealing, lying, and deception."[76] Occasionally Philipp appears to be dabbling in sorcery, too, however: "He toys with sorcery, and more than once the Devil has tried to snatch him out of bed with his hooked talons."[77] Once Philipp gains Julius's trust, he drives away the duke's good advisers and fills the court with "Flaccianer" (followers of the controversial Lutheran Flaccius) and other rogues, thus sullying Julius's reputation as a level-headed broker in the contentious landscape of Lutheranism.[78]

Anna, on the other hand, appears in the *Spottlied* mostly as the gossip mill had long portrayed her (Anna's comment years before about the chattering "mockingbird" comes to mind). She is Heinrich's "whore," who had "murdered a child" in Meissen and simply laughs at acts of perfidy, despite her friends' warnings. She is also a fraud of sorts: "Anna Maria wanted to be pregnant, [so] she placed a pillow above her legs and puffed out for a long time, until finally the telltale circle disappeared."[79] Finally, she is an adulteress with dangerous access to Duke Julius. According to the ballad, she had a key made so that she could enter into Julius's private chambers, "to advise on what happened before bed." And yet, the balladeer also allows Philipp to voice a very different narrative about Anna. "He spoke. Gentle Anna Maria once had a good knight," the *Spottlied* relates, referencing Anna's story about her first marriage. Born of noble rank, according to Philipp she was also learned, "a daughter of philosophy, pious, young, beautiful, German"; she refused even the king of Denmark's hand in marriage. She "burst into the light" after lying in the womb "entirely spotlessly" for only twenty weeks. Although the *Spottlied* does not comment explicitly on any special qualities that may have resulted from Anna's birth, the very next verse notes that she has a very special kind of expertise: she knew how to incorporate magic characters into shirts, and made such a shirt with "planets" on it (presumably astrological signs) for Duke Julius.[80]

Although this *Spottlied* focuses mostly on Philipp Sömmering, it also nicely encapsulates not only the publicity of the entire affair, but also the conflicting narratives that swirled around Anna Zieglerin. Clearly some regarded her as sexually available, murderous, and fraudulent; and yet others saw her as not only virtuous and pious, but also endowed with a nearly preternatural ability to combine domestic arts such as sewing with helpful magic. Both narratives seem to have had some traction, although the fact that the ballad attributed the latter narrative to Philipp, whom it discredited as a sorcerer, liar, and politically manipulative rogue, left little doubt about which version the author(s) preferred. Anna Zieglerin's identity was still profoundly unsettled, not only at court, as Duke Julius mulled over what to do, but in the streets of Braunschweig as well.

Sorcery, Poison, and the Devil

Even as several different interpretations of who Anna Zieglerin was and what she had done vied for primacy in Braunschweig and Wolfenbüttel, one charge never emerged: witchcraft.[81] Hedwig's charges of sorcery continued to swirl around Anna, to be sure, but before her arrest she was never accused of collaborating with demons, attending the witches' sabbath, or signing a pact with the devil—that is, of being a witch. This is an important reminder that, despite our modern, post-Enlightenment habit of tossing witchcraft, alchemy, and magic into a single "waste bin" category of "the occult," "superstition," or "pseudoscience," early modern Europeans often drew careful distinctions among them.[82] Many early modern scholars insisted on a fundamental difference between natural magic, which relied on hidden forces in nature for efficacy, and demonic magic, which drew on the assistance of demons and thus was obviously more problematic.[83] Demonologists, meanwhile, identified witches primarily through their pact with the devil and participation in the witches' sabbath, although sometimes witches were thought to resort to demonic magic as well.[84] These carefully constructed and maintained distinctions were easily muddied, but in general, early modern Europeans understood alchemy to be intellectually and theologically distinct from both witchcraft and demonic magic. Certainly, alchemists had nothing to do with the devil (unless they were also necromancers), and few alchemists would have claimed that their transformations of nature required magic of any sort. They relied instead on their understanding of the qualities and behaviors of natural materials to carry out their processes successfully.

Moreover, while German princes and princesses clearly *were* willing to identify both sorcerers and poisoners in their midst, they were averse to accusing elite men and women of the proximate crime of witchcraft.[85] Julius knew this well, for even as he was considering the accusations against Anna and Philipp, he was also closely involved in managing another scandalous case of sorcery and poison that did involve witchcraft. The trial, involving Julius's cousin Duke Erich II of Braunschweig-Lüneburg-Calenburg (1528–1584) and his wife, Sidonie of Saxony (1518–1575), came to a head at exactly the same moment as the campaign against Anna Zieglerin and her companions. The entire incident laid bare simultaneously the fear of poison and sorcery and the reluctance to advance witch accusations against women of elite status in the princely courts of central Europe.

Erich and Sidonie had been estranged since Erich had abandoned his Protestant faith and returned to Catholicism in 1547, only two years after their marriage (Sidonie remained Lutheran). The couple did not officially divorce, but they lived separately and fought bitterly for nearly a decade over finances, Sidonie's residence in the Calenberg castle, and even the silver and jewels she had been given upon their marriage. Erich essentially placed Sidonie under house arrest for a period, but she was not easily defeated. She marshaled the support not only of her brother, Elector August of Saxony, but also Duke Julius and even Holy Roman Emperor Maximilian II, who in 1569 appointed Julius to serve as an imperial commissioner in the ongoing attempts to resolve the matter.[86]

As rancorous as Erich's dispute with Sidonie over religious and financial matters was, what made it particularly ugly—and relevant to the fate of Anna Zieglerin—was a poison accusation that mushroomed into a series of witch trials between 1568 and 1574. When Erich fell ill in 1564 with a strange malady, his physician concluded that it was poison. In response, Erich initiated an investigation, accusing four women in Neustadt am Rübenberge, close to Hannover, of both trying to poison him and using sorcery to disrupt his marriage, keep him away "from his land and people," and make Sidonie barren. Under torture, some of the women confessed to not only poison and sorcery, but also to witchcraft, that is, to entering into a pact with the devil and attending the witches' sabbath. They also implicated a fifth woman

(Godela Kuckes), who even without torture confessed to attempting to poison Sidonie five years earlier. Three of these women (Frau Hart, Frau Timme, and Frau Bardelen) were executed by fire in 1568, while Kuckes died in prison. The final suspect, Gesche Role, was released because she refused to confess, even under torture. Four years later, however, Erich decided to bring Role in again for renewed interrogation. This time she yielded to torture, implicating an additional five people (Gesche Herbst, Annecke Rotschröder, Margarethe Ölfin, Annecke Lange, and her husband, Hans Lange) in a plot to poison Duke Erich. At the end of March 1572, Herbst, Rotschröder, and Annecke Lange were sentenced to execution by fire (Hans Lange had died in prison).[87] Two months later, an additional six women were also executed by fire in Neustadt, although we know very little about their names or trial; in total, as many as sixty people may have been swept up in the witch trials in and around Neustadt and executed by fire.[88]

Up until this point, as far as is known, all of the people who were caught up in this witch hunt were either villagers or of middling status at best. Ölfin, for example, was a commoner (*Frau aus dem Volke*), Gesche Role was the widow of an official (*Vogt*), and Hans Lange was a barber surgeon.[89] The 1572 interrogations, however, also implicated a cluster of noblewomen (Anna von Rheden, Katharina Dux, and Margaretha Knigge[90]), and it was not long before Duke Erich's estranged wife, Sidonie, herself was accused of directing the poison plot against her husband, purportedly because of his relationship with his mistress, Katharina von Weldam.[91] This escalation of the trial as it reached into the nobility proved to be too much, apparently, even for Duke Erich II, who halted the trial before the noblewomen were sentenced and sought counsel from jurists and other authorities all over the empire before proceeding. Sidonie, meanwhile, sought refuge at the imperial court in Vienna, all the while lobbying Emperor Maximilian and Julius to intervene.

On December 18, 1573, the proceedings began again in Halberstadt. The assembled luminaries give a sense of the visibility and importance of this trial, as well as how many parties had a stake in it. In addition to the imprisoned noblewomen, Sidonie, Duke Erich II, and their advocates, the attendants included Erich's cousins, Duke Julius (as imperial commissioner in the matter) and Duke Wilhelm von Lüneburg, as well as representatives from Elector August of Saxony, the Archbishops of Magdeburg and Bremen, Markgrave Georg Friedrich of Brandenburg-Ansbach, and the Landgrave of Hessen.[92] During the Halberstadt proceedings, it became clear that Duke Erich himself had been present at the previous year's interrogations and had coerced the accusations against Sidonie. When the imprisoned women were questioned again by imperial commissioners, they all retracted their earlier statements and were released. Sidonie, too, was cleared of all charges, and spent the final year of her life in Kloster Weißenfels, which her brother Elector August of Saxony had granted her as a refuge.[93]

Between 1569 and 1574, therefore, Erich and Sidonie's ever-intensifying conflict and the witch hunt with which it was so closely intertwined were a major preoccupation for Duke Julius. Erich's scenario—that Sidonie collaborated with witches to procure and use poison against him—was troubling and plausible, given their acrimonious relationship. Nevertheless, equally troubling was the possibility of a miscarriage of justice. Throughout the affair, Julius was cautiously supportive of Sidonie, working continuously to contain the escalating charges of poison and sorcery emerging from Erich's dispute with the duchess, particularly once they targeted elite women. Julius not only worked officially as an imperial commissioner to resolve the matter; he also corresponded regularly with Sidonie and provided her refuge in April 1572 when the ongoing interrogations turned against her, hosting her at Schleunigen Palace with his widowed stepmother, Sofia.[94] In the summer of 1573, Julius, as a kind of neutral party, also housed the imprisoned noblewomen while Duke Erich and Emperor Maximilian continued their negotiations.

Sidonie and the prisoners, meanwhile, also found strong advocates among their fellow noblewomen, including some of the same duchesses who worked so hard to condemn Anna Zieglerin and Philipp Sömmering: Duchess Hedwig and Julius's sister Duchess Margaretha of Münsterberg-Oels, as well as Julius and Margaretha's stepmother, Duchess Sofia. The three duchesses wrote directly to Duke Erich to advocate for leniency, clearly skeptical of the charges against the noblewomen.[95] Some of these women were linked through the different branches of the Duchy of Braunschweig, of course; like Sidonie, Hedwig and Sophia had married dukes, and Margaretha was Julius's sister. The imprisoned noblewomen were not family, but they were linked not only to Sidonie, but also to the Wolfenbüttel court via patronage networks.[96] These privileged networks, combined with a more general resistance to witch trials in the upper echelons of German society, encouraged a sense of solidarity with Sidonie and her confidantes. In a sense, these social and family ties worked to close ranks *in support of* Sidonie and the other noblemen in the same way that they worked *against* Anna Zieglerin during her time in Wolfenbüttel. In both cases, however, there seems to have been resistance to charges of witchcraft at court, even as accusations of poison and sorcery swirled around these women, and this may account for the fact that Anna never was accused of witchcraft.

*   *   *

With sorcery, poison, fraud, and dramatic judicial proceedings hanging over everything, 1573 was a remarkable year at the Wolfenbüttel court. The man who had supposedly murdered Hedwig's father, Lippold, was executed in January for poison and sorcery. Over the summer, Julius continued to try to resolve another poison and sorcery accusation, this time against his cousin Erich's wife, Sidonie. In November, Jobst Kettwig raised the specter of fraudulent alchemists in his letter to Duke Julius, only to escape his own confinement. And in December, Duke Erich's charges against his wife, Sidonie, were shown to be not only baseless, but a willful perversion of justice. All of this gave context to the letter Duke Julius received the following February from his brother-in-law, Elector Johann Georg of Brandenburg, warning him to beware. Sorcery and poison were serious concerns, to be sure. Two of Julius's fellow princes, Duke Johann Friedrich II in Saxony and his own father-in-law, Joachim II, had both fallen victim, albeit it in different ways, to political sabotage via sorcery and poison. Yet Julius was not naïve; the conflict between Erich and Sidonie taught him that charges of poison or sorcery could be hurled not only for legitimate reasons, but sometimes also as a way to denounce one's enemies.

Julius surely had plenty to think about in 1574. Should he stand by his loyal adviser Philipp Sömmering, perhaps even stake some of his reputation on affirming his support publicly for the pastor-alchemist? Anna Zieglerin was just getting started on the lion's blood, which still had many secrets to yield; if he let her go, he would never know the full potential of the golden oil. On the other hand, Julius's own family—his wife, his sister, and his brother-in-law—clearly wanted Anna Zieglerin, Philipp Sömmering, and Heinrich Schombach gone from Braunschweig-Lüneburg and the ducal court, and, haunted by fears of sorcery and poison, they warned of dire consequences if Julius kept the alchemists around. Sömmering had even offered to leave, so Julius could easily have let them slip away quietly, giving up on his alchemical projects, and returning to the weighty business of establishing Braunschweig-Lüneburg as a strong, modern Protestant state. But would he regret this if the trio offered their considerable skills to another princely court, where they then bore fruit?

Meanwhile, the satirical ballad continued to circulate in Braunschweig and Wolfenbüttel. The final verses were ominous, suggesting that it would not end well for Anna and the others: "He [the balladeer] wishes the prince well, / [and] that he will not sell out his heart, / Yet the knaves belong on the gallows / farewell until I sing again."[97] As it happened, the balladeer did not have to wait for long. In June 1574, Julius ordered Anna, Heinrich, and Philipp's arrest in Goslar, and they were returned to custody in Wolfenbüttel for questioning, thus launching a seven-month process that, just as Anna feared, eventually led to the scaffold.

CHAPTER 7

Toad Poison and Other Fictions

Philipp Sömmering hoped to leave Wolfenbüttel with Julius's blessing by Trinity Sunday, 1574, which fell in early June that year. Instead, he Anna, Heinrich, and others found themselves in Wolfenbüttel Palace, having been taken into custody in Goslar the week before. Their reversal of fortune came mostly during the month of May.[1] The immediate cause of their arrest is obscure and confused today by the avalanche of new accusations that emerged later during their trial. Clearly, however, the mounting pressure Duke Julius faced, both privately from his family and publicly from sentiments like those in the Braunschweig ballad, played a role in eventually convincing him to investigate further.

Initially, the only real charge seemed to be that Philipp had intended—indeed, attempted—to flee Braunschweig-Lüneburg without fulfilling the contract he had signed with Duke Julius in 1571.[2] The day after the alchemists' arrest and transfer back to Wolfenbüttel, Julius's Hofgericht (the highest legal court in the duchy, headed by the duke himself) heard testimony from Anna's servant, Anna Wegemann, as well as a man named Bernd Hübener, who testified that Philipp Sömmering had been making plans to flee over the Harz Mountains and out of Duke Julius's jurisdiction entirely.[3] Reneging on a contract with a prince was a serious crime, and Philipp would not be the first alchemist to face the charge of fraud in this context.[4] Julius's councillors also questioned Philipp directly, who, unsurprisingly, contested the claim that he intended to flee.[5] Anna or Heinrich had not yet been charged with any particular crimes, although they, too, were in custody and were certainly under suspicion, perhaps for assisting Philipp. Unable to resolve the question of Philipp's flight with this initial evidence and hoping to clarify the situation, Julius's Hofgericht conducted further interrogations throughout the month of July 1574.

As ducal officials questioned Anna, Heinrich, Philipp, and several others, the concern about Philipp's flight quickly mushroomed into a much larger and very serious set of charges that engulfed them all: fashioning duplicate keys and using them to access the ducal chambers; various types of magic (including love magic to win Julius's affections and harmful magic to cripple Hedwig and sow discord between the duke and duchess); poison; murdering an envoy whom Julius sent to the Duke of Saxony; financial fraud involving the purchase of a book for Julius; and slander. Anna and her colleagues would have been unsurprised by some of these claims, to be sure. They must have known that the long-standing rumors—involvement in the Grumbach Feud, for example, or even the very generic suggestion that they were involved in some sort of "magic" or moral impropriety—would shape their interrogations. As they were questioned repeatedly, however, these vague charges gained precision; there were now details and specific instances, as if the unspecified unease they had long provoked in people like Duchess Hedwig had now been transformed into particular, actionable crimes. Moreover, there were striking new allegations that surfaced for the first time only during the trial: poison, murder, duplicate keys. Anna Zieglerin and her companions, apparently, had finally committed the abominable acts that their enemies always knew they would, and then some. By November of 1574, the interrogations, helped along by torture, had yielded fruit, and Duke Julius had in hand confessions to forty-six crimes by eight individuals. Anna Zieglerin confessed to eleven, and Philipp Sömmering confessed to twelve, far more than any of the other six individuals who also faced punishment; ultimately, Anna and Philipp were understood to be the chief offenders.

The 1574 trial carefully refashioned Anna's image, transforming her from expert, holy alchemist into sorceress and poisoner. It is not enough simply to say that Duke Julius deployed the juridical power of the state to secure confessions to predetermined crimes. The process was far more complex than this. There is no evidence that Julius began the investigation with a particular outcome in mind, and his ambivalence about the alchemists before their arrest may well have extended into the interrogations themselves. Moreover, unlike modern observers, patrons like Julius did not begin from the premise that alchemy's promises were false—quite the opposite. At stake for him was his own reputation as a duke and ability to establish the truth about his alchemists in the face of competing narratives from his family, the alchemists themselves, and popular rumors. Philipp Sömmering had begged Julius to use his power to declare Philipp's innocence and clear his name. In the end, Julius decided not to take anyone's word—neither Philipp's nor Hedwig's nor his brother-in-law, Johann Georg's—and instead to rely on a legal process to determine the truth about who his alchemists were and the validity of their claims. Anna and the others would not be condemned without evidence, and Julius and his officials proceeded with care, mindful of following legal procedures and inclined to seek juridical expertise where they were unsure.

Inside the chambers where Anna and the others faced interrogation and torture, however, a more complicated battle of the wills took place. Anna held fast to her own self-image, insisting on her expertise and purity, even while enduring torture, for much longer than expected. Meanwhile, the ducal councillors, motivated by the internal dynamics of the Wolfenbüttel court, and the executioner, influenced by Lippold's recent trial in Berlin, exerted their own power to shape the trajectory of the trial by deciding what questions to ask, which to follow up on, and which utterances to ignore. The July 1574 interrogations were a contest among all of these parties. Despite obvious asymmetries in power, Anna still found ways to make her voice heard, even if, in the end, her interrogators used the legal process to produce new "facts" about her life, practices, and ideas.

Procedural Matters: Torture

During the month of July 1574, Julius's Hofgericht worked hard to produce these truths. The proceedings were overseen by Julius's counselor Erasmus Ebener, who had first come into the service of Julius's father, Heinrich, years before. The old duke had granted Ebener a privilege to establish a zinc mine and develop the production of brass in the duchy, and he continued on to serve Duke Julius as well.[6] Certainly Ebener's metallurgical expertise would have been useful to Julius in understanding the alchemists' activities. Joined by a handful of Julius's other advisers, Ebener conducted interrogations that July nearly every day of the week except Sunday, questioning not only Anna, Heinrich, Philipp, and, once they had been captured, Jobst Kettwig and Sylvester Schulvermann, but also several other more peripheral witnesses. Philipp, especially, would have known his interrogators well, for until his arrest he had been their peer, a fellow adviser and confidant of the duke. Each defendant made an initial statement describing his or her biography, especially how she or he came to Wolfenbüttel and into Julius's service, and addressing an initial list of topics.[7] The topics on these lists were not formal legal charges per se, but they do give a sense of the matters of interest, or perhaps suspicions or assumptions that Julius's councillors brought to the interrogations. The Hofgericht then followed up with subsequent interrogations, sometimes with breaks for torture sessions, probing further on particular points, inquiring about new issues that had come up in other witnesses' or defendants' testimony, or attempting to resolve inconsistencies.

Before the bulk of the interrogations could begin, however, the Hofgericht had to settle an important matter. Faced with the inconsistencies among Hübener's, Wegemann's, and Sömmering's initial statements and a lack of any additional evidence to settle the matter, Julius's councillors had to decide whether the evidence presented thus far warranted the use of torture in further interrogations of Philipp. They urged the duke to proceed carefully and according to the 1532 imperial legal code, the *Constitutio Criminalis Carolina* (in German, *Peinliche Gerichts- oder Peinliche Halsgerichtsordnung*), which the duchy had adopted in 1568 under Julius's father, and which Julius had affirmed in 1570.[8] Julius's chancellor, Mutzeltin, in particular, expressed concern that the case for torture was weak, and that Julius risked the appearance of tyranny if he proceeded.[9]

The debates about whether to use torture in settling the question of Philipp Sömmering's fraud reveal the limitations of Duke Julius's juridical infrastructure in 1574. Legal experts played a crucial role in early modern sovereigns' consolidation and realignment of power in the wake of the Reformation.[10] Duke Julius recognized this, and, like many other German princes in the sixteenth century, he was committed to replacing the old Saxon law with Roman law, but he still had work to do in cultivating the expertise he needed.[11] More immediately, he also lacked some of the local resources that he needed to ensure that the case against Anna, Philipp, and the others was handled professionally and efficiently. He had not yet established a university in the duchy, for example (the University of Helmstedt would not be founded until 1576), and thus he lacked a local legal faculty to consult about questions such as torture. More practically, Julius also does not appear to have had an executioner in his employ to conduct the interrogations, with or without torture.[12] The 1574 trial, therefore, was a test of Julius's leadership and resources at a moment when his rule was still fragile. Julius was under pressure from his family to dispatch the alchemists swiftly, restoring the safety of the ducal house, putting rumors to rest, and allowing Julius to restore his reputation as a prince capable of selecting wise and honorable advisers. On the other hand, as Julius's own advisers reminded him, acting recklessly or with impunity risked the appearance of tyranny; he must proceed according to the law.

Julius's brother-in-law, Elector Johann Georg of Brandenburg, guided the duke through the entire process and exercised enormous influence over the course of events that summer in Wolfenbüttel. Julius frequently turned to him for advice in the early years of his reign, so it was unsurprising that his brother-in-law would play this role again as Julius faced this crisis.[13] On June 17, 1574, Johann Georg wrote to Julius to say that he was pleased to hear that Philipp Sömmering and his entourage had been taken into custody. Moreover, he continued, we "do not doubt that Your Dearest will find out about all kinds of great knavery that they [Philipp et al.] have carried out against lords and other people. In order to aid your investigation, we are sending Your Dearest our executioner, who will know how to defrock the pastor (so that Your Dearest does not assault a member of the clergy) and interrogate him and the company that stands behind him. We request that you do not keep him [i.e., the executioner] long, because we have a number of evil knaves sitting here, for whom we also need him."[14] Presumably, Julius needed to "borrow" this executioner from Berlin to assist with interrogations and to be ready if and when he and his councillors decided that the case warranted torture according to the *Constitutio Criminalis Carolina*.

If at this point Julius was still open-minded about the alchemists' crimes, his brother-in-law had a more certain outcome in mind. Johann Georg continued, "In addition, we request that you do not ask us anything else about the pastor's art of gold; because now even more than during the life and rule of our Lord Father of pious memory such adventurers are approaching us, bragging about gold- and silver making and accomplishing nothing more than making things go up in smoke. It is good, however, that for once one of these will be punished, so that the others will have more fear of trying to undertake such arts with princes and lords and to defraud them."[15] Johann Georg saw an opportunity to use Philipp's interrogation to address the "problem" of fraudulent alchemists more generally. Duke Julius, he argued, should make an example of Sömmering, and in so doing, perhaps he could deter fraudulent alchemists (*Betrüger*) from taking advantage of other princes and patrons.

Julius was less concerned in the short term about the nuisance of alchemical fraud than about the threat of political sabotage.[16] As he pondered how to proceed with Sömmering's case in late June of 1574, Julius requested that

Johann Georg send him copies of the trial dossiers from the recent Lippold
case in Brandenburg, making it clear that he was considering its possible
relevance, as his family had long suggested.[17] Meanwhile, the executioner
arrived from Berlin, ensuring that the interrogations would proceed swiftly.
Soon thereafter, Julius asked Johann Georg if he might be able to secure
documentation from the legal record of the Grumbach Feud as well. Johann
Georg wrote immediately to his fellow elector August of Saxony, who was
happy to assist. "Upon receiving Your Dearest's letter," August replied, "we
had carried out a search for the confessions [*Urgichten*] from the Grumbach
[case] and other cases prosecuted at that time, but we did not find anything
more about Schielheintz and the Goldmaker [i.e., Heinrich and Philipp] than
what is included here. It would please me if Your Dearest wishes to send it
to Duke Julius."[18] As Anna, Philipp, Heinrich and others languished in the
Wolfenbüttel Palace jail, in other words, Duke Julius did his research. He
collected documentation from other recent trials that might serve as a model
(Lippold) or shed light on his alchemists' past in Gotha (Grumbach). The
Grumbach case did not uncover much that was useful, in fact. It revealed
nothing at all about Anna Zieglerin, only Grumbach's brief comment that
he did not think much of Heinrich Schombach, and another courtier's
remark that the ducal secretary had tried to sabotage Sömmering's alchemical
work and that "the duke [Johann Friedrich II of Saxony] was bewitched."[19]
The Lippold case, however, ultimately would offer Julius a way forward that
would profoundly shape the direction of the 1574 interrogations and have
devastating consequences for Anna and the others.

Eventually, Julius and his advisers decided that additional interrogations
would be necessary to determine the truth of the matter and that the imperial
legal code allowed for torture in this situation. During the summer and fall
of 1574, the ducal Hofgericht proceeded to conduct multiple interrogations
of Anna, Philipp, Heinrich, and several others about their activities over the
previous years, and they carefully documented the statements of the accused,
as well as the questions put to them. The Brandenburg executioner assisted
and was still in Wolfenbüttel at the end of July, despite Elector Johann
Georg's multiple entreaties that Julius return him to Berlin.[20]

Xenexton and Toad Poison

Anna Zieglerin endured four interrogation sessions during the month of July.
She was tortured at least twice. In November, she was questioned again in

order to confirm her confessions in light of new testimony from Kettwig and Schulvermann, who had just been arrested. During Anna's dramatic encounters with the machinery of justice, the executioner and Julius's councillors confronted her with questions on a huge range of topics, many of which emerged from the testimonies of others linked to the trial, as well as her own statements. The power differential in these encounters was and is clear. Faced with judicial violence, Anna had to decide whether she could shape her own testimony to meet her interrogators' expectations, and if she even wanted to. Given these circumstances, it is difficult to know what to make of the "truth" of her testimony.[21] Nevertheless, in Anna's utterances and in the councillors' questions, one can locate space between Anna and her interrogators, moments when it is possible to glimpse two different interpretations of a particular detail vying for primacy. It was in this space that the executioner collaborated with Julius's court to transform Anna's knowledge of nature's medical and material virtues, to which she freely admitted, into a sinister ability and intent to use nature to harm.

Over the course of the investigation in July, Anna and her interrogators returned again and again to one particular topic, salamanders, which makes the distance, negotiations, and tensions between the adversaries especially visible. During her first interrogation on July 8, Anna responded to a prompt about salamanders, perhaps surprisingly, by describing a Paracelsian remedy for plague: "The thing is Xennex and from Theophrastus; one takes it from a salamander and hangs it around the neck for pestilence."[22] This remedy, known as Xenexton, Zenechton, or Xenextum, was associated with Paracelsus in the sixteenth century, and seems to have first appeared in print in 1559 in Paracelsus's *Two Books on the Plague*, edited by Adam von Bodenstein.[23] The discussion of Xenexton that appeared there was brief, mentioning only "the Xenexton, which hangs around the neck," as well as a sapphire amulet, as examples of the power of certain substances to attract pestilence from the surrounding air and thus prevent it from entering the wearer's body. *Two Books on the Plague* did not include a recipe, but elaborate recipes appeared later when Xenexton took on a new afterlife in the seventeenth century, when physicians and natural philosophers such as Oswald Croll, Johannes Baptista von Helmont, and Athanasius Kircher debated the ingredients, preparations, and efficacy of the plague amulet.[24] Decades before Xenexton became a topic of learned curiosity for such prominent scholars, however, Anna offered her own interpretation to Duke Julius's councillors and the executioner brought in from Berlin to interrogate her.

Although later recipes for Xenexton almost always included toads, often as many as eighteen or more, Anna's key ingredient was salamanders. During her first interrogation, remarkably, she had enough presence of mind to offer the court her recipe, which was duly recorded. "Put the salamanders in a pot," the transcript reads, "leave it for 3 weeks, then add hard wine vinegar, and a slime comes out, from which one makes a little cake, and hangs it around the neck. She did this in Gotha. The salamanders creep along the earth, and nourish themselves as do other animals. The shepherd takes care of the salamanders. In the forest [?], these things creep along in the fields; they are black and yellow. She didn't know who brought Philipp these worms."[25] If Anna Zieglerin had shared this recipe in a different context, perhaps in a letter to Duke Julius offering him protection against the plague, its historical significance would lie in the surprising fact that this self-taught woman possessed such cutting-edge medical knowledge, a recipe that would continue to interest better-known natural philosophers many decades later. One could note, too, her likely familiarity with Paracelsian treatises, both the 1559 *Two Books on the Plague* and the 1570 *On Worms*, where Paracelsus discussed the many positive uses for animals otherwise considered to be poisonous, including salamanders, spiders, toads, and snakes. "Rendered snake fat and also the flesh contains a great cure," *On Worms* explains, "not only for fresh wounds, but also for all poisonous bites."[26] Salamanders also had a special significance for alchemists, in part because of their reputed ability to withstand (even extinguish) fire, just as gold, too, did not degrade in fire. "The alchemists made many trials on salamanders," *On Worms* continues, "until they discovered in them how to make good gold and a tincture for metals."[27] Anna's recipe for Xenexton is clearly an engagement with this Paracelsian tradition, combining the alchemically significant salamander with Paracelsus's broader interest in extracting medicines even from poisonous animals to combat the most devastating illnesses.

And yet Anna repeated her recipe in the context of an interrogation that she well knew could involve torture. She was treading on very dangerous ground. Salamanders may have been associated with gold and alchemy, but they were also known to be poisonous. Anna followed Paracelsus in arguing that "the dose makes the poison"—that is, anything could be poisonous if given in the wrong dosage—and that the proper preparation turned even poisonous substances into beneficial medicines.[28] Still, poisonous animals, especially toads, as well as elementals associated with fire, were as closely associated with malicious acts as with Paracelsian medicaments in Anna's day.[29] By sharing her knowledge of how to extract "slime" from salamanders as a remedy for plague, Anna offered her interrogators, however inadvertently, the opportunity to associate her knowledge of these medically useful but also lethal animals with criminal deeds. Did Anna know that only the day before, her husband Heinrich had testified that Philipp had collected salamanders and extracted their poison to use against Duchess Hedwig and Julius's two sisters, Katharina and Margarethe, all of whom had made their opposition to the alchemists known? Heinrich also claimed that Philipp planned to use the poison against Julius's councillors, and, shockingly, "finally also [against] His Princely Grace himself, if the Duke no longer supported them, but rather became ungracious to them."[30] If Anna had known this, she might have put even more distance between herself and the salamanders. Instead, perhaps unwittingly, she emphasized her expertise with Xenexton. Although she first claimed that Philipp was the one who was engaged with making it, she insisted that "she taught Philipp and showed him the process."[31]

Anna's second interrogation came five days later, on July 13, and once again, she was asked about salamanders, among many other things. She began with a chilling statement to Julius's advisers: "She says that she wants to confess everything just as she would confess it before Christ's Seat of Judgment. And she asks to be spared, and not to be declared guilty. Because if one does not want her to stand by the truth, then she will denounce all of the councillors before God's Seat of Judgment."[32] She then answered one more question, before there was a break in the questioning. Only the brief note in the subsequent record—"*post Torturam*"—tells us why. The executioner, then, was present, and Anna was tortured. When Anna's interrogators continued, "*post Torturam*," they asked again about the salamanders. They were not interested in what she had to say about Paracelsian plague remedies, but rather wanted to pursue Heinrich's claim that Philipp was making poison. Given Anna's professed expertise with salamanders, surely she must have been involved with Philipp's poisons? She refused to admit any involvement with such a nefarious act, however, responding only that "she does not know how to make poison out of salamanders."[33]

Julius and his councillors were surprised at Anna Zieglerin's tenacity. Indeed, they said as much: "Illustrissimus [Julius] was given a report about what happened yesterday and was amazed that she did not want to confess" begins the record of the third session on the following day, July 14. They continued with an unsubtle suggestion that Anna cooperate: "Between such pain and death [there is] little difference, [but] she could be spared, if she wants to help herself."[34] Only during the interrogation that followed (and that was interrupted once again by another torture session) did Anna finally give in. The list of substances she confessed to making or using expanded dramatically, extending far beyond the Xenexton to include poisons, magic to bind affections, and a salt made of absinthe wood, among other things. The devil also made an appearance for the first time in Anna's third testimony, egging her on to malicious acts against Hedwig, Julius, and his councillors.

The crimes to which Anna confessed in this third interrogation session were a mix of medicine, magic, and poison, both demonic and natural, and all designed to kill, cripple, manipulate affections, and even strengthen one's resolve not to confess evil deeds. For example, she now admitted to making a poison made from dried, powdered toads. She and Philipp "poured water over the toads and then crushed them in the devil's name; one does it in the devil's name so that it will do damage," she stated. "It was meant for the Duchess, but [the plan was] not set in motion."[35] She said that she got the recipe from "a book," and that she persuaded someone named Bartold Taube to use duplicate keys to enter into Hedwig's chambers and pour the poison everywhere "in the devil's name, so that all who pass over it may waste away and weaken."[36] At first, she said, Taube did not succeed in carrying out this plan, so the officials called for torture again, taking another break in the proceedings. "*Post Torturam*" Anna continued, explaining that eventually she persuaded Taube to poison the duchess by promising him the philosophers' stone. "When it was in [Hedwig's] goblet, she was supposed to become ill," Anna added.[37] Her description of what actually happened is quite confused, in fact, suggesting that she was struggling mightily to thread the needle between giving what the councillors and executioner wanted and standing her ground.[38] Anna's questioners came back to the topic later, however, and then she gave a very clear description of how Taube took the toad poison and copied keys to Hedwig's chambers, following her clear instructions to pour it "in the devil's name."[39]

At the very end of Anna's confession about the toad poison, we see a little glimpse of the Xenexton again. She turned the conversation back to salamanders, saying that "she got the salamander powder from Philipp. One must extract the virtues of the salamander powder with vinegar when one wants to hang it around the neck."[40] The questioning moved on to other topics, including the preparation of several different *philtra*, special concoctions designed to secure favor, and their placement in food and drink for

Duke Julius, his councillors, and Philipp.[41] But then the interrogators came back to ask about salamanders again. "They wanted to extract the Xenex of Theophrastus [Paracelsus] from them," Anna began. "Xenexton means poison. . . . If one leaves any poisonous animals to rot, and then makes a juice out of them, then one can do many evil things."[42] The plague amulet had become a poison. The interrogation briefly drifted off to another topic, but then it soon returned to Xenexton once again. According to the transcript, Anna said: "She never did anything with the salamanders other than extracting the Xenextum from them. When one wants to make poison out of them, one must putrefy them and place them in the earth. She saw salamanders sitting in Philipp's laboratory, and asked his *Laborant* what they were doing there, and he answered, that he lacked a fire, and the *Laborant* wasn't yet finished. That's why they were sitting there for so long. . . . With putrefaction, she believes, one can poison anyone with it. But she doesn't know of any recipes for this."[43] Remarkably, even after at least two rounds of torture and clear pressure from the court to admit to making and attempting to use a poison made from toads and/or salamanders, Anna continued to assert her expertise and to insist that her only interest in salamanders was in making the Paracelsian plague remedy Xenexton.

One could follow other threads through Anna Zieglerin's interrogations and find a similar pattern, as she and her interrogators struggled to close the space between them, and each attempted to pull the truth in her or their own direction. Despite the Hofgericht's threats and use of torture, Anna's voice still comes through, and her desire to keep the interrogation on her own terms is striking. In the end, of course, the Hofgericht won this battle. Anna Zieglerin's final confession included eleven crimes. But if, in the end, these were taken as the facts of the case, it is worth noting that they were not produced easily. Rather, they emerged only through a prolonged, and violent, confrontation between Anna Zieglerin and the executioner and interrogators who questioned her.

Lippold's Example

One additional detail in the case stands out, suggesting that Anna's testimony may have been shaped not only by broad cultural forces linking alchemy, sorcery, and poison and by the use of judicial torture in her interrogations, but also by the particular executioner who carried them out. Might this executioner, whom Julius had "borrowed" from his brother-in-law, Johann Georg, to help conduct the interrogations in the summer of 1574, have been the very same executioner who oversaw Lippold's interrogation in Berlin the year before, eliciting confessions to a damning list of crimes that led to the court Jew's execution in January of 1573? The similarities between Lippold's confessions and those of the Wolfenbüttel alchemists are remarkable, suggesting the intriguing possibility that the Brandenburg executioner on loan indeed had been involved in the Lippold trial the year before and that his experience shaped the way that he directed the interrogations in Wolfenbüttel.

Whether or not the executioner who had interrogated Lippold in Berlin was the same person who came from Brandenburg to interrogate Anna Zieglerin and the others, it is clear that the two trials were linked in the minds of Julius's court, underscoring deep cultural connections at work that made it possible for a cluster of charges deployed against a court Jew in Berlin to also work against the alchemists in Wolfenbüttel. Both cases shared broad similarities, involving controversial courtiers who were accused of trying to secure their sovereign's favor in part through alchemical promises, and in part by recourse to magic and collaboration with demons. It is the details of Lippold's and Anna's confessions, however, that suggest that Lippold's confessions may have influenced more directly the kinds of crimes Anna was encouraged to confess to as well.[44]

For example, both Lippold and Zieglerin confessed to owning books that contained dangerous arts. Lippold said he possessed a book ("in the Hebrew language") purchased in Prague from "two Jews, one named Smolle and the other named Jacob," in which "were written all kinds of things both good and evil,"[45] including alchemical recipes and methods for conjuring a demon, sowing discord between spouses or winning the love of young women, and using "devil's herbs" to kill. In her fourth and penultimate interrogation on July 16, Anna Zieglerin also confessed to possession of "a book, which contained all of her arts, which Philipp wanted to attain."[46] Moreover, Lippold and Zieglerin also eventually confessed, after torture, to expertise in magic to cause harm to individuals, sow discord between spouses, cripple, and murder, as well as love magic to secure affections. Both Anna and Lippold were said to have relied on duplicate keys to carry out their crimes as well. Finally, and most troublingly, both confessed, ultimately, to poison. We know already about Anna's salamanders; Lippold, too, finally admitted that "he himself sought to poison the blessed Lord His Electoral Grace with a drink as he was lying in bed."[47] Poison, magical trespasses, and duplicate keys: the constellation of crimes from Lippold's trial appeared again the following year in Wolfenbüttel.

The particular focus on bodies, thresholds, and keys is striking. The act of sewing magical objects into clothing or placing them under a threshold in order to harm the wearer or inhabitant emerged in both trials. Thresholds were a particular site of concern in early modern culture, and Europeans frequently relied on rituals, charms, mezuzahs, and apotropaic signs and objects to protect these vulnerable boundaries from trespass by malevolent spiritual forces (conversely, those with malicious intent were believed to exploit this vulnerability by burying harmful objects or inscriptions beneath them).[48] Lippold confessed that he knew how to make someone waste away and die, either by sewing magical objects into their clothes and then burying them under their threshold, or by writing down "demonic characters" and the name of the person one wanted to harm, then running nine threads through the tongue of a black rooster, and burying both the paper and the rooster "in the devil's name" under the person's threshold.[49] Anna eventually also related a series of boundaries that she asked Taube to cross to enter Hedwig's chambers: he crossed a bridge, entered a gate to the palace, walked through the library, where he met Philipp and got the duplicate keys; then, finally, he entered Hedwig's chambers.[50] Some of the magical acts to which Anna confessed even traversed the boundary of the body, traveling into Julius's and his advisers' bellies via food in order to secure their affections.

Anna also confessed (again, "*Post Torturam*") to making various *philtra* for this purpose. The first involved nutmeg placed into a roasted bird, but the second was far more elaborate, made of a rutting deer, "cross of pike and black caraway," kidneys of hares, and the "nature" (semen) from a man, all of which were combined in an eggshell, which was then glued back together and put into warm ashes. After nine weeks, it could be removed and placed in a retort until it glowed, then painted with rose or lavender water and fed to someone whose favors one wanted to secure. She also confessed to adding nutmeg and her own menstrual blood to a crab soup she served to Julius "so that he would hold her dear."[51] So, too, were both Lippold and the alchemists accused of having a duplicate key, known as a "Dietrich," to their patrons' chambers so that they could come and go as they pleased (Lippold allegedly stole his, whereas Philipp and Anna had theirs made).[52] These allegations underscore early modern anxieties about the porosity of boundaries

and the permeability of the court, the fear of easy access to the sovereign's
body, rooms, and secrets. It must have been terrifying to think that someone
could copy keys and thereby enter the chambers of the duke, duchess, or
elector at will, or even turn the very items that clothed and nourished the
ruler's body into a weapon against him or her.[53]

With this very recent "example" of Lippold's treachery in mind, both
the executioner and Julius's councillors may have been inclined to look for
something similar in Anna's case; in the end they produced similar confessions. And yet there was an important difference, for in Wolfenbüttel a
woman, not a man, sat before the executioner, and this shaped expectations
about the particular types of magic and poison she used. The contents of the
poison that both Zieglerin and Lippold were forced to confess to making and
using makes this clear. When Lippold was asked to describe the contents of
the "drink" he used to poison the elector, he confessed "that in addition to a
nutmeg and long pepper,[54] he used oil, smoke from a metalwork, and sublimated mercury"; the latter two substances were strongly associated with mining, minting, and alchemy. When questioned about how he secured the
sublimated mercury, Lippold added that although apothecaries did not usually sell mercury to Jews, "because he was a minter, and handled silver, he
obtained it from the old apothecary under the ruse that he wanted to smelt
real gold with it."[55] Lippold's supposed expertise with poisons and access to
the materials to make them rested on his knowledge of alchemy, in other
words, and on his official role in minting coins. It was his knowledge of
metals and important role in the administration of Brandenburg that made
him dangerous. Anna Zieglerin, on the other hand, was allowed little such
authority by the Wolfenbüttel court. Her interrogators ignored her repeated
assertions of familiarity with Paracelsian remedies and denied her alchemical
and medical expertise, reducing her knowledge of nature to the common
criminal's malevolent proficiency with toad poisons.[56]

Day of Justice

Julius's Hofgericht concluded its interrogations in November of 1574. In the
end, the councillors questioned not only the eight core defendants (Sömmering, Zieglerin, Schombach, Schulvermann, Kettwig, the Braunschweig bailiff
Hans Hoyer, the Wolfenbüttel court pastor Ludwig Hahne, Bernd Hübener,
who supposedly helped the trio plan their escape over the Harz from Goslar,

and Dr. Kommer, a jurist and one of Julius's councillors who allegedly collaborated with the alchemists to slander Julius[57]), but several additional witnesses as well. At the beginning of the trial, when Elector Johann Georg of Brandenburg offered to send his executioner to Wolfenbüttel to help carry out the investigation against the alchemists, he assured Julius that all kinds of crimes would be revealed.[58] And, indeed, the Brandenburg executioner did produce a long list of confessions. The threat and use of torture eventually wore down not only Anna, but also Philipp and the others, who all sooner or later shaped their statements to conform to the court's expectations.

The question of appropriate punishments remained, however, and because Duke Julius did not have local legal expertise to guide him, on December 18, 1574, he did what many German princes did in this situation: he sent copies of the confessions to *Schöffenstühle*, or legal experts, in Magdeburg, Wittenberg, and Brandenburg, seeking their advice about the proper punishments for the crimes to which the eight courtiers had confessed.[59] In the sixteenth century, as more princes founded universities in their territories, *Schöffenstühle* were increasingly replaced by new cadres of scholars at the law faculties of local universities. Julius's University of Helmstedt was still in formation, however, and so the duke looked beyond his borders for additional expertise.[60] Within a few weeks, Duke Julius received legal opinions (*Schöffensprüche*) from Brandenburg and Magdeburg, listing each crime and the corresponding recommended punishment.[61]

In the end, then, what were the crimes to which Anna Zieglerin confessed? There is not a clear answer to this question. The extant archival record for the end of the trial is as sparse as the interrogation records are voluminous. By the nineteenth century, all that remained were three documents: Julius's secretary Wulf Ebert's summary of the Brandenburg *Schöffensprüche* for all eight of the condemned; the original Brandenburg and Magdeburg *Schöffensprüche* regarding Philipp Sömmering only (and even these documents were "in private hands" until they returned to the archive "recently"[62]); and the transcript of the proceedings on the "Day of Justice": the final, formal confessions that took place on February 4, 1575, in the buildings of the *Lustgarten* at Duke Julius's Wolfenbüttel Palace.[63] The record, then, is frustratingly incomplete.[64] And yet, however imperfect, the extant sources are revealing.

The ducal secretary Wulf Ebert's internal summary of the Brandenburg *Schöffensprüche* on all eight of the convicted criminals contains a list of crimes, with the *Schöffenstuhl's* recommended sentences, for each person, including Anna Zieglerin.[65] Ebert listed eleven crimes for Anna (only Philipp

Sömmering had more, at twelve), which fell into four different categories of punishment: flogging and exile, and execution by sword, water, and fire. She was condemned to flogging and exile for "deliberate alchemical fraud and [the fact that] she confessed that she was ignorant of Philipp's process, and that she was just as certain that never in her godless life could she hope to receive God's grace";[66] slander against Duke Julius; fabrication of stories about her own "purity" and a Saxon healer named Mutter Eyle; the preparation of *philtra* in crab soup and cake; infanticide; and various counts of adultery.[67] Anna faced drowning for stealing water from Philipp and for committing a scam involving an alchemical book called the *Testament of Hermes*,[68] and she was condemned to death by the sword for making false letters and seals from "Count Carl," as well as participating in the murder of a messenger involved in the *Testament of Hermes* scam.[69] Finally, the *Schöffenstuhl* suggested death by fire for using toad poison on both Hedwig and Philipp's wife, and salamander poison on the two men who allegedly helped Kettwig escape from his jail, Peter Dussel and Jacob Finning, in order to cover up Philipp's role in the episode.[70] Taking all of this into account, the Brandenburg *Schöffenstuhl* recommended that Anna's punishment was to be "led around to the place of punishment, six pinches with tongs, and fire, before her final death."[71]

This list is overwhelming. One can trace carefully any one of these charges through the archive to identify its origin, locating the first utterance, act, or shred of gossip and then following its transformation into a confession of crime, just as her knowledge of the plague amulet Xenexton evolved into several counts of poison using toads and salamanders. Some of the crimes to which Anna Zieglerin confessed, for example, are easily recognizable as versions (or inversions) of her own self-image. The offense listed first—the double and intertwined fraud of knowingly pretending to possess alchemical expertise that she did not have and claiming to operate with God's blessing that she could never hope to receive—maintains Anna's own claim that her alchemical and spiritual work were integrally linked. The cleansing, healing, generative lion's blood disappeared during the interrogation, replaced by the menstrual blood that she admitted using to make binding *philtra* for Duke Julius and others; likewise, Anna's early claims to be "pure," both bodily and spiritually, dissolved into her confession that she was in fact thoroughly polluted by numerous adulterous liaisons. The healer named Mutter Eyle and, more importantly, Count Carl, her two tutors in natural knowledge, were dismissed as mere fabrications. In short, Julius's Hofgericht and the executioner brought in from Berlin dismantled Anna's long-standing claims about herself—what she knew, how she knew it, and who she was. And then they added new charges to top it all off, including murder and slander.

What is not clear, however, is how deliberate or systematic this process was. The similarities between the Lippold and Zieglerin cases are striking enough to assume that the executioner played a significant role in shaping Anna's confessions. And perhaps he did arrive from Berlin with instructions from his employer, Elector Johann George, about just what those confessions should look like. But there was also something more chaotic about the long list of crimes to which Anna Zieglerin confessed, as well as their range: fraud, murder, theft, poison, magic, slander, and adultery. What else was there to confess to, frankly, other than witchcraft and heresy (two charges that would have been extremely dangerous for Duke Julius, who was still in the process of establishing the Reformation in his lands)? Perhaps, in the end, not even the trial could resolve the tensions surrounding Anna Zieglerin's and her companions' knowledge, actions, and character. Unable to settle on a clear, single narrative and confronted with Anna Zieglerin and the others' many different activities, Julius's Hofgericht may have simply brought the full force of the law down upon her.

The transcript of the *Rechtstag* on February 4, 1575—the "Day of Justice" on which Anna, Philipp, Heinrich, and the others were sentenced—presents a completely different picture, one that is difficult to reconcile with the jumble of confessions that, according to Julius's secretary, Ebert, the *Schöfftenstühle* were asked to evaluate. The *Rechtstag*, as outlined in the *Constitutio Criminalis Carolina*, was in principle a formal and ritualized affair, where the sentence (which had previously been determined) was read out in the presence of the criminal, executioner, officials, and witnesses. The role of the condemned was only to confirm his or her confession and, if desired, to plead for mercy; otherwise, the *Rechtstag* was a performative ritual of justice meant to enact publicly a verdict and sentence that had already been determined privately.[72] The *Rechtstag* in February of 1575 conformed to this model and was indeed a dramatic affair.[73] Julius formally invited the landed estates of the Duchy of Braunschweig-Lüneburg, which meant that over fifty prelates, nobles, and cities, in addition to notaries, official witnesses, Julius's councillors, and other courtiers assembled in Julius's *Lusthaus* on that winter day.[74] Notably, Julius himself did not preside over the *Rechtstag*; his ten-year-old son and heir, Heinrich Julius, officiated instead.[75]

Chancellor Mutzeltin began the proceedings by asking Julius's secretary, Wulf Ebert, to read out the recommendations from the Brandenburg and

Magdeburg *Schöffenstühle* for each of the eight condemned in turn.[76] Then
he asked each of the assembled constituents of the estates—the lords, knights,
prelates, and cities—to comment in turn on the recommended sentences. All
lamented the crimes, but understood the need for justice. The abbots, for
example, noted that they were "loathe to hear that the devil had seduced
these poor sinners" and begged to be excused from weighing in so that the
spiritual estate might be spared from shedding blood; nevertheless, they noted
that they "held that such evil could not go unpunished."[77] The knights and
cities asked that in cases where the Brandenburg and Magdeburg *Schöffenstühle* did not agree, Julius take the milder of the two options.[78]

After a break for lunch, the assembled returned to continue. A jurisdic-
tional dispute between Duke Julius and the Braunschweig city councillors
(not to mention a history of tension between the Wolfenbüttel dukes and the
city more generally) complicated matters and focused the proceedings on
Kettwig's flight while in the custody of the Braunschweig city council.[79] Perhaps for this reason, therefore, for the remainder of the proceedings Chancellor Mutzeltin focused his attention on the people who had confessed to
participating in Kettwig's escape—namely, the Braunschweig bailiff Hans
Hoyer, Philipp Sömmering, Jobst Kettwig himself, and, finally, Anna
Zieglerin.

One by one, each of the confessed criminals was led into the chamber,
read her or his testimony, and asked to confirm it before the assembled group.
This was a crucial part of the ritual, and all but one played the part. Kettwig,
for example, simply affirmed his confession, promised not to hold anything
against the Braunschweig city council, and stated that he was prepared to
die.[80] Philipp Sömmering, however, was not yet resigned to his fate, and used
the opportunity to make his case to Duke Julius one last time. After affirming
that he participated in poisoning Dussel and Finning to cover up his role in
Kettwig's escape, Philipp then turned to alchemy as his saving grace, or so,
at least, he hoped.[81] He begged for mercy, pleading that the court should
"grant him his life and pardon him, because he spoke with His Grace about
many important things (which he does not think it is necessary to repeat),
and then Dear God had granted him a gift and he has learned how to make
the philosophers' stone, which he also wanted to make for Johann Friedrich."
After asking again that he be given his life, "if at all possible," so that he
might share this divine gift, he boldly staked his life once more on his reputation among Paracelsian scholars and the profits he could offer Duke Julius.
"Men as learned as Adam von Bodenstein and other Paracelsians will attest

that he has some ability," he declared, "and that he wants to make not just one, two, or three tons of gold, but many millions of gold, which should free this land of all financial concerns [*Steuer und Schätzung*]." Finally, Sömmering added that "he had written a report about a mine, from which His Grace could get twice as many profits, and for this reason hopes that they will intercede on his behalf." Until the very end, then, Sömmering hoped that his utility to the state would be enough to save him. Unfortunately, Philipp misjudged this situation. He threw himself at the young prince's feet, but Heinrich Julius clearly was not moved. "He shouldn't prostrate himself," Julius's ten-year-old son commented; "he should beg for forgiveness of his sins." Philipp replied that he "was begging for mercy, not privilege," but this final plea fell on deaf ears.[82] Philipp was about to leave behind eight "motherless and perhaps soon fatherless, poor, inconsolable children," the youngest only seven weeks (Philipp's wife, Catharina, must have died in or shortly after childbirth) and the eldest eighteen.[83]

Anna Zieglerin was the last of the condemned to be brought in to the chambers. Chancellor Mutzeltin spoke first: "So that it is made known to her why she was brought before Illustrissimus Heinrich [Julius], the Estates, and Councilors, her prior testimony about Jacob Vennig [Finning] and [Peter] Dussell was produced; therefore, she should not endanger her soul, but rather listen to her testimony, and state the truth and neither more nor less." As unwavering as Anna had been during the interrogations during the summer six months before, on this February day she was notably resigned. Anna confirmed that, as her testimony stated, "she poisoned both people, Dussel and Vennig [i.e., Jacob Finning]," thus ensuring that the ritual of justice would proceed smoothly. Even her final plea was part of the script: "She asks only that, God willing, she will be forgiven for committing sins against God and the world, and that Illustrissimus [i.e., Duke Julius] and the Duchess will intercede on her behalf."[84]

The proceedings concluded with a speech from Heinrich Julius that reveals why Duke Julius asked his young son to officiate in his stead, and hints at the role that the entire affair may have played in the subsequent politics of the duchy. After acknowledging his father's request that he preside over the *Rechtstag*, the boy set out his understanding of the role of the state in promoting justice. "We are of no other opinion," he began, "than that from childhood on we should learn to think about justice, that is, to practice law and justice in all matters, to protect the pious and to defend against the wicked, so that fear of God, an honorable Christian life, and good doings and dealings may be planted, cultivated, and managed in this land, just as the Holy Scriptures asked of all Christian authorities."[85] Indeed, Heinrich Julius already had been studying law at the school his father had founded in Gandersheim, so he spoke as a young jurist. He continued his speech with an announcement that his father was embarking on a new venture; the school was about to be given an imperial privilege. In fact, it would open as the University of Helmstedt in the fall of the following year, and Heinrich Julius would serve as the university's first rector, beginning his term at age twelve. The alchemists' day of sentencing, therefore, was also the young prince's debut, an opportunity to demonstrate his power and learning, and to proclaim his understanding of the law and the role of justice in the lands he would eventually rule. Heinrich Julius has been celebrated as a "baroque prince" for his own learning, and for the rich court culture he cultivated. He also, however, oversaw one of the most brutal witch hunts in early modern Germany, as well as the expulsion of Jews from Braunschweig-Lüneburg.[86] One wonders what effect this early experience with alchemists convicted as sorcerers and poisoners may have had on his subsequent pursuit of piety and justice in his lands.

*　　*　　*

Anna and the others would not live to see the young Heinrich Julius's reign in Wolfenbüttel. Indeed, Anna's plea for mercy, in front of the assembled authorities of Julius and Hedwig's duchy that winter day in 1575, is the last trace of her voice that we have. At the end of Anna's life, between the interrogations in July 1574 and *Rechtstag* in February of 1575, the officials of the ducal court collaborated to replace one narrative with another. Anna related one set of complex and sophisticated stories, describing herself as an alchemist, a new Virgin Mary, and Paracelsian healer capable of restoring a fallen world. We do not know which, if any, elements of Anna's stories were true, and which were fantasies or strategic falsehoods, designed to curry favor with Duke Julius. Perhaps Anna's stories are better understood as something in between, aspirational narratives designed to embellish or make sense of an otherwise mundane or tragic life, an example of that characteristically Renaissance practice of self-fashioning.[87] Regardless, Julius's court effaced those narratives, producing a new Anna Zieglerin and then substantiating it with new facts and confessions. Above all, this new Anna was a poisoner and sorceress who used her knowledge of nature to undermine the state that Duke Julius and Duchess Hedwig were so carefully trying to build.

Early modern rituals of justice typically were carefully choreographed, every detail meant to establish a clear narrative about the nature of the crime(s) and to reinscribe the power of the state and the integrity of the community in the wake of their violation. Executions, especially, usually were not only public events, but also carefully documented for posterity, as the broadsheet printed by Thurneisser in the wake of Lippold's execution illustrates.[88] It is all the more curious, therefore, that there appears to be no contemporary documentation of the executions of Anna Zieglerin, Heinrich Schombach, Philipp Sömmering, Jobst Kettwig, or Sylvester Schulvermann. If indeed the *Schöffensprüche* guided their final punishments, the residents of Wolfenbüttel would have witnessed spectacular and grisly executions by fire and sword that winter. For all the effort that went into constructing this new image of Anna and using the institutions of justice to produce the confessions that could sustain it, it is striking that Julius's court did not follow through to document it in print, crafting and propagating the message that Anna's and her companions' deaths were meant to convey to the community and to the empire. Instead, this task would be left to future generations.

Afterlives

The first printed account of Anna Zieglerin's and her collaborators' time in Wolfenbüttel did not appear until nearly ten years after their deaths, in Heinrich Bünting's 1584 *Braunschweig-Lüneburg Chronicle*. Bünting was succinct, as chroniclers usually are:

> In the year 1575 on the 7th of February in Wolfenbüttel, Duke Julius had the following people (who threatened him and his beloved consort life and limb, committed a great alchemical fraud, poisoned several people, and also carried out other evil deeds) justly executed and killed. Cross-eyed Heinz was beheaded and quartered. His wife Anna Maria was burned to death as a sorceress. D[r]. Kummer was beheaded. *Magister* Philipp was pinched 5 times with hot tongs, and afterwards quartered. Sylvester and Kettwig were broken on the wheel, quartered, and then the pieces were hung on the gallows.[1]

Bünting reduced the alchemists' complex role in Wolfenbüttel to the grisly moment of their executions. He also simplified the nature of their offenses. Out of all of the numerous charges that swirled around the alchemists leading up to their 1575 executions, Bünting settled on a familiar trio of crimes: alchemical fraud, poison, and sorcery. Anna Zieglerin died by the fire as a sorceress, his readers learned, while the men were condemned to equally spectacular deaths, each punishment carefully tailored to communicate the nature of the individual's crimes.

The delay between Anna Zieglerin's death and this first account in print is striking, even puzzling. Perhaps there was a printed broadsheet commemorating the triumph of justice or some other kind of contemporary comment, and these documents simply have been lost to time. On the other hand, perhaps Duke Julius did not wish to publicize the affair, or Wolfenbüttel was still too insignificant to attract much attention from observers. Nor did Anna publicly document her own life or ideas, by publishing her recipes or visions of the Last Days, for instance, or even leaving behind a corpus of texts that could circulate in manuscript and inspire study and commentary. Ordinarily such a paucity of sources would ensure that an individual would be forgotten. The silences around Anna's death, however, have proven to be productive.

Anna has had a stubborn afterlife indeed, especially in the region around Braunschweig and Wolfenbüttel where her life came to an end. Generations of historians, novelists, and local residents have imagined what happened and drawn their own lessons, ensuring that Anna Zieglerin's memory endures. Anna's public image has undergone many transformations in the intervening centuries, however. If the 1574–1575 trial refashioned her self-professed expertise in alchemy and medicine into a dangerous proficiency in sorcery and poison, in the nineteenth century, historians and artists transformed her yet again, this time into a witch and a charlatan. As modern Germany aspired to new forms of rationality, justice, and political unity, Anna offered a foil from the distant past, while simultaneously satisfying a darker, more lurid, "gothic" sensibility. She also offered antiquarians and historians an opportunity to gather remnants of the distant past scattered in personal collections, archives, and Wolfenbüttel Palace, and weave them together again into a coherent narrative that the archives otherwise failed to provide. Along the way, myths found their way into popular and scholarly accounts of Anna Zieglerin's life, ideas, and death, and new sources and artifacts emerged. Today she is mostly fodder for novels and local histories.[2]

The myths about Anna Zieglerin and her companions should not be confused with rigorous scholarly accounts, for they do not rest on the same standards of evidence. At the same time, it is imperative to recognize that all of Anna's many afterlives, including this book, serve important purposes. Following Anna's story through the centuries after her death reveals the blurred lines between myth and scholarship, making visible the fact that modern historical scholarship, too, can sometimes rest on very unstable foundations. Before drawing our own final lessons about Anna Zieglerin's life and death, therefore, we must consider how the modern Anna Zieglerin came into focus, and the powerful continuities, as well as the ruptures, between modern accounts and those that were constructed while Anna was alive.

The Making of a Myth: A "Report," a Chair, and *Schlüter Ilsche*

Bünting's 1584 account of Anna Zieglerin, Philipp Sömmering, and Heinrich Schombach's demise in Wolfenbüttel created a fairly stable narrative that persisted for the next two hundred years. In 1722, when the Braunschweig historian and pastor Philipp Julius Rehtmeyer (1678–1742) assembled a new chronicle of Braunschweig-Lüneburg-Göttingen, he incorporated Bünting's account explicitly, ensuring that it would reach a new eighteenth-century audience.[3] Rehtmeyer did introduce an error that would remain a point of confusion, referring to Anna Zieglerin as "Anna Maria Hedewig Schulver-mann," thus giving Anna Sylvester Schulvermann's name, rather than Heinrich Schombach's, and, curiously, merging Anna and Hedwig. Otherwise, he followed Bünting's account from the previous century, albeit with a bit more editorializing: "Anna Maria was burned at the stake; . . . thus this mob came to an end. These evil people did not only commit a great fraud with alchemy, but apart from the above-mentioned deeds, they threatened the duchess life and limb, and poisoned the Secretary of the City of Braunschweig, Jacob Tonning [Finning], and carried out more evil deeds besides."[4] Like Bünting's, Rehtmeyer's version of events sidestepped the issue of Julius's serious support for alchemical projects and reduced Anna, Philipp, and the others to criminals whose evil deeds struck at the heart of the duchy. These accounts were fairly matter of fact; they were not overly dramatic, nor did they dwell on the personal motivations or personalities of the alchemists and their accomplices.

When the eighteenth century turned to the nineteenth, the alchemists' story resurfaced repeatedly in a cluster of printed books that tapped into a growing interest in regional history.[5] As Anna's and Philipp's demise in Wolfenbüttel was recounted again and again and became a standard episode in the history of Braunschweig and its environs, Bünting's and Rehtmeyer's brief accounts were supplemented by newly discovered, purportedly sixteenth-century sources. These sources—documents and an artifact—gave the episode new life and offered observers the opportunity to draw fresh lessons.

Two published accounts were especially instrumental in creating the nineteenth-century myth of Anna Zieglerin and her collaborators. Zieglerin first reappeared in her modern guise in a surprising location: the German economist and technologist Johann Beckmann's *History of Inventions* (1780–1805), a sprawling account of everything from butter, carp, and ribbon looms to cochineal and water clocks.[6] In a chapter on zinc, Beckmann devoted quite a bit of space to Duke Julius's mining industries. At the end of his discussion of whether Julius issued a ban on the sale of zinc (and almost as an aside), Beckmann added a footnote: "Just how much Duke Julius, who otherwise did a great service to his land, allowed himself to be taken in by goldmakers, is proven by the story in Rehtmeier, in connection to which I have obtained an old handwritten report from Herr Obercommissioner Ribbentrop, which one cannot read without astonishment. At the palace in Wolfenbüttel, one can still point to the iron chair in which the *Betriegerinn* Anna Maria Zieglerin, known as Schlüter Ilsche, was burned to death on 5 February 1575."[7] With this 1792 footnote, Beckmann introduced three new details that were thereafter inextricably linked to Anna's life: a "handwritten report," an iron chair, and a nickname, Schlüter Ilsche. The tantalizing "handwritten report" supposedly came to Beckmann from the Braunschweig jurist Philipp Christian Ribbentrop (1737–1797), who published several books of his own in the 1770s on the history of Braunschweig and thus, presumably, lent credibility to this "report."[8] Notably, Beckmann's footnote is also the first account of Anna's execution *in an iron chair*, a previously unreported detail of her death more than two centuries earlier. Alongside a new manuscript, then, Beckmann also incorporated an artifact into her legacy: the iron chair "to which one can still point" in Wolfenbüttel Palace. Finally, he added a curious nickname, Schlüter Ilsche, or "Lock-Up Ilsche,"[9] suggesting that Anna held a post at court that gave her access to keys, and thus the ability to transgress thresholds at will. Beckman did little more than mention these provocative new "facts," however, leaving it to later generations to elaborate.

Thirty years later, Beckmann's cryptic references could be filled out in much more detail, when, as part of the celebration of the founding of the University of Helmstedt, the Braunschweig jurist Friedrich Karl von Strombeck (1771–1848) published a biography of Duke Julius written by "a contemporary of this prince," the duke's jurist (*Landesfiskal*) Franz Algermann (1548–1613).[10] Algermann's account was written in 1598, nine years after Julius's death, and was then revised in 1608, but it was not printed in his lifetime. Nevertheless, Strombeck explained, it had survived in ten different manuscript copies, housed in the princely archives, regional library, and collections of various private persons in Braunschweig. Unfortunately, Strombeck had access only to later copies, which were "more or less defective or incomplete." Moreover, the oldest copy at Strombeck's disposal was "hardly 120 years old," which would date it from only about 1700, long after Anna Zieglerin's execution. The reconstruction of Algermann's valuable text, therefore, took a bit of scholarly work. As Strombeck explained,

"The circumstances require a precise comparison of all ten manuscripts, in order to establish a single text, as is related here."[11]

The 1822 edition of Algermann's two-hundred-year-old text may have been itself an assemblage, but the advantages of a new, eyewitness account of the life of the founding duke of modern, Protestant Braunschweig outweighed any uncertainty about its origins. Algermann's biography of Julius was not all celebratory, for he did include the somewhat embarrassing episode of the alchemists. In Algermann's account, Julius's support for Philipp Sömmering could be explained by the duke's long-standing problems with his legs.[12] Philipp promised Julius a cure in the form of the philosophers' stone, Algermann said, which, if prepared and used according to Paracelsus's instructions, "was supposed to remove all ill health from a person and restore him in such a way that he appears to be a youth of 18 to 20 years old."[13] Anna Zieglerin appeared in Algermann's account only as Heinrich Schombach's wife. Indeed, perhaps following Rehtmeyer's mistake about Anna's name, Algermann says that it was "Anna Maria Hedwig Schilffermann," not Anna Zieglerin, who was burned to death in an iron chair.[14] Nevertheless, like Beckmann, Strombeck turned to a footnote to editorialize, commenting, "Anna Marie Ziegler . . . was an utterly cunning swindler and proof that the vice and folly of humanity never changes. She pretended to be as pure as an angel and comparable only to the Mother of God. A man who partook of this purity would live to be a hundred years older than a normal person. She was supposedly married to a Count from Oettingen, who was the son of Theophrastus Paracelsus, and she was destined to create children with her husband from whom would emerge a new world of innocence and purity." Strombeck concluded by offering his readers a more recent point of comparison, the Italian swindler Alessandro Cagliostro (1743–1795): "In sum, not even a Cagliostro could have spoken as boastfully as she, and she was believed just as he was, until she found her death in the flames."[15] These new details about Anna were not in Algermann's 1608 biography. In 1822, however, Strombeck clearly thought they merited mention, if only in a footnote, and cited "an old manuscript" as his source. Perhaps this source was the "report on Anna Zieglerin" that Beckmann had included in his 1792 footnote? Or had Strombeck found something else in the archive in the meantime?

The new sources about Anna Zieglerin that Beckmann and Strombeck added around 1800—a manuscript report, an artifact, and a nickname—worked their way into several different cultural spaces in the nineteenth century. The iron chair and the nickname Schlüter Ilsche or Schlüterliese became a standard part of the history of chemistry, for example, from Johann Friedrich Gmelin's 1797 *History of Chemistry* to Hermann Kopp's 1886 *Alchemy in Olden and Modern Times* nearly a century later.[16] For these prominent German chemists writing the history of their own discipline, Zieglerin was only one of many examples of the misguided or outright fraudulent historical alchemists who could safely be condemned as a way of distancing modern chemistry from its (increasingly controversial) alchemical roots.

In the 1840s and 1850s, Julius's court alchemists also became stock characters in popular general histories of Braunschweig and Wolfenbüttel.[17] The essentials of the story were always the same. Philipp and his collaborators were swindlers who pretended to be alchemists to take advantage of the good duke; only Hedwig recognized their treachery, and so they attempted to silence her by making duplicate keys to enter her chambers and murder her. The plot was foiled just in time, however, and the alchemists were delivered to swift justice on the gallows. Sometimes Anna was described as just another member of the group, merely Heinrich's wife.[18] Other authors, however, followed the lead set out in Strombeck's footnote, recounting Anna's own story about her birth, purity, prophecy, and the Count von Oettingen.[19] The newly rediscovered sources played a crucial role in supporting these accounts. The "report" was cited as the source of Anna's stories about her own body and connection with Count Carl. The chair commemorated Anna's just punishment, and her nickname, Schlüter Ilsche, explained her access to the keys central to the murder plot.

Clearly, this brief episode from the remote sixteenth-century past still resonated with nineteenth-century German audiences. It offered these moderns the opportunity to reflect on early modern folly (e.g., belief in alchemical elixirs) and its political consequences, as well as the cruelty of sixteenth-century justice. It also offered lessons on gender, pitting Hedwig, the good wife and clear-eyed moral compass of the ducal court, against Anna, who seduced Julius away from his household with her false promises. But perhaps one of the things that made this old story so riveting is the possibility that the past was not so remote after all. In the mid-nineteenth century, in fact, the iron chair was still to be seen in Wolfenbüttel Palace, hanging in the rafters as a silent relic. Moreover, Anna seemed to be living an afterlife among the folk as Schlüterliese, who, as one local historian put it, "according to legend even now goes about haunting the Wolfenbüttel castle."[20]

Eventually, some turned to art to escape the limits of the available sources and tell a fuller version of the alchemists' story. Fr. Brandes, for example, tapped into his own fantasy to imagine Zieglerin's execution in a painting now in the State Museum of Braunschweig (Figure 16).[21] As a crowd of Julius's well-dressed councillors looks on and someone stokes the fire below, Anna swings in the iron chair from a rickety wooden frame, her head obscured by smoke. The iron chair, hanging in the palace, evidently was not enough to satisfy local curiosity about Anna's demise and so, drawing on his artistic imagination in the absence of contemporary descriptions of the event, Brandes recreated the scene in paint. In doing so, he also addressed a question that must have hovered around the chair, although it was never articulated: how exactly *was* the chair used in Anna's execution? The answer was not, in fact, obvious at all, for criminals executed by fire in early modern Europe were not usually seated.[22] Brandes offered an answer, supplementing the silent object with additional information: the chair was transformed into a swing, he proposed, the loops under the arms used to suspend her over the flames.

If Brandes's painting could capture the moment of Anna's execution and explain the most compelling yet puzzling object associated with it, literature offered a way to give voice to the alchemists' motivations. The prolific novelist Georg Hiltl fictionalized Philipp, Anna, and Heinrich's story in his 1873 *The Devil's Doctor from Wolfenbüttel,* one of his several historical novellas. In his version, Hiltl invoked the long-standing image of the alchemist as fraud (*Betrüger*), depicting Philipp Sömmering, the eponymous "devil's doctor," and his collaborators as unabashed charlatans. They pretended to be alchemists only to conceal their true aim, stealing Julius's treasure. As in Algermann's version of events, their bait was a Paracelsian elixir, which, they convinced Duke Julius, could restore his youth and give strength to his increasingly painful legs (they also promised the philosophers' stone, but this was incidental). Hiltl gave Anna a more important role in the ruse, however, as Philipp presented her, disguised as a mysterious "white woman," as proof of the efficacy of the elixir. She posed as one-hundred-fifty-year-old Countess of Oettingen, restored to the appearance of a thirty-year-old through an elixir administered by Paracelsus himself when she was ninety. She had a "long, spectral form" and presented herself to Julius as "a woman dressed in white and completely veiled" until she invited him to gaze upon her: "As courageous as Duke Julius was, both because of his learning and his innate strength," Hiltl writes, "he nevertheless trembled involuntarily, as he beheld this pale, indeed beautiful, but eerie face. The woman's lineaments were carved out of marble; not a single muscle twitched and only the eyes swiveled

FIGURE 16. Fr. Brandes, "The Execution of Anna Zieglerin," date uncertain (nineteenth century?). Braunschweigisches Landesmuseum, R 1848. Photo credit: Braunschweigisches Landesmuseum, I. Simon.

like black wheels in sockets shadowed by thick brows."[23] This is but a disguise, of course, which Anna casts off later in the novella, "throwing off her veil and removing the color that had lent her a ghostly pallor from her face with a sponge dipped in water."[24] Once again, there is no trace of the sixteenth-century Anna Zieglerin's serious engagement with alchemy and prophecy; she is nothing but an impostor and a swindler, a minor actor in Philipp Sömmering's avaricious scheme.

Hiltl's reimagined tale of Anna's and Philipp's demise smugly proposed that even the most honest and learned early modern men and women made themselves vulnerable to swindlers by placing their faith in philosophers' stones, rejuvenating elixirs, and ghosts. The fictionalized Julius is well intended, but led astray by his desire for the elixir; he is, the novella confides, but "a child of his time," taken in by the idea of the philosophers' stone, the promise of which "overtook many like an illness" in the sixteenth century and went "hand in hand with the mania of witchcraft."[25] Hiltl's version of Hedwig is less gullible, but even she is misled by her belief in ghosts. The real hero of the story is the character of the Juncker Erich von Reden, who assures Hedwig of "the truth"—namely, that "in Paris and London, where I was, there are a large number of enlightened gentlemen, who have revealed all such phantoms to be nonsense." The group's murderous scheme unfolds over the rest of the novella, as Hedwig's and Julius's credulity endangers their happy marriage, their young children, and, most of all, their treasure. In the end, of course, the modern and clear-eyed Erich von Reden sees through Philipp and Anna's ruse, foils the plot, and the criminals are executed; modernity triumphs. And, of course, one of the final images of the novella is Anna's iron chair: "A particularly ingenious and horrifying punishment was devised by the judges, whose verdict the duke was not allowed to alter. She was placed in an iron chair, bound, and then, thus seated, burned alive. All of the pieces of this chair are still visible at the palace."[26]

Not everyone was willing to cede the historical past to Brandes's or Hiltl's flights of fancy, however, or to the popular regional histories that simply repeated the same version of the story again and again. Eventually, Anna and Philipp's story worked its way into more specialized venues for historical scholarship as well. In 1857, over sixty years after Johann Beckmann published his *History of Inventions*, his grandson A. Beckmann published a longer article on the alchemists in the *Zeitschrift für deutsche Kulturgeschichte* (Journal of German Cultural History), devoting sustained scholarly attention to the episode for the first time.[27] The younger Beckmann set out to tell the story properly, he wrote, not as a "scandal," as his predecessors had done; rather, his aim was "to retouch, as appropriate, the places in the historical picture of this interesting prince [Julius] where the truth has worn away a bit—and a few more shadows in the picture should emerge."[28] In order to achieve this more "truthful" and well-rounded image of Julius, Beckmann promised not simply to repeat what was in the usual printed sources, but rather to make use of "mainly old manuscripts that find themselves in the possession of the author and that contain many new things."[29] Duke Julius's engagement with alchemy, in other words, offered A. Beckmann the opportunity to demonstrate how modern historical scholarship should be done.

Despite the fact that A. Beckmann's article was called "Therocyclus in Wolfenbüttel," the heart of the article was a new manuscript: a "report about Anna Zieglerin" by an "indignant unknown contemporary and subject of the Duke." Beckmann quoted the four-page document in full, leaving it to speak for itself.[30] This anonymous "report" related Zieglerin's story about her own birth, subsequent "purity," and Count Carl, echoing the material in Strombeck's footnote to Algermann's edition. The author of this document transformed Anna into a procuress; she is an *Angsthure*,[31] or intermediary, who promised to fulfill Julius's desire, not sexually, but by obtaining the philosophers' stone from Count Carl. Of course, this promise was completely false, too, according to the report, merely a charade to secure Julius's favor: "With this Lord [i.e., Count Carl] the Duke would be shamefully betrayed and made a fool of. . . . The whore pretended devilish and unbelievable things to the pious duke, and enchanted the duke into believing everything, in order to do her will."[32] Was this "report," perhaps, the same "manuscript" that A. Beckmann's own grandfather referenced decades earlier?

A. Beckmann's inclusion of this "report" in his 1857 article highlights an interesting moment at the origins of modern historical scholarship. Today we expect sixteenth-century manuscripts to be securely housed in archives and libraries, authenticated by their institutional affiliation and the promise of a reliable origins story. For the elder and younger Beckmann, however, manuscripts passed more freely through historians' hands, staying for generations in private collections and allowing historians to base their work on sources that could not be verified or consulted by others. The younger Beckmann did not offer further information about his anonymous and unidentified manuscript, so his readers would not have been able to consider its authenticity, origins, or point of view. The originality of his article, in fact, depended on his exclusive access to this "new" source, rather than his interpretation of it. Even his (by then de rigueur) mention of the iron chair and the fact that it "is at this moment still available in Wolfenbüttel" depended on the rhetorical strategy of privileged access. He commented, "In 1787 the author's grandfather [i.e., Johann Beckmann] was shown the iron chair on which Zieglerin suffered the fiery death of a witch, hanging in a hallway of the Wolfenbüttel castle."[33]

As it happens, the mysterious "report" seems to have disappeared with A. Beckmann, for it is no longer extant, and was neither cited nor discussed by subsequent authors. Indeed, one wonders whether it ever existed at all. Twenty-five years after A. Beckmann's article appeared, another historian with a similar interest in reading and locating manuscripts, the Braunschweig judge and legal historian Albert Rhamm,[34] returned to the material once again, this time with an eye to a book-length study. His title, *The Fraudulent Goldmakers at the Court of Duke Julius von Braunschweig-Wolfenbüttel, According to the Trial Records,* signaled that his study, too, was based on new (or in any event previously unused) documents. Rhamm began by setting out the scholarly state of affairs: "At the moment, what we know about Philipp Sömmering's adventures at the court in Wolfenbüttel is limited to the comments that Franz Algermann recorded in his manuscript sketch of the life of Duke Julius von Braunschweig."[35] Rhamm then noted that the trial records and ducal correspondence located in the Herzogliches Landes-Haupt-Archiv challenged Algermann's account. From these new sources, he argued, "it follows that even Algermann is not to be seen as an entirely reliable informant; in some places his account exaggerates, and in others it fails to convey exhaustively the real facts."[36] Fortunately, Rhamm concluded, the archival documents (*Akten*) offer a much more complex and reliable way to approach the alchemists' story and were reason enough to justify telling it once again.

With Rhamm's 1883 book, Anna Zieglerin and the rest of Julius's alchemists finally entered the familiar frame of modern historical practice. His account rested on archival documents that he meticulously cited, making it possible for subsequent historians (including this one) to locate them and read them with new eyes. He evaluated his sources carefully, demonstrating for his readers his expertise as a historian. He rejected some of the stock elements of the tale, for example, the association of Anna Zieglerin with Schlüterliese or Schlüter Ilsche. "But this fact is incorrect," Rhamm declared; "Frau Anne never found employment at the palace nor did she reside there."[37] He simply ignored the "report on Anna Zieglerin" that figured so prominently in A. Beckmann's account, suggesting that he did not lend it much credence. On the other hand, he did include the detail of the iron chair, and noted that it was locked up in a room in the castle, having been stolen in the 1850s, only to have been recovered later.[38] Moreover, Rhamm, too, bemoaned the frustrating absences in the documentary record. Confronted with missing pieces, Rhamm's instinct was to try to track down more documents. Unfortunately, he was mostly unsuccessful. "Only those *Schöffensprüchen* related to Philipp Sömmering (recently arrived in the archive—Fasc. VIII—from private hands) have been kept," he noted in footnote. "In response to inquiries with the relevant states, the Royal State Archive in Magdeburg and the Royal District Court in Brandenburg, I was told that the legal decisions concerning Sömmering and his companions are no longer extant."[39]

In the end, Rhamm had to confess the limits of the available archive. Perhaps it was a desire to protect Julius's reputation, he suggested, that led to the disappearance of some of the documents related to the case. Nevertheless, he was aware that the numerous fantasies about the alchemists were made possible, at least in part, by the fragmentary nature of the archival record. "At a minimum, it allows us to explain how a legend could form so quickly after the event," he concluded, "and how, for example, already Algermann, who came to Wolfenbüttel in the service of the court by 1575, could transform Bartold Taube's attempt to sprinkle poison in Hedwig's chambers into a nighttime assault, by which the collaborators tried to burst into the chambers of the Duchess with murder weapons, stab the princess and plunder the treasury." Rhamm lamented that "this story, along with a string of other inaccuracies, is repeated in all subsequent reports," even most recently Hiltl's novella, which "adds to [the story of] Sömmering and his crew a pair of lovers, the original source of which, however, one could search for in the piles of yellowed documents in vain."[40]

This is a forceful condemnation of the fantasies that had been spun around Philipp, Anna, and the others, and a defense of the archival record as the most authoritative source about the alchemists. It is all the more curious, therefore, that Rhamm ended his book on a more ambivalent note than we might expect, conceding that perhaps there is a role for invention after all. "In the places where these events took place their memory has not entirely faded," he remarked. "Until a few decades ago the chair, on which Zieglerin suffered death by fire, was hanging from chains in the rafters of the palace, across from the old place of justice [*Richtplatz*]," Rhamm continued, reinscribing one of the myths that he apparently was not interested in challenging. "And under the name 'Schlüter-Lise,'" he concluded, "Frau Anne lives on among the folk as a witch and poisoner, even though her personality [*Gestalt*], her crimes, [and] her companions have been woven together out of a colorful mix of truth and fiction."[41] With the final words of his book, therefore, Rhamm acknowledged a version of events beyond the one he had just related, licensing a semifictional Anna Zieglerin who could live on in Wolfenbüttel in memory, fiction, and one very charismatic object.

Albert Rhamm ended his book, therefore, where we began, with the iron chair on which "Frau Anne," witch and poisoner, supposedly was executed. This, too, seems at first to be a puzzling choice. In the book itself, Rhamm demonstrated very little interest in Anna, in fact, like other nineteenth-century authors, focusing his attention mostly on Duke Julius and Philipp Sömmering and leaving the life of "this strange woman" mostly to footnotes. Despite his comparative lack of interest in Anna Zieglerin's life, however, Rhamm nonetheless chose to leave his readers with the image of this ghastly relic that had come to symbolize the nineteenth-century version of her death. The object was, and is, powerful, and it remains an ideal emblem for Anna Zieglerin. It endures because it is concrete enough to send a chill down viewers' spines, and, whether hanging in a palace or displayed in a museum with all of the trappings of institutional authority, seems to bear mute witness to the past. At the same time, it does not overdetermine one's interpretation; its message is vague enough to allow room for fantasy. Viewers can conjure up whatever cruelty, murderous schemes, or irrational beliefs they want to attribute to their early modern forebears—authorizing precisely the "colorful mix of fact and fiction" that Rhamm conceded would continue to live on "among the folk," as indeed it has.[42]

*   *   *

Anna Zieglerin will likely continue her afterlife in the "witch's chair," the protagonist of local histories and fiction. Myths have many uses, and there is no strong reason to dismantle this one, although it is also important to recognize that these modern myths tell us more about what nineteenth-century Germans found interesting about early modern people and events than they do about the early modern period itself. *Anna Zieglerin and the Lion's Blood* has aimed to return to the moment before those modern myths coalesced. We can locate their origins in Zieglerin's 1574–1575 trial, where Julius's instruments of justice produced new "facts" about Anna, transforming her into a criminal condemned to the pyre, the false alchemist, sorceress, and poisoner who would feed modern fantasies about the early modern past.

Even before Julius's Hofgericht constructed its own fantasy about Anna, she put forward myths, stories, and fantasies of her own. Out of the shambles of a difficult, if not traumatic, youth, war, and perhaps also a body that could not produce life in the way typically expected of early modern women, Zieglerin constructed an orderly narrative that made sense of those facts. She rewrote the signs of her body as prophecy, as corporeal confirmation of the special role she was to play in the unfolding story of sacred time. Notably, she chose to articulate this vision through alchemy, drawing on the art's promise of the salvation and generation of matter, to imagine how her body could produce superior kinds of life at the end of time. Whether Anna's stories about her past and future were true or false is beside the point. These were powerful and effective narratives that forced Julius, Hedwig, their families and subjects, to grapple with alchemy's most ambitious claims in a fallen world, and the possibility that a woman with a troubled past and an unusual body might be the one to realize them.

Anna Zieglerin's engagement with alchemy was sophisticated, if not particularly scholarly. As far as we can tell, she was self-taught, learning through limited reading in German and most likely conversations with and observations of other alchemists. Unlike other holy men or women, she did not seek followers, and unlike other scholars, she did not seek to engage a broader intellectual community, whether through print, correspondence, or travel. Her interest in alchemy was largely personal and local, a vehicle for self-invention, and perhaps even self-preservation. Nevertheless, it was surprisingly well informed. She knew about and engaged some of the most cutting-edge debates in a Paracelsian thread of alchemy, considering the alchemical generation of life, for example, or the plague remedy Xenexton. Moreover, she incorporated laboratory practice into her work, through her recipes, which recorded techniques for making and using the lion's blood, and through the work she began at Duke Julius and Duchess Hedwig's court, and ended in a small chamber in Goslar. In all of this, she placed her own body at the center of her work, anchoring even her loftiest aims in the material world and ensuring that only she could take advantage of all that the lion's blood promised.

Anna Zieglerin's alchemical work, however, was interrupted by the mounting accusations against her and Philipp, underscoring the fact that her vision of alchemy, the body, and faith, however well rooted in the religious and intellectual culture of her world, was always also in tension with another, darker view of the court alchemist as a threat to the body and rule of the sovereign. In Anna Zieglerin's immediate orbit, Duchess Hedwig, her brother Elector Johann Georg of Brandenburg, and Duke Julius's sister Margravess Katharina of Brandenburg-Küstrin propagated this view most forcefully, with an eye to the murder, as they saw it, of Hedwig and Johann Georg's father, Elector Joachim II of Brandenburg. These terrifying associations with alchemy had deeper roots, however, in the Grumbach Feud in ducal Saxony, events in which Anna, Heinrich, and Philipp had played no direct role, but which continued to haunt them until their deaths several years later in Wolfenbüttel. From this perspective, Anna Zieglerin's and her colleagues' expertise with nature promised not salvation (or, in Philipp Sömmering's case, profit), but danger, instability, dishonor, and even death. From the moment that Anna, Heinrich, and Philipp arrived in Wolfenbüttel, this sinister vision of who they were and what they could do was intertwined with Anna's understanding of herself. Whatever her own knowledge and intentions, those around her could read her actions quite differently. Duke Julius was optimistic, but he was outnumbered by his many family members who took the dimmer view. Meanwhile, Anna Zieglerin's life hung in the balance.

More than a century ago, Albert Rhamm likened Anna Zieglerin to a spider who, sitting at the center of her web, used her charms to capture Philipp Sömmering, Duke Julius, and others.[43] At a moment in which Germany's own political unity, formalized only in 1871, seemed precarious, Zieglerin seemed to shed light on another anxious and uncertain moment, when the princes of the Holy Roman Empire tried to rebuild their world anew after the ruptures of the Reformations. Today, Anna Zieglerin offers a different kind of opportunity, allowing us to explore questions about how the history of alchemy intersects with Christianity, the body, gender, and politics. If Anna Zieglerin can be likened to a spider, therefore, it is not, as Rhamm suggested, because she lay in wait, hoping to catch unwitting prey, but rather because from her position at the center of a vast web, she could pluck strands that stretched out beyond alchemy to the most consequential matters in sixteenth-century Europe: the fragility of political power, the difficulty of rebuilding Christianity anew in the wake of the Reformation, and the fate of the world as it hurtled toward its final moments.

1541	Paracelsus dies.
ca. 1545	Anna Zieglerin is born.
1546	Luther dies.
1560	Julius and Hedwig marry in Cölln an der Spree, near Berlin, on February 25.
1566	Holy Roman Emperor Maximilian II signs imperial ban against Duke Johann Friedrich II of Saxony in May, entrusting Elector August of Saxony to enforce it. Philipp Sömmering and Abel Scherdinger sign a contract for alchemical work with Duke Johann Friedrich II on November 6.
1567	In January, Elector August of Saxony's troops besiege Gotha, where Duke Johann Friedrich II and his family take refuge. In April, the Gothaners surrender. Wilhelm von Grumbach, Hänschen, and others are quickly executed in Gotha; Johann Friedrich II is imprisoned in Austria, where he remains until he dies in 1595.
1568	First round of witch trials associated with Duke Erich II and Duchess Sidonie of Braunschweig-Lüneburg-Calenburg take place in April, resulting in three executions. Julius and Hedwig take the throne in Braunschweig-Lüneburg and introduce the Reformation when Julius's father, Duke Heinrich the Younger, dies in June.

1571	Elector Joachim II of Brandenburg dies suddenly in January. Philipp Sömmering signs an alchemical contract with Duke Julius and moves to Wolfenbüttel, joined by Heinrich Schombach, Anna Zieglerin, and Sylvester Schulvermann.
1572	In March, three more women are executed for witchcraft in connection with Duke Erich II and Duchess Sidonie of Braunschweig-Lüneburg-Calenburg. When the trials begin to target noblewomen, Duke Erich II pauses the proceedings. Jobst Kettwig arrives at Duke Julius's court in Wolfenbüttel, and Julius's sisters, Margravess Katharina of Brandenburg-Küstrin and Duchess Margarete of Münsterberg, visit Wolfenbüttel in November.
January 1573	Lippold ben (Judel) Chluchim is executed in Berlin.
April 1, 1573	Anna Zieglerin presents her book of recipes to Duke Julius.
November 1573	Jobst Kettwig is imprisoned in Braunschweig after a warrant for his arrest is issued by Heinrich von Rantzau in the name of the Danish crown. Kettwig denounces Anna Zieglerin and Philipp Sömmering.
December 1573	Kettwig escapes his Braunschweig prison.
January 1574	Duchess Sidonie of Braunschweig-Lüneburg-Calenburg and several other noblewomen are acquitted in Halberstadt after witnesses recant previous accusations of witchcraft and poisoning.
February 1574	Elector Johann Georg of Brandenburg warns Duke Julius about Philipp Sömmering.
May 1574	Margravess Katharina of Brandenburg-Küstrin dies.
Late May–June 1574	Anna Zieglerin, Philipp Sömmering, and Heinrich Schombach are arrested in Goslar and returned to Wolfenbüttel for questioning.
July–November 1574	Duke Julius's councillors, with the help of an executioner from Berlin, conduct interrogations of Anna

Zieglerin, Heinrich Schombach, Philipp Sömmering, and many others at Wolfenbüttel Palace.

December 1574   Duke Julius sends copies of the resulting confessions to *Schöffenstühle*, or legal experts, in Magdeburg, Wittenberg, and Brandenburg.

February 1575   After Duke Julius's ten-year-old son, Heinrich Julius, presides over the Day of Justice, Anna Zieglerin, Heinrich Schombach, Philipp Sömmering, Sylvester Schulvermann, Jobst Kettwig, and Dr. Kommer are executed in Wolfenbüttel.

October 1576   Official opening of the University of Helmstedt on the twelfth birthday of Duke Julius's son Heinrich Julius, who becomes the university's first rector.

1577   Formula of Concord establishes a unified Lutheran confession.

1589   Duke Julius dies.

1602   Duchess Hedwig dies.

NOTES

INTRODUCTION

1. Buck, "Angeblicher Hexenstuhl," 120–22. See also Ute Frietsch's comprehensive investigation of this chair, "Leben und Sterben in der Alchemie." I am immensely grateful to Frietsch for sharing this work with me while it was still in progress.

2. "Die Hinrichtung der Anna Maria Zieglerin, genannt Schlüterliese, 1575 war einer der grausamsten und Aufsehen erregendsten Prozesse in Wolfenbüttel." Buck, "Angeblicher Hexenstuhl," 120.

3. "sie hettes alles bekanst [*sic*] aber nit alles gethan, solches were von ir aus furcht geredet worden." "Anna Zieglerin's Verhör," 16. November, 1574, 8am, NLA WO, 1 Alt 9, Nr. 314, fol. 35r.

4. For an introduction to the history of crime and punishment in early modern Germany, see Dülmen, *Theatre of Horror*; Rublack, *Crimes of Women*; Schuster, *Verbrecher, Opfer, Heilige*. On execution by fire in particular, see Dülmen, 91–92.

5. On estimates of the number of people executed during the European witch hunts, see Goodare, *European Witch-Hunt*, 27–29.

6. For an introduction to Anna Zieglerin's life and career in Wolfenbüttel, see Nummedal, "Alchemical Reproduction" and "Anna Zieglerin's Alchemical Revelations." See also Rhamm, *Die betrüglichen Goldmacher*.

7. The core documents related to Zieglerin and her fellow courtier-alchemists' stay at the Wolfenbüttel court can be found in the NLA WO, 1 Alt 9, Nrs. 306–36.

8. "Auszug der brandenburgischen Schöppensprüche gegen Ph. Sömmering, Heinrich Schombach, Anna Maria Ziegler, Jobst Kettwig, Sylvester Schulvermann, Dr. Georg Kommer, Bernd Hueffner, Hans Hoyer," NLA WO, 1 Alt 9, Nr. 311, fols. 23r-35v.

9. See the Conclusion, "Afterlives," below, for a much more detailed discussion of this issue. The earliest histories to mention the 1575 executions are Franz Algermann's 1598 biography of Duke Julius (revised 1608), which remained unprinted until 1822, and Heinrich Bünting's 1596 chronicle of the region. In 1722, Philipp Julius Rehtmeyer essentially combined Bünting and another source in his chronicle of Braunschweig-Lüneburg. All three texts note that Zieglerin was burned to death (*verbrannt*), but none mentions a chair. See Algermann, "Leben, Wandel und tödtlicher Abgang des Herrn Juliussen"; Bünting, *Braunschweigische und Lunebürgische Chronica*; Rehtmeyer, *Braunschweig-Lüneburgische Chronica*.

10. The history of German crime and punishment is richly documented, including the standard forms of punishment by fire, water, sword, hanging, and so on. Standard works on this subject include Dülmen, *Theatre of Horror*. While criminals were sometimes beheaded

while seated in a chair, I have been able to find only a single reference to someone being executed by fire while strapped to a chair: the Hungarian rebel György Dózsa (or Jörg Zeckel), who was executed in 1514. See *Die auffrur so geschehen ist im Vngerlandt*; Erdélyi, "Tales of a Peasant Revolt."

11. Frietsch, "Leben und Sterben in der Alchemie," 17–20; Bettina Zöller-Stock, *Stühle,* 13.

12. Buck, "Angeblicher Hexenstuhl," 120. This theory originated with the masterful archival historian and art historian Friedrich Thöne, who cited as evidence the ornamentation, the swivel seat, and loops, as well as the fact that Julius gave a similar chair to his stepmother, Sofie of Poland, who was also disabled, shortly before Zieglerin's execution. Thöne, *Wolfenbüttel. Geist und Glanz,* 47–48.

13. Rhamm, *Die betrüglichen Goldmacher*; A. Beckmann, "Therocyclus in Wolfenbüttel."

14. Rhamm, "Vorwort," in *Die betrüglichen Goldmacher,* III–IV. See the Conclusion, "Afterlives," below, for a fuller discussion of how the nineteenth-century view of Anna Zieglerin emerged.

15. Standard twentieth-century works on Duke Julius's reign and the Duchy of Braunschweig-Lüneburg in this period include Kraschewski, *Wirtschaftspolitik im deutschen Territorialstaat des 16. Jahrhunderts*; Graefe, *Staatsklugheit und Frömmigkeit*; Lippelt and Schildt, *Braunschweig-Wolfenbüttel in der Frühen Neuzeit.*

16. On the operation of these tropes and the marginalization of alchemy, see Nummedal, *Alchemy and Authority*; Principe, "Alchemy Restored."

17. "Diese Studium—namentlich der Chemie und Physik—in unseren Tagen ein großer Hebel der Aufklärung, war für seine Zeit eine Klippe, an der gar leicht der gesunde Menschenverstand scheiterte." Beckmann, "Therocyclus in Wolfenbüttel," 553.

18. "Es wäre in Wunder gewesen, wenn ein mineralogischen Studien zuneigender Fürst in der damaligen Zeit nicht seine Alchymisten besessen hätte. Kein Hof war damals ohne sie, auch nicht die kaiserliche." Merkel, "Herzog Julius von Braunschweig und Lüneburg (1529–1589)," 35.

19. For the former view of Sömmering, see, for example, A. Beckmann, "Therocyclus in Wolfenbüttel." Albert Rhamm had a more historically sensitive understanding of alchemy. Rhamm, *Die betrüglichen Goldmacher,* 12.

20. "ein schlaues, gewandtes und verführerisches Weib" (Heinemann, *Geschichte von Braunschweig und Hannover,* 420); "blühend und kräftig, liederlich und listig," "ungebildete Courtisane," "Hedwig von Brandenburg, einer liebenswürdigen, frommen, und klugen frau" (A. Beckmann, "Therocyclus in Wolfenbüttel," 556; 557; 552); "eine abgefeimte Intrigantin, von einnehmenden Manieren" (Kopp, *Die Alchemie,* 1: 171).

21. "Sie trieb mit derlei Versprechungen nicht minder, wie durch ihre persönliche Liebenswürdigkeit den armen Philipp immer tiefer in ihre Netze, bis er sich in ihrem Garne gefangen sah und, zu spat, als betrogenen Betrüger erkannte." Rhamm, *Die betrüglichen Goldmacher,* 13.

22. The literature on alchemy's place in the history of science is large and growing. For an introduction, see a cluster of recent synthetic reflections on the field: Garber, "Untwisting the Greene Lyon's Tale"; Moran, "Focus: Alchemy and the History of Science"; Martinón-Torres, "Some Recent Developments"; and Nummedal, "The Alchemist." For an important corrective to an older literature that identified alchemy with magic and "the occult," see the now classic pair of historiographical essays by William R. Newman and Lawrence M. Principe, "Alchemy vs. Chemistry" and "Some Problems with the Historiography of Alchemy."

23. Alchemy's religious resonances in early modern Europe have received relatively little scholarly attention. For a more detailed discussion of this historiographical question, see

Nummedal, "Introduction: Alchemy and Religion in Christian Europe." There are a handful of other important exceptions to the general trend, including Baldwin, "Alchemy and the Society of Jesus"; Shackelford, "Unification and the Chemistry of the Reformation" and "Paracelsianism and the Orthodox Lutheran Rejection of Vital Philosophy"; Crisciani, "*Opus* and *Sermo*"; Forshaw, "'Alchemy in the Amphitheatre'" and "Vitriolic Reactions"; Garber, "Transitioning from Transubstantiation to Transmutation"; DeVun, *Prophecy, Alchemy, and the End of Time*; Janacek, *Alchemical Belief*.

24. In the decades following Zieglerin's death, heterodox Lutheran figures such as Jacob Boehme would articulate these relationships in new ways, forging what we might consider a tradition of "spiritual alchemy." On the term "spiritual alchemy," see Zuber, "Spiritual Alchemy from the Age of Jacob Boehme to Marry Anne Atwood"; Tilton, "Alchymia Archetypica."

25. Hedesan, *Alchemical Quest for Universal Knowledge*; Matus, *Franciscans and the Elixir of Life*.

26. This topic is vast. Particularly influential for this project are Bynum, *Holy Feast and Holy Fast*; Duden, *Woman Beneath the Skin;* Sibum, "Reworking the Mechanical Value of Heat"; Park, *Secrets of Women*; Smith, *Body of the Artisan*.

27. Smith, *Body of the Artisan*. Alchemists knew this well, and their written works commented on how physically taxing alchemical work could be. Only the fit would succeed, and, conversely, alchemical work could disable the bodies of the unfit and unskilled, making them weak, blind, and physically disordered, as stock images of the fraudulent or incompetent alchemist showed. Nummedal, *Alchemy and Authority,* 45.

28. Bynum, *Fragmentation and Redemption*, 184.

29. Brady, *German Histories in the Age of Reformations*.

30. See especially DeVun, *Prophecy, Alchemy, and the End of Time*; Matus, *Franciscans and the Elixir of Life* and "Resurrected Bodies and Roger Bacon's Elixir."

31. See, for example, the following recent studies: Findlen, "Scientist's Body"; Leong, "Making Medicines in the Early Modern Household" and "Collecting Knowledge for the Family"; Pal, *Republic of Women*; Rankin, *Panaceia's Daughters*; Ray, *Daughters of Alchemy*.

32. On courtly alchemy in the Holy Roman Empire, see Moran, *Alchemical World of the German Court*; Smith, *Business of Alchemy*; Nummedal, *Alchemy and Authority*.

33. For a helpful recent overview of the "rise of the state" in early modern Europe, see Capra, "Governance."

34. This book is the beneficiary of many years of scholarship on microhistory. See especially Brown, "Microhistory and the Post-Modern Challenge"; Egmond and Mason, *Mammoth and the Mouse*; Findlen, "Two Cultures of Scholarship?"; Ginzburg and Davin, "Morelli, Freud, and Sherlock Holmes"; Magnússon and Szíjártó, *What Is Microhistory?*; Muir and Ruggiero, eds., *Microhistory and the Lost Peoples of Europe*; Peltonen, "Clues, Margins, and Monads."

35. I took up some of these issues in the context of alchemy in *Alchemy and Authority*. See also Darnton, *Great Cat Massacre*; Duden, *Woman Beneath the Skin*.

36. My thinking on these complicated questions has been shaped especially by Davis, *Return of Martin Guerre*; Snyder, *Dissimulation and the Culture of Secrecy in Early Modern Europe*; Greenblatt, *Renaissance Self-Fashioning*; Martin, *Myths of Renaissance Individualism*.

37. On this middle ground among Renaissance historians, artists, and archaeologists, see Grafton, *Forgers and Critics*; Rowland, *Scarith of Scornello*; Wood, *Forgery, Replica, Fiction* and "Credulity Problem"; Margoszy, "Certain Fakes and Uncertain Facts."

CHAPTER 1

1. Brady, *German Histories*, 45. The classic work on the Grumbach Feud is Ortloff, *Geschichte der Grumbachischen Händel*. For a recent reappraisal, see Starenko, "In Luther's Wake," 12–13. On Moritz of Saxony, see Blaschke, *Moritz von Sachsen*.

2. Peter Starenko places Zieglerin and Schombach at the ducal court before Hänschen Henkel's arrival, although, given that Zieglerin claimed that she and Schombach were married in 1563, they may have arrived at the ducal court shortly after Hänschen instead. Starenko, "In Luther's Wake," 167.

3. In addition to the various German diminutive forms of Hans (Hänsen, Hänsel, or Hänschen), there were also several last names associated with the boy, including Müller, Henkel, and Tausendschön. He was also referred to as *der Engelseher*, or "the angel seer." See Ortloff, *Geschichte der Grumbachischen Händel*, 1: 272; and Winzlmaier, *Wilhelm von Grumbach*, 41.

4. Ortloff, *Geschichte der Grumbachischen Händel*, 1: 272–28; Starenko, "In Luther's Wake," 32.

5. According to Jürgen Beyer, these were "the options available to contemporary theologians." Beyer, "Lutheran Popular Prophets," 64.

6. Starenko, "In Luther's Wake," 53.

7. "Doctor Jonas guttliche aussage 1567," HStA Dresden Loc. 7191 38b, fol. 6r: "welcher es anfegklick vor Zeuberey vnndt Teuffels gespenst gehaltenn." As cited in Starenko, "In Luther's Wake," 48.

8. Starenko, "In Luther's Wake," 55.

9. Ibid., 68–72.

10. Languet, *Historische beschreibung der ergangenen Execution*, Fiiiir-v. In addition, "welcher entweder vom Grumbachen selbst, oder je andern auß derselben Gesellschaft, zu Zauberey, und Abgöttischer gauckelen gewehnet."

11. Stößel and Selnecker "condemned the boy, Grumbach, and the angels as agents of Satan, and denounced the duke for allowing himself to believe in either Hänschen or the 'angels of darkness.'" Starenko, "In Luther's Wake," 240–41.

12. "*QUOD SATAN NUNQUAM UERITA DICIT. Aber dieser Geist hab nun mer als hundert mal die warheit gesagt. ERGO NON EST MALUS SPIRITUS.*" "Doctor Johansen guttliche aussage Grumbachs Bubenn vom Sundthauszenn so man den Engelseher genant belangende. Anno 1566," HStA Dresden, fol. 112v. As cited in Starenko, "In Luther's Wake," 66; see also 59n104.

13. Starenko, "In Luther's Wake," 48–49, 53–55, 63–72.

14. The Elizabethan natural philosopher John Dee was equally sanguine about the possibility of communicating with angels and famously devoted significant energy and resources to developing the tools, methods, and personnel to communicate with them. See Harkness, *John Dee's Conversations with Angels*.

15. Starenko, "In Luther's Wake," 96.

16. Untitled document, NLA WO, 1 Alt 9, Nr. 307, fol. 177v: "So were Schilheintz . . . ein Narr. Den s. Hern Hertzog Johannes Fried unsern beweiß hett wie s. l. sie solt selbst berichten haben." In 1572, Schombach was described as "Hertzog Johaneß Friedrich zu Sachssen Cammer und leib diener . . . und vielleicht etwaß mehr." Protocol of Duke Julius's discussion with Heinrich Schombach about Thangel's slander against Johann Friedrich II, 14. February, 1572, NLA WO, 1 Alt 9, Nr. 313, fol. 1b.

17. Starenko, "In Luther's Wake," 42, 102.

18. Ibid., 122–38.

19. Ibid., 138–54.

20. Ibid., 154–66. On lore surrounding the Springwurtzel, see Bächtold-Stäubli, Hoffmann-Krayer, and Lüdtke, *Handwörterbuch*, s.v. "Springwurtzel," 8: 313–20.

21. Starenko, "In Luther's Wake," 154–70.

22. A catalog of Duke Johann Friedrich II's library does not suggest a serious scholarly engagement with alchemy. It contained only two books on the subject, both popular compendia of alchemical texts: Jacob, *De alchimia opuscula complura veterum philosophorum*; and Gratarolo, *Veræ alchemiae artisque metallica*. StAC, LA E 2405, fols. 241r and 256r, as cited in Starenko, "In Luther's Wake,"168n42. On other German princes with an entrepreneurial interest in alchemy, see Nummedal, *Alchemy and Authority*, chap. 3.

23. On Matthes (or Mattheus Friedrich) and Karle (or Carol), both of whom were working in Reinhardsbrunn in 1566, see Starenko, "In Luther's Wake," 168; and Ortloff, *Geschichte der Grumbachischen Händel*, 3: 268–90.

24. For the claim about the sum of 10,000 gulden, see "Hansen Beiers Bekentnus unnd Peinliche Aussage den 16. Aprilis Anno Dei geschehen," reprinted in Ortloff, *Geschichte der Grumbachischen Händel*, 4: 535. For a discussion of other alchemical contracts, as well as sums of money involved, see Nummedal, *Alchemy and Authority*, 109–18.

25. "Grumbach sei dem alchemisten feindt gewesen." "Verhör der Anna Maria Zieglerin," 16. July, 1574, NLA WO, 1 Alt 9, Nr. 314, fol. 32v. On Rudolf, see Ortloff, *Geschichte der Grumbachischen Händel*, 2: 301.

26. On Thangel's alchemical interests, see Starenko, "In Luther's Wake," 166ff.; and Ortloff, *Geschichte der Grumbachischen Händel*, 2: 30. On David Baumgartner's alchemical experiments, see Ortloff, *Geschichte der Grumbachischen Händel*, 3: 330; "Verhör der Anna Maria Zieglerin," Nr. 314, fol. 28; and Otto Puchner, "Baumgartner, David," in *Neue Deutsche Biographie* 1 (1953): 663.

27. This contract is paraphrased in Ortloff, *Geschichte der Grumbachischen Händel*, 3: 271–72.

28. Cob. Arch. Engelsanz. V. Bl. 159.160. [old notation], as cited in Ortloff, *Geschichte der Grumbachischen Händel*, 3: 272n1.

29. Moritz Hausner, n.p., 15. December, 1562, StAC, LA A 2034, fols. 20r-v, as cited and translated in Starenko, "In Luther's Wake," 198–99.

30. Wilhelm von Grumbach, Coburg, to Johann Friedrich II, Gotha, 17. December, 1562, StAC, LA A 2022, fols. 11v-12r, as cited and translated in Starenko, "In Luther's Wake," 200–201.

31. On this episode, see Starenko, "In Luther's Wake," 201–9.

32. On the delegitimization of imperial and Catholic authority in the sixteenth century, see Starenko, "In Luther's Wake," 202–6, 33.

33. Wilhelm von Grumbach to Johann Friedrich II, Gotha, 12.6.62, StAC, LA A 2023, fol. 109v, as cited and translated in Starenko, "In Luther's Wake," 209.

34. Ibid., 217.

35. [S.d.], n.p., 11.15.62, StAC, LA A 2031, fol. 55r, as cited and translated in Starenko, "In Luther's Wake," 193.

36. Wilhelm von Grumbach to Duke Johann Friedrich II, Würzburg, 9.10.63, StAC, LA A 2017, fol. 118v, as cited and translated in Starenko, "In Luther's Wake," 212.

37. Ortloff, *Geschichte der Grumbachischen Händel*, 1: 453, as cited and translated in Starenko, "In Luther's Wake," 213.

38. N.p., 22. and 23. January, 1564, StAC, LA A 2032, fol. 212r, as cited and translated in Starenko, "In Luther's Wake," 216. See also also Ortloff, *Geschichte der Grumbachischen Händel*, 1: 494.

39. Starenko, "In Luther's Wake," 218–19.

40. Moritz Hausner to Wilhelm von Grumbach, n.p., 15. January, 1574, StAC, LA A 2025, fol. 7r-9r, as cited and translated in Starenko, "In Luther's Wake," 219–20. See also Ortloff, *Geschichte der Grumbachischen Händel*, 1: 489.

41. For the duke's statements of conviction, see Starenko, "In Luther's Wake," 220–21, 35.

42. StAC, LA A 2027, fol. 3v ([MH], n.p.: 2.4.66 [GH 2:501; 3:7, 12]), as cited and translated in Starenko, "In Luther's Wake," 244.

43. Ibid., 246.

44. Ibid., 247.

45. Moritz Hausener, n.p., 14. August, 1564, StAC, LA A 2023, fol. 65r-v, as cited and translated in Starenko, "In Luther's Wake," 222. See also Ortloff, *Geschichte der Grumbachischen Händel* 2: 124 and 27.

46. Starenko, "In Luther's Wake," 224.

47. Ibid., 225.

48. Beck, *Johann Friedrich der Mittlere*, 2: 328.

49. Starenko, "In Luther's Wake," 242.

50. Zechariah 1:8–10 suggests that four horses are sent by God to patrol the earth and execute judgments; Zechariah 6:1–5 describes four different horses that patrol the north and south.

51. StAC, LA A 2027, fols. 56r-59v ([MH], n.p.: 5.6.66 [GH 2:516, 3:96]), as cited and translated in Starenko, "In Luther's Wake," 248.

52. Ibid., 249.

53. Ibid., 250–53.

54. McGinn, *Visions of the End*, 70–79.

55. On the German myth of the "third Friedrich," or *Friedenskaiser*, as Last World Emperor, see Shaw, "Friedrich II as the 'Last Emperor'"; and Struve, "Utopie und gesellschaftliche Wirklichkeit." On the sixteenth-century identifications of this "third Friedrich," see Barnes, *Prophecy and Gnosis*, 47; Starenko, "In Luther's Wake," 250–53; Webster, *Paracelsus*, 220, 231–32.

56. On this series of events, see Starenko, "In Luther's Wake," 12–14.

57. Ortloff, *Geschichte der Grumbachischen Händel*, 4: 96–97.

58. Rudolphi diary, as cited in Ortloff, *Geschichte der Grumbachischen Händel* 4: 96.

59. Ortloff, *Geschichte der Grumbachischen Händel* 4: 97.

60. Ortloff estimates that seven thousand people died as a result of the invasion, more from hunger and disease than from the battle itself. See Ortloff, *Geschichte der Grumbachischen Händel*, 4: 264.

61. Beck, *Johann Friedrich der Mittlere*, 1: 165. Interestingly, Lukas Thangel was a member of this commission.

62. For a detailed description of these executions, see Ortloff, *Geschichte der Grumbachischen Händel*, 4: 155ff.; and Dülmen, *Theatre of Horror*, 93–94.

63. Ortloff, *Geschichte der Grumbachischen Händel*, 4: 176.

64. Beck, *Johann Friedrich der Mittlere*, 1: 382–404; Kirwan, "Empowerment and Representation," 121–23.

65. Hubert Languet, *Historica descriptio svsceptae a Caesarea Maiestate executionis contra S. Rom. Imperij rebelles, eorumqúe Receptatorem: & captae urbis Gothae, solóque aequati castri Grimmenstein: Anno Domini M.D.LXVII, XIII. Aprilis*. [s.l.], 1568. I have consulted the German edition of this text, Languet, *Historische beschreibung*.

66. Languet, *Historische beschreibung*, Aiiv.

67. Ibid., Aiiir-Ciiiv.

68. Ibid., Ciiir-v.

69. SächsHStD, Geh. Arch. Loc. 9161, Verzeichnus der Schriften und Briefe/so in der kleinen Laden von Gothaischen Sachen befunden, 1567, fols. 1r-4v, as cited in Starenko, "In Luther's Wake," 74.

70. Languet, *Historische beschreibung*, Fiiiir.

71. Starenko, "In Luther's Wake," 75.

72. Languet, *Historische beschreibung*, Ciiir-v.

73. Abel Scherdinger, Pastor in Hohenkirchen, and Philipp Sömmering, Pastor in Schönau and Ampt Jorgenthal, to Duke Johann Wilhelm, Gotha, 4. April, 1567. LATh-HStA Weimar, Ernestinisches Gesamtarchiv, Reg. P (Grumbachsche Händel und die daraus entstandenen Gothaischen Händel), fols. 167–168, Blatt 5r-v.

74. Emphasis added. Scherdinger and Sömmering to Duke Johann Wilhelm, Gotha, 4. April, 1567. LATh-HStA Weimar, Ernestinisches Gesamtarchiv, Reg. P, fols. 167–168, Blatt 7r.

75. Scherdinger and Sömmering to Duke Johann Wilhelm, Gotha, 4. April, 1567. LATh-HStA Weimar, Ernestinisches Gesamtarchiv, Reg. P, fols. 167–168, Blatt 7v.

76. Scherdinger and Sömmering to Duke Johann Wilhelm, Gotha, 4. April, 1567. LATh-HStA Weimar, Ernestinisches Gesamtarchiv, Reg. P, fols. 167–168, Blatt 7v, reprinted in Beck, *Johann Friedrich der Mittlere*, 2: 305–6.

77. Christoff Ackermann, "Bephel haber zue Jorgthals," to Johann Wilhelm zu Sachsen, 27. February, 1567. LATh-HStA Weimar, Ernestinisches Gesamtarchiv, Reg. P, fols. 167–168, Blatt 1r-2v.

78. In Protestant territories in sixteenth-century Germany, a superintendent was an ecclesiastical administrator.

79. Melchior Weidemann, Superintendent in Gotha, to Duke Johann Wilhelm, 30. April, 1567, fol. 17r-v.

80. Although they would spend the next several years issuing numerous appeals to the new duke, Johann Wilhelm, the pastors never succeeded in winning back their posts and property in ducal Saxony. For the documents related to Scheringer's and Sömmering's appeals, see Sömmering, Scherdinger, and Hellefeld in Gotha, 1567–1572, LATh-HStA Weimar, Ernestinisches Gesamtarchiv, Reg. P, fols. 167–168 F3, Nr. 4, Blätter 1–64.

81. The two parties' animosity toward one another had its origins in the 1562 controversy over Striegel's Declaration, Duke Johann Friedrich II's attempt to impose doctrinal unity in his lands. Sömmering and Scherdinger both signed the document and then later withdrew their signatures, but Weidemann refused to sign at all. In other words, the two alchemist-pastors' opposition to synergism proved to be slightly weaker than Weidemann's in 1562. Beck, *Johann Friedrich der Mittlere*, 1: 382–404; 2: 153 and 159. Sömmering cited the "hatred and grudge, that he [Weidemann] held against me, on account of my position as a partisan of the Synergists (meiner Confession halb, alß ein *patronus Synergisternn*). Philipp Sömmering to Duke Johann Wilhelm, 20. June, 1567. LATh-HStA Weimar, Ernestinisches Gesamtarchiv, Reg. P, fols. 167–

168 F3, Nr. 4, Blatt 33v. See also Abel Scherdinger to Duke Johann Wilhelm, 18. June, 1567. LATh-HStA Weimar, Ernestinisches Gesamtarchiv, Reg. P, fols. 167–168 F3, Nr. 4, Blatt 30r-31v. On synergism in ducal Saxony, see Dingel, "Culture of Conflict," 15–64.

82. Georg Ernst, Graf und Herr zu Hennenberg to Johann Neuradt, Schosser to Georgenthal, 1. March, 1568. LATh-HStA Weimar, Ernestinisches Gesamtarchiv, Reg. P, fols. 167–168, Blatt 38r-47v.

83. Philipp Sömmering to Johann Wilhelm's *Räthe*, 6. May, 1568. LATh-HStA Weimar, Ernestinisches Gesamtarchiv, Reg. P, fols. 167–168, Blatt 49r-50v.

84. Catharina Sömmering, Schmalkalden, to Johann Wilhelm, Weimar, 17. March, 1569. LATh-HStA Weimar, Ernestinisches Gesamtarchiv, Reg. P, fols. 167–168, Blatt 53r-54v.

85. Philipp addressed a number of rumors circulating about his time in Saxony in a letter to Julius of 9. February, 1572, for example. Philipp Sömmering to Duke Julius, 9. February, 1572, NLA WO, Nr. 306, fols. 88–93.

86. For the specific claim that Sömmering lived off of the money he had received from Duke Johann Friedrich II, see "Was Philip Sommering am anfangs und zum Eingang in der gute berichtet und ausgesagt," n.d. [July 1574], NLA WO, 1 Alt 9, Nr. 311, fol. 15v. In this document, Sömmering also claims that he continued to practice alchemy in Schmalkalden, successfully completing a *solutionem solis*, although this may have been a later attempt to justify financing his sojourn there with Johann Friedrich II's money.

87. LATh-HStA Weimar, Ernestinisches Gesamtarchiv, Reg. P, fols. 5–8, 30–54.

88. Of particular interest to August was the band of nobles, including Anna's brother Hans Ziegler, who had been accused of highway robbery in 1565. See Ortloff, *Geschichte der Grumbachischen Händel*, 3: 39–46.

89. Philipp Sömmering described this in a letter to Herzog Julius, 8. January, 1574, NLA WO, 1 Alt 9, Nr. 307, fols. 202v-203r.

90. Rhenanus had been consulting on Julius's saltworks since 1571. Kolbe, Forche, and Humburg, *Die Geschichte der Saline Salzliebenhalle.*

91. Rhamm, *Die betrüglichen Goldmacher,* 9.

92. It is not clear whether Catharina Sömmering and her children stayed in Schmalkalden or accompanied the others to Wolfenbüttel. "Was Philip Sommering am anfangs und zum Eingang in der gute berichtet und ausgesagt," n.d. [July 1574], NLA WO, 1 Alt 9, Nr. 311, fols. 15r-17r. See also Rhamm, *Die betrüglichen Goldmacher,* 5–8. On Rhenanus, see Nummedal, *Alchemy and Authority,* 87–88; Hans-Henning Walter, "Rhenanus, Johannes," in *Neue Deutsche Biographie* 21 (2003): 494–95, http://www.deutsche-biographie.de/pnd13334150X.html.

CHAPTER 2

1. "Schreckliche Zeichen am Himmel, auch gespenst gesehen, und grewliche treume gehabt, welche sie da selbst außgedeutet, das Ihnen groß ungluck bevorstende und sie alle semptlich schedtlich umbkommen wurden." Regarding the ghost of the woman, Anna said, "Sie habe gedacht, es sei d teuffel. . . . Die leute in d herbergn haben gesagt, es Pflege da so zugehen und viele leut umbzukommen." "Verhör der Anna Maria Ziegler," 8. July, 1574, NLA WO, 1 Alt 9, Nr. 311, fols. 2r-v.

2. I have pinpointed Zieglerin's birth somewhere between 1541 and 1547, but most likely around 1545. In 1574 Philipp Sömmering said that she was twenty-three, which would mean that she was born in 1551. "Gütliche Aussage des Philipp Sömmering," 9. July, 1574, NLA WO,

1 Alt 9, Nr. 308, fol. 62r. However, Zieglerin's purported father, Caspar von Ziegler, died in 1547, so it is probable that she was born before then. In 1574, Zieglerin said that she was twenty-nine—that her previous claim to be twenty-four was a lie—which would put her birth date in 1545. She also claimed that her mother got married when the old duke in Dresden died (Heinrich "der Fromme," who died in 1541), so she was likely born after 1541. "Fraw Anna Maria Zieglerin verhort," 8. July, 1574, NLA WO, 1 Alt 9, Nr. 314, fol. 14v.

3. Anna named "Caspar Ziegler" as her father in her testimony from 8. July, 1574, NLA WO, 1 Alt 9, Nr. 314, fol. 2. On Caspar von Ziegler, see Tentzel, "Vorstellung oder kurtze Beschreibung," 41–42 and 61–62. Tentzel notes that Caspar von Ziegler's wife was Clara von Schönberg, and that they had two sons, Christoph and Hans, although he does not a list a daughter. On the landholdings of the Ziegler family, see Sächsisches Staatsarchiv, 10319 Grundherrschaft Klipphausen bei Meißen, 10590 Grundherrschaft Taubenheim bei Meißen, 10236 Grundherrrschaft Gauernitz, 10560 Grundherrschaft Schönfeld bei Dresden, 10416 Grundherrschaft Nickern, 10276 Grundherrschaft Helfenburg, and 10674 Grundherrschaft Zscheckwitz.

4. Franz Menges, "Schönberg," in *Neue Deutsche Biographie* 23 (2007): 386–87, http://www.deutsche-biographie.de/sfz115183.html. On the Schönberg family, see Sächsisches Staatsarchiv, 12614 Familiennachlass von Schönberg. See also Kobuch, *Die Geschichte der Familie von Schönberg.*

5. "Zu Pilnitz sey sie geboren wie ervor Ir berichtet und zu dresden getaufft, sie hette fursten und ander zugeuattern gahap." "Gütliche Aussage und peinliches Verhör des Heinrich Schombachs," 5. July, 1574, NLA WO, 1 Alt 9, Nr. 314, fols. 69r-v. See also Nr. 313, fol. 26.

6. Ortloff, *Geschichte der Grumbachischen Händel,* 3: 43–46. Hans Ziegler, who eluded arrest in 1566, later asked Anna to convince Duke Julius to intervene with Elector August of Saxony on behalf of Ziegler and his coconspirators Hans Heinrich and Antonius Pflug. "Verhör der Anna Maria Ziegler," 9. January, 1575, NLA WO, 1 Alt 9, Nr. 314, fol. 45r. See also her testimony on 26. January, 1575, NLA WO, 1 Alt 9, Nr. 314, fol. 45r.

7. He sold the estate to August's Rath Christoph von Loß the Younger. Tentzel, "Vorstellung oder kurtze Beschreibung," 42.

8. The Dresden Hofkirche did not keep a *Taufregister* (baptism registry) until 1660. The only church in Dresden that did keep a *Taufregister* during Anna Zieglerin's lifetime was the Dreikönigskirche in Dresden Neustadt, which started a *Taufregister* in 1560 (although the record has a gap from 1572 to 1577), a *Totenregister* (death registry) in 1570, and a *Trauregister* (marriage registry) in 1582. Pillnitz did not keep a *Taufregister* until 1638. Blanckmeister, *Die Kirchenbücher im Königreich Sachsen,* 97, 165.

9. Blaschke, John, Gross, and Starke, *Geschichte der Stadt Dresden,* vol. 1.

10. August, Churfürst zu Sachsen, *Künstlich Obstgarten-Büchlein.*

11. On August of Saxony, see Rößler, "August," in *Neue Deutsche Biographie* 1 (1953): 448–50, http://www.deutsche-biographie.de/pnd119458446.html; Brady, *German Histories in the Age of Reformations,* 239–41; Nummedal, *Alchemy and Authority,* 81–84; Blaschke, *Politische Geschichte Sachsens und Thüringens.* On Anna of Saxony, see Rankin, *Panaceia's Daughters* and "Becoming an Expert Practitioner"; Keller, *Kurfürstin Anna von Sachsen.*

12. On marriage among the ruling classes, see Wunder, *Er ist die Sonn', Sie ist der Mond,* 80–81.

13. Zieglerin testified to this several times and on different days: "Verhör der Anna Maria Ziegler," 14. July, 1574, NLA WO, 1 Alt 9, Nr. 314, fols. 22–23; "Verhör der Anna Maria Ziegler," 16. July, 1574, NLA WO, 1 Alt 9, Nr. 314, fol. 37v; and "Verhör der Anna Maria Ziegler," 16. November, 1574, NLA WO, 1 Alt 9, Nr. 314, fol. 45. For example, "der

Handorff hette sie erst zu falle gebracht, wie sie noch nit virzehen jar alt gewesen," NLA WO, 1 Alt 9, Nr. 314, fol. 37v; "Im Landt zu Meichssen were Sie zu falle gekomen," fol. 45r. The man Anna would later marry, Heinrich Schombach, recounted a more elaborate version of these events, claiming that Anna's mother had betrothed her to Handorf, but that Anna had put him off because she had no interest in him. One day, however, Handorf "snuck up on her and attacked her" (presumably sexually), "but that time he was unsuccessful because she was too young." Finally, Handorf "gave her something in an apple, so that she would do his will and become pregnant." When Anna did in fact become pregnant following this incident, according to Schombach, she informed Handorf of her pregnancy, and then, when she gave birth to the child, had a maid wrap it in a linen cloth and drown it. "Der fal mit Niclas von Hondorf hatt sie also zugetragen, dß Ihre Mutter seine frawen Ihme gelobt, Sie hett aber keine lust zu Ihme gehabt, und Ihn fast ein Jahr aufgehalten, daruber [derselb] sie ein mal geschlich und sich uber sie hergemacht, das mal es Ihne aber nicht angang, aus ursach, daß sie [??] zu jung gewesen. Dernach hett er Ihr in einem Apfel etwas geben, das sie seinen willen gethan und geschwanger werden." "Gütliche Aussage und peinliches Verhör des Heinrich Schombach," NLA WO, 1 Alt 9, Nr. 313, fol. 27r. See also "Aussagen Heinrich Schombach gegen Anna," NLA WO, 1 Alt 9, Nr. 314, fols. 69v-71r. Rhamm repeats Schombach's version of events in *Die betrüglichen Goldmacher*, 13. I have not been able to locate any contemporary evidence of this rape or its fallout, but the incident clearly followed Anna to Wolfenbüttel.

14. "Heinrich [Schombach] hett vol gewist, dass sie zuvorn zu fall bracht, sie habe auch zuvor eine Eheman gehabt, der habe Ihren fall, auch wol gewust, und habe Rottenburg geheiss von Grunberg. Mit dem habe sie auch keine Ehestifftung aufgericht, der habe am 9 tage dem halßentz wie gesturzt Pferdt." "Verhör der Anna Maria Ziegler," 8. July, 1574, NLA WO, 1 Alt 9, Nr. 314, fol. 3; see also "Verhör der Anna Maria Ziegler," 16. November 1574, NLA WO, 1 Alt 9, Nr. 314, fol. 37v: "Ir schwager und schwester hetten dem Rotenburg gessagt, das sie albereidt zufalle gekommen were, hette denselben nur 9 wochen und Heinrich 9 Jahr gehapt. Zu konigeswort hette sie den Rotenburg genomen." Anna said that "Mit dem [Rottenburg] habe sie auch keine Ehestifftung aufgericht," "Verhör der Anna Maria Ziegler," 8. July, 1574, NLA WO, 1 Alt 9, Nr. 314, fol. 3. Königswartha did not keep a *Trauregister* until the seventeenth century in any case. Blanckmeister, *Die Kirchenbücher im Königreich Sachsen*, 131.

15. On widowhood in early modern Germany, see Wunder, *Er ist die Sonn', Sie ist der Mond*, 174–90.

16. For nobles with a patrimony at stake, perhaps there was an additional reason for an older man to marry a much younger woman for a second (or third) wife, particularly if his prior marriage(s) had not yet produced an heir. Indeed, Karl-Heinz Spieß has estimated that 75 percent of the princes in the Holy Roman Empire between 1200 and 1600 left behind widows. Spieß, "Witwenversorgung im Hochadel," 87.

17. Wunder, *Er ist die Sonn', Sie ist der Mond*, 187.

18. Most likely, this was her deceased husband's brother, since he, too, was a Silesian Juncker.

19. According to Rhamm, Schombach was a "Kammerdiener und Hofnarr." Rhamm, *Die betrüglichen Goldmacher*, 13.

20. "Schilheintzen hab er Grumbach allewege fur einen leichtfertigen Lecker gehalten, habe dünne Oren, kan vil erfaren, was er gehort hab er hin und wider getragen, von iren Hendeln und Practicen hab er nichts gewust." "Grumbach's Guthliche Aussage," 15. Aprilis Ao. [15]67, as quoted in Ortloff, *Geschichte der Grumbachischen Händel* 4: 531. Rhamm also

cites this same passage, but slightly modernizes the phrasing: "habe dünne Ohren, könne viel erfahren; was er gehört, trage er hin und wider." Rhamm, *Die betrüglichen Goldmacher*, 13.

21. "Ehr sei zum Prediger der Kirchen gangen, und gebeten, sie Ihme zugeben, für den Pfarrer habe sie einen leiblichen aidt gethan, w.. Sie ihm ehelich gegeben." "Heinrich Schombachs Verhör," 5. July, 1574, NLA WO, 1 Alt 9, Nr. 313, fol. 27r.

22. "Es sei keine ehestiffung aufgericht. Es sein vil leute dabey gewesen allß Joachim Haubitz und ander schlesische Junckern." "Verhör der Anna Maria Ziegler," 8. July, 1574, NLA WO, 1 Alt 9, Nr. 314, fols. 3r-v. Anna also reiterated this story in a letter to Johann Friedrich several years later, where she reminded the duke who she was by saying, "E. f. g. haben sich ohn allen zweiffel gnedigst zubesinnen, das ich durch E. f. g. begern und gnediger vieler vorschrifften meinen jtzigen eheman Heinrichen Schumbach zur ehe gegeben worden bin." (Without a doubt Your Princely Grace recalls that I was given in marriage to my current husband, Heinrich Schombach, as a result of your many gracious requests and instructions.") Anna Zieglerin to Herzog Johann Friedrich II, 19. December, 1573, NLA WO, 1 Alt 9, Nr. 307, fol. 193r. Rhamm seconds Anna's version of events. Rhamm, *Die betrüglichen Goldmacher*, 13.

23. At least one face might have been familiar to her in ducal Saxony, for the man she claimed was her brother, Hans Ziegler, may also have been affiliated with Johann Friedrich II's court. According to Rhamm, Hans Zieglerin served in the retinue of Johann Friedrich II's close adviser and confidant, Grumbach. Rhamm, *Die betrüglichen Goldmacher*, 13. On the other hand, Grumbach did not list Ziegler among his servants (*Knechte*) in his 1567 testimony. Ortloff, *Geschichte der Grumbachischen Händel*, 4: 525.

24. On the "culture of conflict" among Protestants in this period, see Dingel, "Culture of Conflict."

25. Thomas Brady prefers the term Real Lutherans to Gnesio, following the German *Echter*, but Gnesio remains the more common term. Brady, *German Histories in the Age of Reformations*.

26. Particularly in the wake of the Protestant princes' defeat in 1547. On these disputes and the "forming of the Lutheran Confession," see Brady, *German Histories in the Age of Reformations*, 264–68; Barnes, *Prophecy and Gnosis*, 64–65.

27. On these controversies, see Dingel, "Culture of Conflict," 15–64.

28. "sei ehr . . . folgends anno 55 zu Schonaw Pfarrer worden, und da [zu Schonaw] gebliebcn biß Gota belagert worden. Sein Predigerampt habe er da gewarten biß der streit mit dem Victorino [Striegel] furgefallen, da sei ehr an Bergkhandel kommen, und haben Ihme Johan Hoffman und Nickel Crolach in seinen Patria erstlich darauff gefurt, von dem habe her erstlich ein alchimistisch buch bekomen." On the dispute over Viktorin Striegel's controversial 1562 *Declaration* on synergism, see below. Philipp Sömmering described his early life in his initial deposition for his 1574–1575 trial, "Was Philip Sommering am anfangs un zum Eingang in der gute berichtet und ausgesagt," n.d. [July 1574], NLA WO, 1 Alt 9, Nr. 311, fols. 14r-17v. On the first generations of Lutheran pastors in this region, see Karant-Nunn, *Luther's Pastors*. On the dispute surrounding Viktorin Striegel, see Beck, *Johann Friedrich der Mittlere*, 1: 382–404; Kolb, *Bound Choice*, 121–23.

29. Nummedal, *Alchemy and Authority*, chap. 3.

30. "demnach von Ein andern zapfern Pfarrhern am hartz noch eins [i.e., noch ein Buch], darin ehr von dem Raute Lunaria gelesen, daraus sie safft bekommen, und die alchimij damit angefangen nach dem Proceß." "Was Philip Sommering am anfangs un zum Eingang in der gute berichtet und ausgesagt," n.d. [July 1574], NLA WO, 1 Alt 9, Nr. 311, fol. 14v. "Lunaria sey ein kraut gar wol reichend, sei gelb an stengel, die blumen [von zwey?] farben. Wen man

angrifft, gehe der safft darauß. Daraus haben sie safft bekommen, und die alchimey damit angefangen nach dem Proceß." "Verhör und Aussagen Philipp Sömmerings," 9. July, 1574, NLA WO, 1 Alt 9, Nr. 309, fol. 1v.

31. Mondraute, St. Petersschlüssel, St. Walpurgiskraut, Botrychium lunaria or Lunaria annua. Bächtold-Stäubli et al., *Handwörterbuch*, s.v. "Mondraute."

32. Della Porta, *Phytognomonica*; Gessner, *De Raris et Admirandis Herbis;* Bock, *De Stirpium.*

33. Bock, *De Stirpium,* 345r, as cited in Bächtold-Stäubli et al., *Handwörterbuch*, 6: 39–40. For a much later claim that it could be used to transmute base metals into gold, see Jacob, *Die Fruchtbare Boriza.* Interestingly, *Die Fruchtbare Boriza* was printed in Silesia, not far from where Sömmering lived, suggesting that alchemical lore surrounding *Lunaria* had a long life in this mining region.

34. "Das meinet auch Salienus da er redet/von dem Kraut Lunatica oder Bera seine Wurtzel ist ein Metallische Erdt/hat einen Roten Stengel/mit einer schwertze belegt/oder besteckt wessentlich/und nimbt auch leichtlich ab/gewint oder bekombt auch zeitlich Blumen/nach 3 Tagen so man das thut inn Mercurium/so verendert es sich in ein vollkommen Silber/unnd so man das weiter seut so verkert es sich in Goldt/dasselbig Goldt verkert 100 theyl in das aller Feinste Goldt." Salomon [Pseudo-]Trismosin et al., "Splendor Solis," in *Aureum Vellus* 1: 19.

35. "Verhör und Aussagen Philipp Sömmerings," 9. July, 1574, NLA WO, 1 Alt 9, Nr. 309, fol. 1r ff.

36. Sömmering said that he acquired "Librum Isaaci, . . . *darin viel gutes Dinges.*" "Was Philipp Sömmering Anfangs und zum Eingang in der Gute berichtet und ausgesagt," 12. June, 1574, NLA WO, 1 Alt 9, Nr. 311, fol. 15r.

37. "Habe demnach den Librum Isaaci bekommen, darin viel gutes dinges, folgends auch das Hexameron Bernanrdj, welchs man sonst Liber de Creatione nennet, fur 400. Thaler." "Was Philip Sommering am anfangs un zum Eingang in der gute berichtet und ausgesagt," n.d. [July 1574], NLA WO, 1 Alt 9, Nr. 311, fol. 15r. Also "habe ehr auch gern haben wollen Exameron Bernardt, welches sie [meinen?] Liber de Creatione, und deßwegen in Erfurdt geritten. dass sei ein herlich buech sei umb 1000 Thaler gebott, doch nicht darumb zugekaufft. Da habe ehr aufbracht 400 Thaler und d frater geschenckt 40 groschen, viel fraw Ihne auf die [knie?] gefallen und gebetn, Ihm das buch zu uberlassen. Welches ehr auch bekommen." "Verhör und Aussagen Philipp Sömmerings," 9. July, 1574, NLA WO, 1 Alt 9, Nr. 309, fol. 2v. I have been unable to identify which book this might have been, although Rhamm suggests it was something attributed to the fourteenth-century Bernardus Trevisanus, who was known for his rejection of the sulfur-mercury theory in favor of the theory that mercury alone was the basis of all metals. Given the enthusiasm for Trevisanus's writings that Sömmering would later express to Duke Julius in Wolfenbüttel, this makes sense. On Trevisanus, see Nummedal, *Alchemy and Authority,* 19, esp. n10.

38. The Elizabethan alchemist and physician Simon Forman, for example, explored this connection in his notes "Upon the firste of Genesis." Kassell, *Medicine and Magic in Elizabethan London,* 194–96. See also Forshaw, "Vitriolic Reactions"; Crowther-Keyck, " 'Be Fruitful and Multiply.' "

39. In 1567 Scherdinger claimed to have held his "Pfarampt" in Hohenkirchen for fifteen years. Abel Scherdinger to Duke Johann Friedrich II, 18. June, 1567, LATh-HStA Weimar, Ernetinische Gesamtarchiv, Reg. P, fols. 167–168, Blatt 30r.

40. "gesagt, ehr wise die heimblicheit die kein mensch wise." "Verhör und Aussagen Philipp Sömmerings," 9. July, 1574, NLA WO, 1 Alt 9, Nr. 309, fol. 3.

41. Nicolaus (also Nikolaus or Nicolas) Solea was the author of *Ein Büchlien von dem Bergwergk* (1600), first written as a manuscript for Count Wolfgang II von Hohenlohe. It was reprinted in Frankfurt am Main by Antonius Humm in 1618 as *Bergwerckschatz, Das ist, Außführlicher und vollkommener Bericht von Bergwercken, nach der Ruten, und Witterung künstlich zubawen . . .* , but Solea's name had disappeared; the title page explained only that it was "vor niemals an Tag kommen, Auff Begern einer Hohen Fürstlichen Person gestellet, Jetz durch Eliam Montanum Fürstlichen Anhaltischen Leib-Medicum zum Briege an Tag gegeben." See the copy at the Digital Collections of the University Library Erlangen-Nueremberg: urn:nbn:de:bvb:29-bv008957139-3. Solea's name was not reconnected to the book until the eighteenth century, when it was reprinted as "Nicolai Soleae Philosophische Grund-Sätze," in *Drey curieuse bißher gantz geheim gehaltene Nun aber denen Liebhabern der Kunst zum besten An das Tages-Licht gegebene Chymische Schrifften*: Als 1. Nicolai Soliae Philosophische Grundsätze, 2. Herrn C. L. von L. Chymischer Catechismus, 3. CXXX. Grund-Sätze hier . . . (Leipzig: Strauß, 1723). See Weyer, *Graf Wolfgang II. von Hohenlohe und die Alchemie*, 212, 284, 315, 390, and also 284.

42. Sömmering's interest in arsenic evidently continued while he was in Wolfenbüttel. See Philipp Sömmering to Duke Julius, 9. February, 1572, NLA WO, 1 Alt 9, Nr. 306, fol. 91v.

43. In fact, what evidently made Solea change his mind was Johann Friedrich's position on the Lutheran controversies, particularly with respect to the Gnesio, Echter, or Real Lutheran party in ducal Saxony. See "Gütliche Aussage des Philipp Sömmering," 9. July, 1574, NLA WO, 1 Alt 9, Nr. 309, fols. 1r-3v.

44. The duke's sudden interest in alchemy may have been linked to the arrival of Moritz Hausner, Grumbach's secretary and himself a practitioner of the art, but Johann Friedrich II easily could have come to alchemy on his own as well. On Johann Friedrich's alchemical activities in these years, see Ortloff, *Geschichte der Grumbachischen Händel*, 3: 267–74.

45. Abel Scherdinger to Johann Friedrich II, Diestags nach Trinitatis, 1565, StAC LA A 10543, fols. 1r-v; Philipp Sömmering to Duke Johann Friedrich II, 1566, StAC, LA A 10543, fols. 3r-v; and Duke Johann Friedrich II, Grimmenstein, to Philipp Sömmering and Abel Scherdinger, 27. November, 1566, StAC, LA A 2020, fols. 4–5. I am deeply grateful to Peter Starenko for generously sharing these documents with me many years ago.

46. Philipp Sömmering to Duke Johann Friedrich II, the Tuesday after [Vigilii], 1566, StAC, LA A 2020, fols. 141r-142r.

47. Philipp Sömmering to Duke Johann Friedrich II, 1566, StAC, LA A 10543, fols. 3r-v.

48. This contract is paraphrased in Ortloff, *Geschichte der Grumbachischen Händel* 3: 271–72.

49. "Verhör und Aussagen Philipp Sömmerings," 8. July, 1574, NLA WO, 1 Alt 9, Nr. 309, fols. 1r-3v.

50. Until the eighteenth century, "Wolfenbüttel" referred to the castle, not the town, parts of which were known variously as "Zu unswer lieben Frawn," "Newstadt," "Heinrichstadt," "Sophienstadt," "Juliusfriedenstadt," etc. For the history of early names for what is now the city of Wolfenbüttel, see Thöne, "Wolfenbüttel under Herzog Julius."

51. This church was destroyed during Duke Heinrich's war with the Schmalkaldic League in the 1540s, and finally replaced with the Marienkapelle in 1561. Thöne, *Wolfenbüttel*, 43.

52. Ibid., 35.

53. The castle and Newstadt were both damaged by the skirmishes between Duke Heinrich and the Schmalkaldic League between 1542 and 1547. Chiaramella also worked in Antwerp for Emperor Charles V, as well as throughout northeastern Germany, in Spandau, Küstrin,

Peitz, Dönitz, Schwerin, and Rostock. As Friedrich Thöne put it, "Erst von Wolfenbüttel ging er in das ostelbische Gebiet und baute dort Festungen aus." Thöne, "Wolfenbüttel under Herzog Julius," 8.

54. "die Heinrichstadt . . . in etwas zu extendiren, größer bebauen und gefestigen, dadurch auch die Straßen und Häuser legen zu lassen, darmit nach Zeit und Gelegenheit sich darin mehr Leute niederlassen. . . . So befinden wir doch, dass darin allerhand lose Kuffen und kleine böse Feuernester, die woll sonst in andere Wege können geordnet werden . . . dass also die Heinrichs-Stadt mit schnur-richtigen, räumigen Strassen und gleichförmigen Häusern von Weite und Höhe schnur-fehlig sollen . . . gebauet werden, auf dass solche Strassen alle auf unser Veste Wolffenbüttel lauffen und in Besichtigung wesentlich können gehabt werden." "Kurtze Nachricht, Nachdem seither Anno 1661, da die vorstehende Fürstl. Feuer-Ordnung publiciret worden . . . als Anhang . . . gedruckt 1738," as cited in Thöne, "Wolfenbüttel under Herzog Julius," 38 and *Wolfenbüttel*, 3.

55. On Duke Heinrich's and Duke Julius's urban planning, architectural, and infrastructure projects, see Thöne, "Wolfenbüttel under Herzog Julius" and *Wolfenbüttel*, as well as Scheliga, "Renaissance Garden in Wolfenbüttel"; Kraschewski, *Wirtschaftspolitik im deutschen Territorialstaat*. For the estimation of one hundred houses in 1571, see Thöne, "Wolfenbüttel under Herzog Julius," 37.

56. The standard modern works on Julius include Mohrmann, "Vater-Sohn-Konflikt und Staatsnotwendigkeit"; Kraschewski, *Wirtschaftspolitik im deutschen Territorialstaat*; Graefe, *Staatsklugheit und Frömmigkeit;* and Kraschewski, "Julius," in *Neue Deutsche Biographie* 10 (1974): 654–55, http://www.deutsche-biographie.de/pnd118558714.html.

57. According to Julius's sixteenth-century biographer Franz Algermann, Julius's nursemaid let him fall off a table, and his feet never healed properly from the injury. Algermann, "Leben, Wandel und tödtlicher Abgang des Herrn Juliussen, Herzogen zu Braunschweig," 6.

58. "Nach diesem [i.e., his stay in Cologne] ist S. F. Gn. nach Löwen zum Studieren geschickt worden, da Dieselben auch so viel gelernet, das Sie etwas Latein verstehen konnten." Algermann, "Leben, Wandel und tödtlicher Abgang des Herrn Juliussen, Herzogen zu Braunschweig," 8. As Wolf-Dieter Mohrmann has pointed out, narratives of Julius's early years, particularly of his relationship with his father, have been strongly influenced by the ducal historian, court official, and poet Franz Algermann (1548–1613), his first biographer, and subsequent Protestant historians, particularly the evangelical pastor of St. Michael's in Braunschweig, Philipp Julius Rehtmeyer (1678–1742), who framed Julius and Heinrich's relationship in confessional terms. Mohrmann summed up this confessional view concisely as follows: "Herzog Heinrich hielt am alten Glauben fest, bekämpfte die neue evangelische Lehre, wo immer sie sich zeigte. Und da sie im Sohn Julius einen glaubensstarken Anhänger fand, bekämpfte er auch diesen, bis derselbe vor den grausamen väterlichen Nachstellungen in höchster Gefahr außer Landes floh." As Mohrmann notes, this polemical confessional version of Julius's history continues to be influential today. Mohrmann, "Vater-Sohn-Konflikt und Staatsnotwendigkeit," 64. For an overview of the Schmalkaldic Wars, see Brady, *German Histories*, 220–27.

59. On the use of the term *Meister* (with no additional name) to refer to an executioner in early modern Germany, see Angstmann, *Der Henker in der Volksmeinung*.

60. Algermann, "Leben, Wandel und tödtlicher Abgang des Herrn Juliussen, Herzogen zu Braunschweig," 7.

61. On executioners in early modern Germany, see Harrington, *Faithful Executioner* and *Executioner's Journal*. On executioner medicine in particular, see also Stuart, *Defiled Trades and Social Outcasts*, es chap. 6, "The Executioner's Healing Touch."

62. Algermann, "Leben des Herzogs Julius zu Braunschweig und Lüneburg," 7.

63. "damit die zeit seines abwesns auch ehrliche und gutte verhaltung eine vergeslichkei-tund minderung des vatterlichen zorns villaicht mit bringen werde." Watzlaw von Wegesowitz an den böhmischen König Maximilian, Teplitz 1556 Mai 30: Beh. Ausgertigung HHStA, Staate-nabteilung Brunsvicensia, Fasz. I, as cited in Mohrmann, "Vater-Sohn-Konflikt und Staatsnot-wendigkeit," 67n14.

64. The location of Schloß Hessen is not to be confused with the territory of Hessen. Julius and Hedwig renovated this palace around 1564. Thöne, *Wolfenbüttel*, 45.

65. Mohrmann, "Vater-Sohn-Konflikt und Staatsnotwendigkeit," 67.

66. For an overview of Julius's implementation of the Reformation in Braunschweig-Lüneburg, see Mager, "Die Einführung der Reformation." See also Reller, "Vorreformatorische und Reformatorische Kirchenverfassung"; Mager, *Die Konkordienformel im Fürstentum Braun-schweig-Wolfenbüttel*. The text of the various *Kirchen-, Schul-,* and *Armenordnungen* that accom-panied the Reformation in the duchy can be found in Sehling and Sprengler-Ruppenthal, *Die Evangelischen Kirchenordnungen des XVI. Jahrhunderts*. The University of Helmstedt has also been the subject of several recent studies, including Maaser, *Humanismus und Landesherrschaft;* Kirwan, "Empowerment and Representation"; Triebs, *Die Medizinische Fakultät der Universität Helmstedt*.

67. Beck, *Johann Friedrich der Mittlere*, 1: 382–404; Kolb, *Bound Choice*, 121–23.

68. Mager, "Die Einführung der Reformation," 27–28.

69. "Was Philipp anfangs und zum Eingang in den gute berichtet und aufgesagt," n.d. [June 1574], NLA WO, 1 Alt 9, Nr. 311, fols. 14r-17v; and Rhamm, *Die betrüglichen Gold-macher*, 11.

70. This contract is now lost, unfortunately, but Philipp Sömmering described it in his 1574 testimony. "Was Philipp anfangs und zum Eingang in den gute berichtet und aufgesagt," n.d. [June 1574], NLA WO, 1 Alt 9, Nr. 311, fols. 14r-17v and "Gütliche Aussage des Philipp Sömmering," 9. July, 1574, NLA WO, 1 Alt 9, Nr. 308, fol. 47r.

CHAPTER 3

1. "es ist heutte dato mein lieber man zu mir alher k[in] braunschweick komen und mir angezeiget wie weit yr auch mit dem durchleuchtigen hochgebornen fursten und herren her-zogk yulio von braunschweick einglassenn." Anna Zieglerin in Braunschweig to Philipp Söm-mering [in Wolfenbüttel?], 1. September, 1571, NLA WO, 1 Alt 9, Nr. 306, fol. 14r.

2. On the payment Philipp, Heinrich, and others received for their alchemical work, see "Gütliche Aussage des Philipp Sömmering," 9. July, 1574, NLA WO, 1 Alt 9, Nr. 308, fol. 51v-52r.

3. Anna Zieglerin to Philipp Sömmering [in Wolfenbüttel?], 1. September, 1571, NLA WO, 1 Alt 9, Nr. 306, fol. 14r. On Thangel, see Georg Müller, "Tangel, Lukas" in *Allgemeine Deutsche Biographie* 37 (1894): 370, https://www.deutsche-biographie.de/gnd123326575 .html#adbcontent. Note that although *Deutsche Biographie* lists Thangel's date of death as 1566, the Deutsche National Bibliothek lists his date of death as 1590, which is correct. "Tangel, Lukas," Indexeintrag: Deutsche Biographie, https://www.deutsche-biographie.de/gnd123326575 .html and http://d-nb.info/gnd/123326575.

4. On Thangel's proposals, see Ortloff, *Geschichte der Grumbachischen Händel* 3: 272n1 and Chapter 1 above.

5. Beck, *Johann Friedrich der Mittlere,* esp. 2: 165.

6. "Georg von Harstall zu Meila, Amtmann in Creuzburg und Gerstungen, an s. Schwä-gerin, die Witwe Dorothea von H[arstall], wegen Warnung des zur Zeit in Creuzburg weilen-den braunschw. Rates Dr. Lukas Thangel vor Sömmering und seinen alchimistischen Komplizen," 16. July, 1571, NLA WO, 2 Alt, Nr. 8080, fols. 7–8. Thangel reportedly denounced Philipp, Heinrich, and Sylvester Schulvermann as "drey Landtbetrieger." He described Philipp as "der gewesene Pfarher zu Schonaw, der Alchimist." See also "Schreiben des Doktors der Rechte Lucas Thangel aus Frankfurt/M. an Herzog Julius betreffend die des Salzsiedens kundi-gen Fachleute," 8. August, 1571, NLA HA, BaCl Hann. 84a, Nr.6/72.

7. "Nuhn weist yr wie es Meynem armes man und mir armen weybe an herren hoffe gegangen wie seyne getraw diesnt, uns mit so beteubttenn, herzeleidt schmach schandt gefer leybes und lebens vergelden ist worden. Auch weyl wir albereidt ohne alle schult von Dockter Daniel an furstlicher Durgleuchtigkeit hoffe geschmet vor schelmen und losse leutte gescholden sindt worden[,] kont yr leichtlich erachten, wan ich zu wolfenbutt, solt, warhafftig sein wie ich solde das kreutze tragen mussenn." Anna Zieglerin in Braunschweig to Philipp Sömmering [in Wolfenbüttel?], 1. September, 1571, NLA WO, 1 Alt 9, Nr. 311, fol. 14r.

8. On the payment Philipp, Heinrich, and others received for their alchemical work, see "Gütliche Aussage des Philipp Sömmering," 9. [July] anno 7[4], NLA WO, 1 Alt 9, Nr. 308, fol. 51v.

9. The literature on court culture and patronage is immense. For a starting place, how-ever, see Moran, *Patronage and Institutions;* Biagioli, *Galileo Courtier;* Ash, *Power, Knowledge, and Expertise;* Duindam, *Vienna and Versailles;* Rankin, *Panacea's Daughters.*

10. For an excellent overview of historiography of the European court and its intersection with women's history, see Akkerman and Houben, *Politics of Female Households,* esp. the intro-duction. See also Keller, *Hofdamen;* Schulte, *Body of the Queen.* On the medical expertise in particular of German noblewomen, see Rankin, *Panacea's Daughters.*

11. On how such networks were constructed and fostered, see Hannah Leah Crummé, "Jane Dormer's Recipe for Politics: A Refuge Household in Spain for Mary Tudor's Ladies-in-Waiting," in Akkerman and Houben, *Politics of Female Households,* 51–76.

12. Basilius Satler, *Ein Predigt gethan bey der Begrebis . . .* (Wolfenbüttel, 1602), fol. Jiiiv., as cited and translated in Jill Bepler, "Posterity and the Body of the Princess in German Court Funeral Books," in Schulte, *Body of the Queen,* 129.

13. On noblewomen as medical practitioners, see Rankin, *Panacea's Daughters.*

14. Scheliga, "Renaissance Garden," 7.

15. On Lippold, see Franz Menges, "Lippold" in *Neue Deutsche Biographie* 14 (1985): 667f., https://www.deutsche-biographie.de/gnd13708384X.html#ndbcontent; Ackermann, *Münzmeister Lippold,* 3–33; Jütte, *Das Zeitalter des Geheimnisses,* 143–44.

16. Before the investigation was even concluded, the new elector, Johann Georg, took revenge on the Jewish community of Brandenburg, authorizing the destruction of the syna-gogue in Klosterstraße, as well as plundering and violence. On this episode and Joachim's II massive debts, see Ackermann, *Münzmeister Lipopld,* 37–39.

17. Ackermann also reproduced many of the archival documents related to the case in *Münzmeister Lippold.*

18. "den gnediger Furst und Herr e. f. g. geliebtter gemahl yhren grymmigen thorn auf uns geworffen, doch wieder gott, und alle verdienst, das da yr f.g. bey E.f.g. nicht hatt erhalten [?] konnen, uns zu ungenaden zubewegen, das yr f.g. andere hohe personem an e.f.g. abfertigen,

ist zu besorgenn." Anna Maria Zieglerin to Duke Julius, 12. December, 1571, NLA WO, 1 Alt 9, Nr. 306, fols. 4r-v.

19. Anna Zieglerin to Duke Julius, s.d. [1571–73?], NLA WO, 1 Alt 9, Nr. 306, fol. 9r.

20. See Chapter 2 above, for more information about Chemnitz's work for Duke Julius, as well as Mager, *Die Konkordienformel*, 165–75. On de Vries's work in Braunschweig-Lüneburg, see Scheliga, "Renaissance Garden"; Thöne, "Hans Vredeman de Vries in Wolfenbüttel."

21. On the "expert mediator," a middling figure who drew on knowledge and networks to help provide expertise to expanding states in early modern Europe, see Ash, *Power, Knowledge, and Expertise*. On brokers, see Biagioli, *Galileo Courtier*.

22. For example, Weyer, *Graf Wolfgang II. von Hohenlohe*.

23. For an overview of Julius's interests in mining, alchemy, and related economic activities, see Nummedal, *Alchemy and Authority*, 80–81; Kraschewski, *Wirtschaftspolitik im deutschen Territorialstaat*, esp. 157–60.

24. For some examples of other contemporary German princely laboratories, see Nummedal, *Alchemy and Authority*, 119–46.

25. In addition to the specific alchemical work Philipp promised in his 1566 contract with the Saxon duke, he also had helped Johann Friedrich II assess his existing alchemical enterprises by identifying incompetent and deliberately deceptive alchemists already in the duke's employ. He claimed to have helped Johann Friedrich stop wasting his money on inept alchemists and "appropriately punish the fraudulent ones" (die betrieger in gebührene straff genohmen). Scherdinger and Sömmering to Duke Johann Wilhelm, Gotha, 4. April, 1567, LATh-HStA Weimar, Ernestinisches Gesamtarchiv, Reg. P, fols. 167–168, Blatt 7r-v.

26. "Matz Rodtermund habe Ihme einen sonderlich aidt gethan, dass ehr Illmo. und Ihme wille traulich diener und die sachen verrichten, und was ehr horr und sagt, nicht nachschwetzen. Die Copie des aids sei unter seinen anderen briefen noch wol zubefinden." "Gütliche Aussage des Philipp Sömmering," 9. July, 1574, NLA WO, 1 Alt 9, Nr. 308, fol. 50r. On the variety of types of work in alchemical laboratories, see Nummedal, *Alchemy and Authority*, 122–33.

27. One of the other men involved in this robbery, Jobst Kettwig, would later come to work in Wolfenbüttel as well.

28. On Schulvermann, see his testimony of 9., 10., 12., and 18 November, 1574, NLA WO, 1 Alt 9, Nr. 315, fols. 12r-50v.

29. For an overview of Brun's artistic oeuvre, see Veronika Braunfels, "Brun, Franz," in *Grove Art Online, Oxford Art Online*, http://www.oxfordartonline.com/subscriber/article/grove/art/T011754. See also Nummedal, "Alchemist in His Laboratory," 1–4.

30. See, for example, Pamela Smith's work on Wenzel Jamnitzer in *Body of the Artisan*, 74–80.

31. According to Brun, he and Philipp also discussed their mutual interest in natural magic, "that is, a natural exploration of hidden things" (das sei ein natrürliche Ausforschunge verborgener Dinge), which, Brun said, the biblical three Magi had also used. "Verhör von Franz Brun," 20. November, 1574, NLA WO, 1 Alt 9, Nr. 315, fols. 2–8. Friedrich Thöne also reproduces this testimony at length, albeit with some unacknowledged elisions, in Thöne, *Wolfenbüttel*, 258–59.

32. "Von Franzen Brun habe ehrs nichts gelernen, der habe gar einen andern Proceß. Der hett wollen in sechs wochen tingieren. Ehr habe nicht auf des laboranten, sonder seine eigene Kunst gesetzt. Den laboranten hatt ehr Illmo. wol gerhumt, und gesagt, dass der tinctur gemacht. Ehr hett nichts gewis, das die Materii, so weit kommen, den alß ehr negst gesehen. Wie

man in seinen Laboratorio in seinen besein inmentirt. Dar Laborant habe einen andern wegk mit dem Proces und Spiritu vini. Ehr woll ein Viol oder Philosophisch faß darzu gebrauchen. Ein Philisophisch faß nennen ehr ein instrument die violen. Er habe Ihme ein ganz Jahre an den gelesen gemangelt. [D]essich glaß sei das allerbest." "Gütliche Aussage des Philpp Sömmering," 9. July, 1574, NLA WO, 1 Alt 9, Nr. 308, fols. 47r–v.

33. "Illimo. habe Ihne selbst verhindert und mit andern henden beladen." "Gütliche Aussage des Philipp Sömmering," 9. [July] 1574, NLA WO, 1 Alt 9, Nr. 308, fol. 47r.

34. Bruce Moran coined the extraordinarily useful term "prince-practitioner" to describe this kind of activity. See, more recently, Alisha Rankin's suggestion of the more inclusive term "court experimentalism," which incorporates the experimental activities of noblewomen as well. Moran, "German Prince-Practitioners"; and Rankin, "Becoming an Expert Practitioner."

35. "so bin ich soliches zutrits halber zu E. f. g. hie und anterswo in verdacht komen, als solte ich etwa E. f. g. . . . zu was Chimishen werken, probationibus und kunsten anreitzen" (90v-91r). Es gleuben E. f. .g nicht was schrecklicher vergiftigung, der geber den mineralischen geist noch in den rohen mineralibus stecken, derohalben wollen E. f. g. gott nicht versuchen üben über trewhertzige vormanung. . . . Es haben E. f. g. nicht allein Ihres leibes und lebens, daran Jtziger Zeit der gantzen Christlichen Kirchen gelegen ist, deßfals zu scheuen sondern auch Ihre Junge herlein und frewlein zu bedencken. . . . Unnd zu diesen allen, sollen auch billig E. f. g. meiner vorschonen, damit die Suspition und vordacht, wan E. f. g. etwa durch sollicher vorgiftiger ding versuchungs beschediget wurden |: do godt vor behuete :|" Philipp Sömmering to Duke Julius, 9. February, 1572, NLA WO, 1 Alt 9, Nr. 306, fols. 91v-92r. On the alchemical understanding of "arsenic," see Priesner and Figala, *Alchemie*, s.v. "Arsen."

36. The most polarizing issue in these debates was the theological concept of synergism, which posited a role for both human will and God's grace in the context of conversion and original sin. The clearest introduction to the synergist controversy can be found in Dingel, "Culture of Conflict," esp. 45–54. On Flaccius in particular, see Ilic', *Theologian of Sin and Grace*.

37. Beck, *Johann Friedrich der Mittlere*, 1: 382–404; Kirwan, "Empowerment and Representation," 121–23.

38. See the Introduction, above.

39. Mager, *Die Konkordienformel*, 142–64.

40. In 1571, before Anna and Philipp arrived in Wolfenbüttel, Duke Julius also joined several other Protestant princes in an appeal to Emperor Maximilian II and Elector August of Saxony at the Reichstag of Speyer to free the imprisoned former duke Johann Friedrich II. When this attempt was rejected, the group petitioned the emperor and elector again the following year (1572), but again, their request was firmly denied. Duke Julius made one last effort in the winter of 1572–1573 before giving up the cause. The other princes involved were Elector Johann Georg von Brandenburg, Elector Friedrich Pfalzgraf of the Rhein, Duke Wilhelm of Jülich-Cleve-Berg, and Landgrave Wilhelm of Hessen. For the attempted intercessions between 1571 and 1574, see NLA WO, 1 Alt 9, Nrs. 220–22.

41. Rhamm, *Die betrüglichen Goldmacher*, 18–19; Krusch, "Die Entwickelung der Herzoglichen Braunschweigischen Centralbehörden," 80–81. Rhamm's and Krusch's suggestion that these appointments were merely part of Philipp's attempt to amass power at Julius's court, however, fails to recognize what was expected of him as *Kirchenrat*, as well as Philipp's own very real commitments on matters of politics and religion. There is a large and growing literature on expertise. For an overview of the role of expertise in the early modern period, see Ash, "Introduction" and the other essays in the *Osiris* volume.

42. "demnach ferner auf sein [Herzog Johann Friedrich II's] und unser, seyner Collegen, geschene unterteniges bitten in gnedigen schutz wider gewalt gnedigs aufgenomen, so haben wir und sunderlich ich gehoffet, wir woltten in wolchem e. f. g. gnedigem schutz unser schweren erliedtnem creutz etwas lind[e]runge befinden. Aber got erbarmes wan ich die Recht worheit sagenn, sohl so ist mir in den vorigen gantzen funff yare nicht erger ergangen." Anna Maria Zieglerin to Duke Julius, Wolfenbüttel, 7. September, 1572, NLA WO, 1 Alt 9, Nr. 306, fol. 6r.

43. Anna Zieglerin to Duke Julius, s.d. [1571–73?], NLA WO, 1 Alt 9, Nr. 306, fol. 9r.

44. "Nuhn werde ich globwurdigk berichtet wie das die herzogin e. f. g. geliebtes gemahel mir solches zum aller obelsten und schreckstlichsten wendet als solde ich solches durch zauberey und schwarzkonsten thun, sich auch zum obelsten, mit weinen und schreihen, kegen e. f. g. uber mich beschweret sol haben." Anna Maria Zieglerin to Duke Julius, Wolfenbüttel, 7. September, 1572, NLA WO, 1 Alt 9, Nr. 306, fol. 306, 6v.

45. In her testimony from 1574, Anna attributed the magical recipe for making these shirts to Paracelsus. "Verhör der Anna Maria Zieglerin," 8. July, 1574, NLA WO, 1 Alt 9, Nr. 314, fol. 4r. See also "Verhör der Anna Maria Zieglerin," 16. July, 1574, 6pm, NLA WO, 1 Alt 9, Nr. 314, fol. 32v.

46. "was aber die furstin darzu treibett, das yr f. g. ohn eynigtts aufhoren, ein gantzes yar langk mich armes weyb gehasset geschmehet und ohn under lag bey e. f. g. verunglimbt [= verunglimpft] is mir verborgen." Anna Maria Zieglerin to Duke Julius, Wolfenbüttel, 7. September, 1572, NLA WO, 1 Alt 9 (Acta publica des Herzogs Julius), No. 306, fol. 6v.

47. "iyr f. g. ewer f. g. berichtet also solden des orts leutte gewessen die mit augen gesehen, her Philip und mich in eynem bette als ehe leutte schlaffen, wider got und alle bylligkeit unzucht mit einander treyben, welches durchaus die lautter unwarheit ist." Anna Maria Zieglerin to Duke Julius, Wolfenbüttel, 7. September, 1572, NLA WO, 1 Alt 9, Nr. 306, fol. 6v.

48. On adultery in early modern Germany, see Roper, *Holy Household*, 194–205; Rublack, *Crimes of Women*, esp. 198ff.

49. "yhr f. g. wehrn mir so feindt das sie mich nicht mit augen sehen konden, sie spritzten mich an, ich solde mich der hoffkirchen entholden, und betzeuget solches, mit vien trehnen." Anna Maria Zieglerin to Duke Julius, Wolfenbüttel, 7. September, 1572, NLA WO, 1 Alt 9 (Acta publica des Herzogs Julius), No. 306, fol. 7v.

50. Further study of Hedwig's correspondence networks, of course, may well reveal additional participants in this discussion about Anna Zieglerin during these years. See, for example, Hedwig's correspondence with various women in her family, "Gemahlinnen der regierenden Herzöge d. mittleren und neueren Lin," 1572–1592, NLA WO, 1 Alt 23, Nr. 129.

51. As numerous studies have noted, morality, including sexual morality, was increasingly seen as the purview of the state in the sixteenth century, although it was policed through gossip networks and more informal modes of discipline as well. Roper, *Holy Household*; Rublack, *Crimes of Women*, esp. chap. 4. Correspondence networks among noblewomen (and men) were, of course, a central part of court life in the early modern world. For the German case, see Rankin, *Panaceia's Daughters*; Keller, "Zwischen zwei Residenzen."

52. See especially Keller, "Kommunikationsraum altes Reich"; Rankin, *Panaceia's Daughters*; Wunder, *Dynastie und Herrschaftssicherung*; Hirschbiegel and Paravicini, *Das Frauenzimmer*; Schulte, *Body of the Queen*; Schattkowsky, *Witwenschaft in der frühen Neuzeit*.

53. "Ehr, Phlipp sey ein verlauffener Pfaff, der sein [ampt?] undt Ehlich weip verlauffen, und sich am den Losen S[ack] die Ziglerin gehenckt. Derselbe verfuhre und vorblende S. L. das S. L. sich aller hern und freunde gentzlich eussertten, vorhetze auch S. L. wid. derselben

altte Rhette, vom adel und *auch ehenlichere* diener dadurch dieselben beungnadet und verur-laubet, Umb desselbigen willen, wurde auch entlichen khein redlich Juncker bey S. L. pleiben." This conversation was reported in a letter from [Hedwig?] to Katharina, s.d., winter 1572/73, NLA WO, 1 Alt 9, Nr. 307, fols. 39r-40v. Philipp Sömmering recorded the same details (nearly word for word, in fact) in a description of a conversation between Katharina, Julius, and Hedwig, on November 5, 1572. Philipp overheard this conversation because he was waiting outside Julius's rooms to speak with him. See Philipp's notes of Katharina's visit to Wolfenbüt-tel, 3–7. November, 1572, recorded 17. July, 1573, NLA WO, 1 Alt 9, Nr. 307, fols. 177r-178v.

54. Indeed, a few weeks after Katharina left Wolfenbüttel and at her urging, Julius wrote again to Duke Johann Wilhelm in Saxony to follow up on the accusations against Philipp, Heinrich, and Anna, but it is not clear whether he got a response. Julius to Herzog Johann Wilhelm, 13. December, 1572, NLA WO, 1 Alt 9, Nr. 307, fol. 181r.

55. "gehortt, wie sie es soltte getrieben habenn Wehre [ric]htig, bey Chur. und Fursten Im gantzen Reich, Sie khontte Zaubern, und alle böse Stucke treiebenn, [?] hette sie ein Eins-pennig knecht beschlafffen, und beschwengert und hette sie das Kindt ermordet. . . . Zu deme wehre Schiel Heinze Ihr Man ein Narr, der seinem Hern Herzogk Johanns Fridrichen untreu beweiset. . . . SL solcher dinge alzumahl viel zu milde berichtet worden wehren . . . S L solche lose Leutte von sich thun, etzo brachte S. l. bey allem Chur: und fursten nachteil." [Hedwig?] to Katharina, s.d., winter 1572/73, 1 Alt 9, Nr. 307, fols. 39r-40v.

56. Katharina's late husband, Margrave Johann of Brandenburg-Küstrin (1513–1571), was the younger brother of Hedwig's father, Elector Joachim II of Brandenburg (1505–1571).

57. "derselben Jungen herschafft zu schaden, schimpf oder nachteil." Katharina in Küs-trin to Hedwig in Wolfenbüttel, 1. April, 1573, NLA WO, 1 Alt 9, Nr. 307, fols. 47r-48v.

58. "Ich mag In warheit woll sagen das ich zum hochsten und eussersten darumb betrübt und bekummert sein, Das Ich schier nicht weis, was Ich von Herzleidt und [wunndt] beginnen und thun soll, Es muß Ja Gott Im Himmel erbarmen, das solch Gott unnd ehren vergessenen Leuthen sollen, furstlich Personen Ire liebe kinder, hab und gutt und alles was Inen von Gott und der nature wegen billich lieb und als Ihr eig. Herz befolen sein solt, vertrawet und under-geben sein. Man wis Ja woll und ist landrüchtig, wie sie sich bey Ihren vor Herren gehalten haben." Katharina in Küstrin to Hedwig in Wolfenbüttel, 18. June, 1573, NLA WO, 1 Alt 9, Nr. 307, fols. 55r-58v.

59. "Was wolt eillig daraus werden, was wolte E. L. vor schimpfliche nachrede daraus erfolgen." Katharina in Küstrin to Hedwig in Wolfenbüttel, 18. June, 1573, NLA WO, 1 Alt 9, Nr. 307, fol. 55v.

60. "Der lieb Gott sey Ja mit gnaden dafur das der leidige Teuffel durch diese Leuthe als sein werckzeug seinen wollen, nicht schaffen müge, Den die erfahrung hats gegeben, wie es wol ehrmals denn herren mit Juden und dergleichen Leuthen gegang und was es vor einen bosen ausgang genomen, Und da Ja E. L. Irer selbst und Ires furstlich gutten gerüchts nicht zuscho-nen bedacht weren, So wolten doch E. L. Ire liebe Junge Herrschafft Ja auch leib und leben, Landt und Leuthe In acht und zu Hertzen nemen, Wie gar geferlich und ergerlich es sein wolte, dieselbig mit solchen leichtfertig unverschembten Leuthen umbgehen und bey Inen erziehen zulassen. Weis Gott Ich meins treulich und von Herz gutt, wollen auch nicht aufh-oren, den lieben Gott zubitten, das ehr E. L. und meinen lieben Bruders Herz und gemuth erleuchten und von diesen vorfluchtig bosen Gottlosen leuthen abwenden wolle, Dann bose geselschafft kann bisweilen woll etwas verursachen, das sonsten nimermehr geschehe, und man Pflegt gemeinlich zusagen, wie die Räthe und Diener sein, also sein die Herrn auch gern, Bitten derwegen E. L. oder mein viel Herzliche schwester lautter umb Gottes willen, E. L. wolte der

leuthe mussig gehen, und sich umb E. L. lieben kinder willen derselben entschlag und nicht etwas thun andern zugefallen, das E. L. dem nachteilig und verweislich sein müchte." Katharina in Küstrin to Hedwig in Wolfenbüttel, 18. June, 1573, NLA WO, 1 Alt 9, Nr. 307, fols. 56r-v.

61. Rublack, *Crimes of Women*, 158–62.

62. "Ehr, Phlipp sey ein [ver]lauffener Pfaff, der sein [ampt? hole in paper] undt Ehlich weip verlauffen, und sich am den Losen Sack die Ziglerin gehenckt." Hedwig to Katharina (?), s.d. [late 1571/early 1572], NLA WO, 1 Alt 9, Nr. 307, fols. 39v-40r.

63. On the elements of early modern honor generally, see Schreiner and Schwerhoff, *Verletzte Ehre*, especially the essays by Schreiner and Schwerhoff, Dinges, and Nowosadtko; Kollmann, *By Honor Bound;* Gowing, *Domestic Dangers;* Cohen, "Honor and Gender."

64. "den gnediger Furst und Herr e. f. g. geliebtter gemahl yhren grymmigen thorn auf uns geworffen, doch wieder gott, und alle verdienst, das da yr f.g. bey E.f.g. nicht hatt erholden konnen uns zu ungenaden zubewegen, das yr f.g. andere hohe personem an e.f.g. abfertigen, ist zu besorgenn." Anna Maria Zieglerin to Duke Julius, 12. December, 1571, NLA WO, 1 Alt 9, Nr. 306, fols. 4r-v.

65. "denn zeuber und schwerzkunstriger gehoren hie ins zeydtliche und dort ins ewige fewer. . . . Ich sage noch wie vor, ich habe doch wunderbarliche schrekung gottes und trewer [leut?] herz." Anna Maria Zieglerin to Duke Julius, Wolfenbüttel, 7. September, 1572, NLA WO, 1 Alt 9, Nr. 306, fols. 7r-v.

66. Sabean, *Power in the Blood*, 46.

67. "die furstin wil derumb auffs neue wieder mich zue ungenaden bewogen"; "ich höre das yr f. g. gern in gottes wort lessen, wort vileicht eynes wiles got, der herlige geist, yr f. g. hertz erleuchtenn, das yr f. g. von yhrem unbefugten zorn von mir ablassen." Anna Maria Zieglerin to Duke Julius, Wolfenbüttel, 7. September, 1572, NLA WO, 1 Alt 9, Nr. 306, fol. 7v.

CHAPTER 4

1. The booklet itself exists only in this copy, which includes a note at the end explaining that "His Grace received it to copy out on the first of April in the year [15]73 at the Wolfenbüttel castle; it was handed to His Grace by his personal valet with the name Rubrecht Lobrinne, and His Grace copied it out with his own hand and in his own writing beginning on the 19th of this month and ending on the 20th." (dieß hadtt s.f.g. bekommen zu [aus]schreyben um den ersten apprillyß anno 73 zu der vestunge wolffenbuttel. Ifg ist gehandigett worden dur[ch] sfg geheymsten leybe und kammer dyenner mytt nahmen heyste der rubrechte lobrinne. Und ifg haben dißer auch auß geschrieben mytt eygen handt und angefangen den 19 huiuß und den 20 gecundigett ihn eynnygem schryben.) Anna Maria Zieglerin, "die Edele und Tewere Kunst Alchamia belangende," 1. April, 1573, NLA WO, 1 Alt 9, Nr. 306, fols. 52r-70r.

2. "Im nahmen Gottes, des vatterß und Sohnes unnd deß heylyger geystes Amen. Waß ich Zieglerin Maria Zygellerinne itzo Heynnerich Schumbachs Ehe Weyb von dem edelen und wolgebornen Herren Herren Carrolo Graf und Her zue Ottynge myt meiner Augen gesehe auch mytt meyner eygener handt selber vollbracht die Edele und Tewere Kunst Alchamia belangende habe ich nycht unterlassen wollen, meyne lyben Hauß herrn und Mann hinter myhr zu laßen und zu offenbaren." Anna Maria Zieglerin, "die Edele und Tewere Kunst Alchamia belangende," 1. April, 1573, NLA WO, 1 Alt 9, Nr. 306, fol. 52r.

3. Anna Maria Zieglerin to Duke Julius, 12. December, 1571, NLA WO, 1 Alt 9, Nr. 306, fol. 4v. This letter is the only documentation of Anna's use of this symbol, so it is not clear

how much significance she attributed to it. Alchemists regularly incorporated symbols into their writings as a kind of shorthand for common materials, but these symbols sometimes could take on more elaborate meanings as well, as, for example, in the famous "hieroglyphic monad" on the title page of Anna's contemporary John Dee's *Monas Hieroglyphica*. Anna's symbol also calls to mind the pseudo-Paracelsian *Archidoxis magica*, the first four books of which were published in the same year of Anna's letter to Julius as *De signis Zodiaci et ejus mysteriis*. I am grateful to Didier Kahn for his consultation on this issue. Interestingly, Philipp Sömmering also included the same symbol (albeit in mirror image) in the address of his letter to Duke Julius the following year. See Philipp Sömmering to Duke Julius, 10. May, 1572, NLA WO, 1 Alt 9, Nr. 306, fol. 33v.

4. In a letter to Duke Julius at the end of March, Anna mentioned another letter that Julius had evidently written and sent to Count Carl. Anna Maria Zieglerin to Duke Julius, 21. March, 1573, NLA WO, 1 Alt 9, Nr. 307, fols. 77r-78v.

5. In September of 1573, Anna sent Duke Julius "this small nugget, the greatest little stone, which I have placed in wine so that the air does not dissolve it," that is, turn it back into a liquid. Anna then reported that she was about to begin a work with mercury. Anna Maria Zieglerin to Duke Julius, 3. September, 1573, NLA WO, 1 Alt 9, Nr. 307, fol. 72r. Philipp Sömmering mentioned "Frau Anna's laboratory" several times in his testimony. See "Gütliche Aussage des Philipp Sömmering," 10. July, 1574, NLA WO, 1 Alt 9, Nr. 308, fol. 111r, 112r, and 115r.

6. See Anna Maria Zieglerin to Duke Julius, 3. September and 17. October, 1573, NLA WO, 1 Alt 9, Nr. 307, fols. 72r-73v.

7. See the interdisciplinary collection of essays on early modern blood, Lander Johnson and Decamp, *Blood Matters*.

8. "laß eß neuhn Stundt wal er warmen darnach Ihn eyn lynde kol fewer leyne da wirtze erstlych oben in dehr kolbh eynen leyb farbe nebel ßehen, der ferbet ßych ohn underlaß byß der helme braun rodt wirdt. . . . So wirdt derßelbe leuchte schone blutte ßych rutzen dyß behalte ihn dehr vorlage dass heyst und nent man den Rotten leuwen ßeyne schonneß her-lycheß reyches blutte außzyhen. Darnach ßol man wan dass glaß wieder weyß wirt ßo laß dem fewer abgehen. Den der lewe hadte von ßych gegeben ßo vielle er ßeynnes bluts ohnne vorletz-unge hadt entrotten konnen."Anna Maria Zieglerin, "die Edele und Tewere Kunst Alchamia belangende," 1. April, 1573, NLA WO 1 Alt 9, Nr. 306, fol. 54r.

9. "Nymme den deyn lewen bludt ßo du von dem leven entpfang[en] hast und nymme wyder frische bley thu auffe firre # [i.e., pfund] bley eyn achzehenne troppen deß schonnen blutzen laß eß wolle myt eyn ander ßych er weßellen auf eyne starcke kol feuer biß du oben ßihest alß eyn fest olle guß eß dehn wyder in eyn Inguß feyll eß fortt zeuge dem leuen widehr daß blute auß ßo hastu eynne ewige erbeyt." Anna Maria Zieglerin, "die Edele und Tewere Kunst Alchamia belangende," 1. April, 1573, NLA WO, 1 Alt 9, Nr. 306, fol. 54v.

10. "ßo hastu eben ßowol dehn rechtten labiden alß festen den kanstu ßo hoche und ville myt dyße [ink smeared—pölff?] dyngiren." Anna Maria Zieglerin, "die Edele und Tewere Kunst Alchamia belangende," 1. April, 1573, NLA WO, 1 Alt 9, Nr. 306, fols. 55r-v.

11. On this distinction, see Priesner and Figala, *Alchemie*, s.v. "Lapis philosophorum."

12. "gehe zu eyn schonnen meyen nachte auf eynne schonne herlych wisen dar vielle schoner welche reychenn? blumen stanne laße den taue auf dem merckurio fallen." Anna Maria Zieglerin, "die Edele und Tewere Kunst Alchamia belangende," 1. April, 1573, NLA WO, 1 Alt 9, Nr. 306, fols. 56r-v.

13. "und durch scheynet daß glaß alß eyn Carbunckel dß du denckest dß gantze glaß ßey lautter rubyn." Anna Maria Zieglerin, "die Edele und Tewere Kunst Alchamia belangende," 1. April, 1573, NLA WO, 1 Alt 9, Nr. 306, fol. 60v. This entire process stretches over fols. 56r-63v.

14. "Ist dehr steynne rechtte und fulle kommen."Anna Maria Zieglerin, "die Edele und Tewere Kunst Alchamia belangende," 1. April, 1573, NLA WO, 1 Alt 9, Nr. 306, fol. 60v.

15. "gißen in gestaltte eynß leven lyndtwurmer trachen kopffe oder waß dier gefelte ist eben gleyche." Anna Maria Zieglerin, "die Edele und Tewere Kunst Alchamia belangende," 1. April, 1573, NLA WO, 1 Alt 9, Nr. 306, fol. 63v.

16. "ßo bluhet der baum oder aste und trege rechtte gutte naturlyche und allemass wolle schmeckende fruchte. Daß habe ich ßelber offt verßuchet." Anna Maria Zieglerin, "die Edele und Tewere Kunst Alchamia belangende," 1. April, 1573, NLA WO, 1 Alt 9, Nr. 306, fol. 65r.

17. "Item wiltu eyn geßunt kynt zeugen: So nymme deß obgedachtten olleß 3 tage nach eynander abentz und morgentze eyne 9 tropfen eyn gebeß auch deynen weybe deß gleyche, und ob syhe je tage werre unfruchtbarre geweßen, ßo entpfehet ßyhe auff dehr stedt [= empfängt sie auf der statt?], weyl aber myt dem kind schwanger ist ßo gyb ehre eynne und alle tage nychtte merre dann 3 tropfen deß olleß eyn. . . . so laße daß kyndt keynne mutter milche schmecken, gibe Imme wedehr zu eßen noch zu tryncken sonder allezeyt jdeß tageß drey malle alß abent morgentze und zue myttage und zu abentze doch jdeß malße nychtte mehre dan 3 dropfen Im munde. laß eß also lygen ßo schleffet zie eß ohn underlaß feste und herte ohne weynnen und allem unfhalle. Du kznst eyn kyntte alßo erszyhen ohnne ander eßen und dryncken 12 gantzer jarlange dß eß keynen tropen begertte denn aber waß zu eßen. dyße kynder werden auch nych ßo lange in mutterleybe getragen alß ander den von wegen der großen wermen wirtte die fruchte inne mutter leybe dardurche desto eher zeyttig ßeyner kreyffte aller ßeyner geburlychen gestalltten." Such a source of nourishment, of course, would have been convenient when food was scarce, as it no doubt was during the seige of Gotha. Anna Maria Zieglerin, "die Edele und Tewere Kunst Alchamia belangende," 1. April, 1573, NLA WO, 1 Alt 9, Nr. 306, fols. 64r-v.

18. "Hadt ehr hochgedachtter ungefer drey tage nach ßeynner ahnkunfft gefordertt, man ßol Ihmme zweyhe pfundt bley geben und wye baltte ehr dass bleyhe bekhamme, nam ehr eynen schmeltztygel, darinnen zerlyße ehr dß bleyhe und gantze schmeltzygke warm werdenn, da gabe ehr eyn kleyn Beuttelleyn dareynne war eyn braunn polffer, da warffe [ich] 2 erbysen große in daße bley, da kugelyette syche dass bley und stieck [= stieg] auff und nyder, wolle ßych gar nycht geben; letzlch da die Tinktur daß bley uberwandt, darnach ßchwame eß oben wie eyn olle war rotter und schonner alß klyn rubyn; dyß lyß ich auf deß Herrn graffen underweyßunge waß kalt werden, doch nychte zue ßer kalt, goße eß in eynen Inguß, da war eß golt drey zehen gradt hoher gradiertte den daß aller beste Arrabysche gol111." Anna Maria Zieglerin, "die Edele und Tewere Kunst Alchamia belangende," 1. April, 1573, NLA WO, 1 Alt 9, Nr. 306, fols. 52v-53r.

19. Zieglerin utilized this style to relate other recipes as well, including one for making emeralds. Anna Maria Zieglerin, "die Edele und Tewere Kunst Alchamia belangende," 1. April, 1573, NLA WO, 1 Alt 9, Nr. 306, fols. 68r-v. For this use of *historia* as an epistemological tool in early modern medicine, see Pomata, "*Praxis Historialis.*"

20. "Nuhn Wie man wytter darmytt arbeytten kan. Sol man nemmen deß ßelbygen tingiertten goldes Feylle eß ganße kleyn den es ist ßo gantz 1 hanyd dyck daß du [e]ß schneyden kanst wie kleyn du wilst, thuet man daß ßelbyge gefeyltte goltte zu eyn kleynne kolbelleyn."

Anna Maria Zieglerin, "die Edele und Tewere Kunst Alchamia belangende," 1. April, 1573, NLA WO, 1 Alt 9, Nr. 306, fols. 53v-54r.

21. For example: "Take your lion's blood that you received from the lion, take some fresh lead again, and add four pounds lead, [and] eighteen drops of the lovely blood; let it all mix together [?] in a strong coal fire until you see oil congeal on top. (Nymme den deyn lewen bludt ßo du von dem leven entpfang[en] hast und nymme wyder fristhe bley thu auffe firre # bley eyn achzehenne tropfen deß schonnen blutzen laß eß wolle myt eyn ander ßych erweßellen [erwälzen?] auf eyne starcke kol feuer biß du oben ßihest alß eyn fest olle.) Anna Maria Zieglerin, "die Edele und Tewere Kunst Alchamia belangende," 1. April, 1573, NLA WO, 1 Alt 9, Nr. 306, fol. 54v. It is worth noting that Zieglerin *never* would have addressed Julius in writing with the simple pronoun "you," but rather with the honorific "Your Grace." Her use of "you" in the booklet, therefore, should not be taken as something directed at Duke Julius, but rather as the common formulation one finds in early modern recipes.

22. "To continue, I will divulge as much as has been reported to me about how to come to this powder, but I have not done this work myself." (Weytter wil ich vermelden wie ville ich berichtes habe zu dyßen polffer zu kommen. Aber Ich habeß nycht gearbeyt.) Anna Maria Zieglerin, "die Edele und Tewere Kunst Alchamia belangende," 1. April, 1573, NLA WO, 1 Alt 9, Nr. 306, fol. 56r.

23. "philoßophischen feuwer dß ßelbige ich nycht schreyben ijeze malß, dawie eß zu gerychtte muß werden den Ich muß Ifg ßelber weyßen." Anna Maria Zieglerin, "die Edele und Tewere Kunst Alchamia belangende," 1. April, 1573, NLA WO, 1 Alt 9, Nr. 306, fols. 58v-59r. On the difficulty of putting craft knowledge into writing, see Smith, *Body of the Artisan*.

24. Roger Bacon, Raymond Lull, Arnald of Villanova, and John of Rupescissa have each generated a significant literature of their own. For a concise overview of this period in Latin alchemy, see Moran, *Distilling Knowledge*, 11–25; DeVun, *Prophecy, Alchemy, and the End of Time*, chap. 5. On the quintessence in particular, see Darmstaedter, "Zur Geschichte des Aurum potabile"; Priesner and Figala, *Alchemie*, s.v., "Quintessenz." Sixteenth-century printed editions of these medieval authors include *De Alchimia* (Nuremberg: J. Petreius, 1541); Lullus, *Künstliche Eröffnung aller Verborgenheyten vn[d] Geheymnussen der Natur* and *De Secretis Naturae sive Quinta Essentia*. Works on distillation from sixteenth-century authors include Brunschwig, *Liber de arte distillandi de simplicibus* and *Liber der arte distulandi simplicia*; Ryff, *Das new groß Distillierbuch*; and Ulsted, *Coelum philosophorum*.

25. John of Rupescissa, to offer only one example of the ubiquitous red and white stones, described them in his *Liber lucis*. See DeVun, *Prophecy, Alchemy, and the End of Time*, 58–59.

26. The process of ceration was necessary to soften the *lapis* enough that it could penetrate the metals it was supposed to transmute. See Newman, *Summa Perfectionis*, chap. 18.

27. Pereira, *Alchemical Corpus Attributed to Raymond Lull*, 6–9; Principe, *Secrets of Alchemy*, 72. Pseudo-Lull's *Testamentum*, the earliest text in what became the Pseudo-Lullian corpus and a touchstone for the addition of gemstones to the elixir's medical and transmutational virtues, first appeared in print in Cologne in 1566: Lullus, *Testamentum*.

28. On the vernacularization of alchemy, see Pereira, "Alchemy and the Use of Vernacular Languages."

29. Nummedal, *Alchemy and Authority*, chap. 3.

30. The literature on Paracelsus is vast and growing. For an introduction, see Webster, *Paracelsus;* Benzenhöfer, *Paracelsus;* Weeks, *Paracelsus.*

31. Anna Maria Zieglerin, "die Edele und Tewere Kunst Alchamia belangende," 1. April, 1573, NLA WO, 1 Alt 9, Nr. 306, fols. 52r-53r. "Sagt ja, das sey der Graff Carl von Otingen,

darvon sie [Anna] gesagt und seine goltkunst gewust, Theophrastus soll der mit einer Graffin von Otingen gezeuget haben, er der Graffe soll die warliche Kunst der tinctur haben und soll so gar Kunst[r?]eich sein als kheiner auf d welt." "Verhör des Heinrichs Schombachs," 6. July, 1574, NLA WO, 1 Alt 9, Nr. 314, fols. 71v-73v. Also Carl "sol Theophrasti Sohn sein." "Verhör der Anna Maria Ziegler," 8. July, 1574, NLA WO, 1 Alt 9, Nr. 306, fol. 13r.

32. Paracelsus, *Bücher und Schriften*. On the various historical and modern editions of Paracelsus's work, including the Neue Paracelsus-Edition (New Paracelsus Edition) of Paracelsus's theological works, still in progress, see the Zurich Paracelsus Project at the University of Zurich: http://www.paracelsus.uzh.ch.

33. For an excellent (and succinct) overview of Paracelsus's innovations in medicine and philosophy, see Moran, *Distilling Knowledge*, 67–98.

34. Classic book-length studies of Paracelsus's medicine and natural philosophy include Pagel, *Paracelsus*; Debus, *Chemical Philosophy*. On his theology, see Webster, *Paracelsus*; Weeks, *Paracelsus*; Cislo, *Paracelsus's Theory of Embodiment;* Gause, *Paracelsus;* Daniel, "Invisible Wombs" and "Paracelsus's *Astronomia Magna* (1537/38)." Additionally, there are numerous collections of essays on Paracelsus, including Classen, *Paracelsus im Kontext der Wissenschaften seiner Zeit;* Schott and Zinguer Schott, *Paracelsus und seine Internationale Rezeption;* Grell, *Paracelsus*; Jütte, *Paracelsus Heute;* Telle, *Parerga Paracelsica* and *Analecta Paracelsica*; Dilg and Rudolph, *Resultate und Desiderate der Paracelsus-Forschung;* Dilg-Frank, *Kreatur und Kosmos.*

35. *Coelum philosophorum, sive Liber Vexationum, Das Thesaurus Thesaurorum Alchemistarum*, and *De tinctura physicorum* all appeared in print as authentic Paracelsian texts the sixteenth-century Huser edition, Paracelsus, *Bücher und Schriften* 6: 363–401.

36. Sudhoff, *Sämtliche Werke*, 14: 391ff. It is impossible to do justice in translation to Joachim Telle's own colorful description of this process of the "frühparacelsistischen Reformbewegung, daß man Paracelsus, einen kraftvollen Anwalt der Alchemia medica (Chemiatrie), seines schlichten Laborantenkittels entkleidete und spätestens seit den 1550er Jahren im fabelbunten Flittergewand eines deutschen Hermes Trismegistus paradieren ließ, eines Wundermannes, dem in der Alchemia transmutatoria metallorum das Höchste gelungen sei. Statt fand ein bald recht ausmäßlicher Glorifikationsprozeß, der nun keineswegs nur den tüchtigen Pharmazeuten Paracelsus zu einem triumphalen Goldmacher entstellte (und das frühneuzeitliche Paracelsusbild maßgeblich prägte), sondern—vielleicht bedeutsamer noch—sich mit zahlreichen Aufgriffen und aktualisierenden Paracelsifizierungen nichtparacelsischer Alchemica verband, namentlich mit Rückgriffen auf Schriften der (vom historischen Paracelsus pharmazeutisierten) Alchemia transmutatoria metallorum." Telle, "Der *Splendor Solis* in der Frühneuzeitlichen Respublica Alchemica," 432. See also Telle, *Rosarium philosophorum*, 2: 167–68.

37. [Pseudo-]Paracelsus, *Büchlin von der Tinctura Physica; Archidoxa Ex Theophrastia; De Lapide Philosophorum; Archidoxa Philippi Theophrasti Paracelsi Bombast*. See Sudhoff, *Sämtliche Werke*, 14: xii-xxv (with caution) for further editions and translations.

38. On the tradition of the elixir in medieval and early modern alchemy, see Moran, *Distilling Knowledge*, 11–35; DeVun, *Prophecy, Alchemy, and the End of Time*, esp. 158–63.

39. [Pseudo-]Paracelsus, *Archidoxa Philippi Theophrasti Paracelsi Bombast*, Zv r-v.

40. Another recipe for a lion's blood appeared in print in 1598 in the alchemical compendium *Aureum Vellus*, where it was associated with Salomon Trismosin, a legendary figure who was purportedly Paracelsus's teacher. Again, however, the recipes are not close enough to make a direct link. Trismosin et al., *Aureum Vellus*. Martin Ruland's *Lexicon alchemiae*, published in 1612, long after Anna Zieglerin was dead, claims that the lion's blood refers to red sulfur that

has been joined with mercury: "Leo rubeus, der rote Löw/ist roter Schwefel/welcher in mercurium resoluirt wirdt/genennet sanguis leonis, wirdt auch Gold gemacht." Ruland, *Lexicon alchemiae*, 303, http://digital.slub-dresden.de/id277095204. The most likely scenario, therefore, is that multiple recipes for "lion's blood" were circulating in sixteenth-century Europe, none of which I have been able to match exactly to the one that Zieglerin mentions in her 1573 book of recipes.

41. Cadden, *Meanings of Sex Difference*, 184.

42. Ibid., 11–39; Wood, "Doctor's Dilemma," esp. 719.

43. DeVun, *Prophecy, Alchemy, and the End of Time*, 73–74, 116–27.

44. Bynum, *Fragmentation and Redemption*, 205–14.

45. Bynum, *Wonderful Blood*, esp. chap. 3.

46. Ibid., 155–56. See also Smith, "Vermilion, Mercury, Blood, and Lizards," esp. 42–45.

47. Bynum, *Wonderful Blood*, 64–66.

48. Ibid., 73–75, 226–28. Christ's blood as an emblem of sacrifice more generally continued to be a powerful force in Protestant soteriology as well. Ibid., 226–28.

49. Anna Maria Zieglerin, "die Edele und Tewere Kunst Alchamia belangende," 1. April, 1573, NLA WO, 1 Alt 9, Nr. 306, fols. 64r-v (to conceive a child), 65r (for the tree), and 54v-55v (for the philosophers' stone).

50. "laß daß wercke [unverweckt], biß ahn 20 dagen stehen, ßo kombt der frucht dyßehelben"; "Wan man auch zu winters zeyt wille gutte reyffe zeytige fruchtte." Anna Maria Zieglerin, "die Edele und Tewere Kunst Alchamia belangende," 1. April, 1573, NLA WO, 1 Alt 9, Nr. 306, fols. 59v and 65r.

51. Aristotle, *Generation of Animals* 2.4, on the contributions of the male seed, menstrual blood, and the womb in generating life. Galen, *On Semen* 1.9. See also Bynum, *Wonderful Blood*, 158; Cadden, *Meanings of Sex Difference*, chap. 1. On Paracelsus's views, see Cislo, *Paracelsus's Theory of Embodiment*, 21–42.

52. Crowther-Heyck, "'Be Fruitful and Multiply,'" 913, 917–19. See also Principe, "Revealing Analogies"; Moran, *Distilling Knowledge*, 27–29; Fissell, "Gender and Generation," 435–36.

53. On alchemy as a technology and debates about these claims, see esp. Newman, "Technology and Alchemical Debate" and *Promethean Ambitions*.

54. On seeds, see Hirai, *Le concept de semence*; Moran, *Distilling Knolwedge*, 27–29.

55. For a particularly insightful commentary on the function of metaphors in alchemical texts, see DeVun, "'Human Heaven.'"

56. Principe, "Revealing Analogies," esp. 216–17; DeVun, "Jesus Hermaphrodite"; Kavey, "Mercury Falling." On the translation of alchemical metaphors into images, see esp. Obrist, *Les débuts de l'imagerie*.

57. Stavenhagen, *Testament of Alchemy*, 29; Stavenhagen, "Original Text of the Latin Morienus."

58. Jacob, *De alchimia opuscula*. For a modern facsimile, with German translation and commentary, see Telle, *Rosarium philosophorum*.

59. There are numerous theories about the dating and authorship of both the original *Rosarium* and the Sol and Luna series as well as the relationship between the two. See Telle, *Rosarium philosophorum*, 2: 162–71, 80–86 and *Sol und Luna*, 34–37; Principe, *Secrets of Alchemy*, 74–75.

60. Telle, *Rosarium philosophorum*, 2: 175.

61. Telle, *Rosarium philosophorum*, 1: 73–74: "Sol est masculus, Luna foemina, & Mercurius sperma. Sed ut fiat generatio et conceptio, oportet ut masculus iungatur foeminae, & ultra hoc requiritur semen, et sic ante fermentationem debet fieri conceptio & impregnatio, & cum multiplicatur materia dicitur quod infans crescit in utero matris et cum fermentatur."

62. John 12:24 NRSV.

63. "Wiewol in gemein darvon zureden, möchte man sagen, das von natur alle ding würden aus der erden geboren mit hilf der putrefaction" [Pseudo?-]Paracelsus, *De natura rerum neun Bücher*, in Sudhoff, *Sämtliche Werke*, 11: 312.

64. "Dan die putrefaction große ding gebiert, dessen wire in schön exempel haben im heiligen evangelio, da Christus sagt: es sei dan, das das weizenkörnlin in den acker geworfen werde und faule, mage es nicht hundertfeltige frucht bringen. dabei ist nun zuwissen, das vil ding in der putrefaction gemanigfaltiget werden, also das sie ein edle frucht geberen; dan die putrefaction ist ein umbkerung und der tot aller dingen und ein zerstörung des ersten wesens aller natürlichten dingen, daraus uns herkomet die widergeburt und neue geburt mit tausentfacher besserung." [Pseudo?-]Paracelsus, *De natura rerum neun Bücher*, in Sudhoff, *Sämtliche Werke*, 11: 312.

65. "Daher denn volgetßdes ein allen gewachsen in allen thieren und menschen die Vermerung geschieht zu seines selbsten gleichen. . . . das alle metalla von einer wurtzel nach der ersten bescheffung gotts in den Adern der Erde wüchsen. Und demnach sie die wachsung der metalle gantz gewiß befunden, haben sie die selbige dem spiritu Generativo zugemessen. Das ist, sie haben klar befunden, das ebensowohl ein gebärende und vermerende krafft und geist in den metallen als in andern ding sey. . . . so haben sie's dafur gehalten, diesse wechsende und vermernde craft der metallen sey so viel Edler, den in denn andern gewächsen, so viel der metalle Leichnam im feuer vor andern Substantia unverändert und bleiblich od. fix seint." Philipp Sömmering to Duke Julius, 25. August, 1573, NLA WO, 1 Alt 9, Nr. 306, fol. 74r-75v.

66. Bynum, *Wonderful Blood*, 1–2.

67. Ibid., 156, 212–13.

68. Wandel, *Eucharist in the Reformation* and *Companion to the Eucharist*, esp. Volker Leppin's essay, "Martin Luther," 39–56.

69. On executioner medicine in early modern Germany, see esp. Stuart, *Defiled Trades and Social Outcasts*, chap. 6. See also Pliny, *Natural History* 28.2.

70. For an overview, see DeVun, *Prophecy, Alchemy, and the End of Time*, 109–18.

71. John of Rupescissa, *Liber lucis*, as cited and translated in DeVun, *Prophecy, Alchemy, and the End of Time*, 113.

Ibid., 118.

72. John of Rupscissa, *Liber lucis*, as cited and translated in DeVun, *Prophecy, Alchemy, and the End of Time*, 118.

73. Ibid.

74. As cited in Hoheisel, "Christus und der philosophische Stein," 72.

75. DeVun, *Prophecy, Alchemy, and the End of Time*, 114.

76. "Den der lewe hadte von ßych gegeben ßo vielle er ßeynnes bluts ohnne vorletzunge hadt entrotten konnen." Anna Maria Zieglerin, "die Edele und Tewere Kunst Alchamia belangende," 1. April, 1573, NLA WO, 1 Alt 9, Nr. 306, fol. 54r.

77. "Wylttu auch schonne au[ch] wolle eyn hymlische vorgleychunge machen ßo nyme eyn kleynneß waltte fogelleynne ßetz zeß In eyn korbchen. . . . den der ßon gottes am stamme deß heyllygen creutzes unß arme ßundern alle myt ßeynen aller heyllygeste Roßen farben blutte dyngiertte und teylhafftig gemacht alß wirdt dieß fogeley der yrdischer tyncktur teylhafft."

Anna Maria Zieglerin, "die Edele und Tewere Kunst Alchamia belangende," 1. April, 1573, NLA WO, 1 Alt 9, Nr. 306, fols. 55r-v.

78. On the emergence of "blood piety" in fifteenth-century Germany, see Bynum, *Wonderful Blood*.

79. "Erstlych wen du eynen außßetzygen wilste reynne machen, ßo gib Ihmme neunner gantzer tage und eyn jdeß tage [marre] 3 troppen deß olleß ßo du von dem dingertten golden getzogen amme 9 tage laß Inne eyn ader schlagen, ßo loffte dehr außßatze von Imme oder auß den Aderlaßeloche mytt herrauße alß ßandte körnner." Anna Maria Zieglerin, "die Edele und Tewere Kunst Alchamia belangende," 1. April, 1573, NLA WO, 1 Alt 9, Nr. 306, fol. 63v. The resemblance between leprosy and grains of sand was common. The fourteenth-century surgeon Henri Mondeville, for instance, wrote that the practitioner could diagnose a patient by handling his or her blood: "If, when you rub this blood between your fingers and the palm of your hand, you find grains like grains of millet, sand or gravel, this is a sure sign of incipent leprosy." E. Nicaise, *Chirurgie de Maître Henri de Mondeville* (Paris: F. Alcan, 1893), 550, as cited in Pouchelle, *Body and Surgery*, 73.

80. Emblem XIII in Michael Maier's *Atalanta fugiens*, for example, depicted this scene, and explained its alchemical parallel with the following motto: "The Philosophers brass is hydropical, and desires to be seven times washed in a river, as Naaman the leper in Jordan." (Aes Philosophorum hydropicum est, & vult lavari septies in fluvio, ut Naaman leprosus in Jordane/ Das ertz der Weisen ist wassersüchtig, und wil gebadet seyn siebenmal im Fluß, wie der aussetzige *Naaman* im Jordan.) Maier, *Atalanta fugiens*, 60–64.

81. Bynum, *Wonderful Blood*, 156. Bächtold-Stäubli et al., *Handwörterbuch des deutschen Aberglaubens*, s.v. "Blut"; Yuval, "Jews and Christians in the Middle Ages," 90.

82. Matthew 8:3, Luke 5:13, and Luke 17:13–14. Leviticus 14:49–53 (NRSV), which discusses how a priest should cleanse the house of a leper, suggests another possible source for Anna's little bird: "For the cleansing of the house he shall take two birds, with cedarwood and crimson yarn and hyssop, and shall slaughter one of the birds over fresh water in an earthen vessel, and shall take the cedarwood and the hyssop and the crimson yarn, along with the living bird, and dip them in the blood of the slaughtered bird and the fresh water, and sprinkle the house seven times. Thus he shall cleanse the house with the blood of the bird."

83. On this point, see Smith, "Vermilion, Mercury, Blood, and Lizards."

84. "Demnach sie den Dingen alle also nachgetrachtet und gewiß worden seint, daß in denn metallen, die vermehrende krafft verborgen und mit dem harten, zusammengefügten leibe fest beschlossen seyn, So haben sie wohl ermessen können, das derselbige metallische vermerende spiritus zu keiner Wachsung oder Vermerung kommen könnet, es würde denn das leichnen aufgelöst Solt es aber aufgelost werden, so mußte es durch eine naturlich bequem mittel geschehen, Damit der geist des metalles vom leib gescheiden und den der leib kont Inerlich und Eusserlich gereiniget werden. Dieses nun ist das ganze geheimnus der Naturlcih hohen Kunst alchimie." Philipp Sömmering to Duke Julius, 25. August, 1573, NLA WO, 1 Alt 9, Nr. 306, fols. 75r-v.

85. What exactly this analogy meant varied enormously from antiquity to the early modern period; it took on different meanings in Gnostic or Christian contexts, for example. Zuber has explored one important thread of this tradition in "Spiritual Alchemy from the Age of Jacob Boeme to Mary Anne Atwood." For an overview, see Priesner and Figala, *Alchemie*, s.v. "Seele"; Kämmerer, *Das Leib-Seele-Geist-Problem*.

86. Telle, *Rosarium philosophorum*, 1: 74: "Ortulanus & Arnoldus: quod infundatur anima corpori, & nascitur rex coronatus. Item in libro turbae Philosophorum habetur haec doctrina: Solvite corpora & inhibite spiritum."

87. Ryff, *Des aller fürtrefflichsten, höchsten unnd adelichsten Gschöpffs aller Creaturen*, v, as quoted and translated in Crowther-Heyck, "'Be Fruitful and Multiply,'" 913.

88. Crowther-Heyck, "'Be Fruitful and Multiply,'" 906.

89. Ibid., 908.

90. Ibid., 926.

91. Anna Maria Zieglerin, "die Edele und Tewere Kunst Alchamia belangende," 1. April, 1573, NLA WO, 1 Alt 9, Nr. 306, fol. 66r: "Wie man dye edelen steynne zue rychtte, dihe mytt dem lapidem philoßophorum dingiertte werden welche steynne alle steyn Ine krefftten und tugentte ßo wolle in arzneyhen alle ander steynne dunne uber dreffen und auch uber alle ander steyne zu zielle wege ßeydt zu gebrauchen den menschen zu gutten." The discussion of gemstones runs from fol. 66r to fol. 68v.

92. On alchemy as a way not just to imitate, but to improve nature, see Newman, *Promethean Ambitions*.

93. On the distinction between *sanguis* (blood) and *cruor* (gore, or bloodshed), see Bynum, *Wonderful Blood*, 17–18.

94. Stuart, *Defiled Trades and Social Outcasts*.

95. Smith, "Vermilion, Mercury, Blood, and Lizards," 43–44.

96. As the Mother of God, of course, the Virgin Mary was always seen as an exceptional woman. As we shall see in the following chapter, theologians and physicians alike took pains to distance Mary's body from those of other women.

97. This text was attributed to the theologian and natural philosopher Albertus Magnus, but was probably written by a "German follower." On this text and the tradition of the "secrets of women," see Park, *Secrets of Women*, esp. chap. 2. Pliny's *Natural History* also articulated an influential and detailed negative view of "the mysterious and awful power of the menstruous discharge itself." Pliny, *Natural History* 28.23. See also Pouchelle, *Body and Surgery,* 73.

98. Paracelsus, *Das Buch von der Gebärung der Empfindlichen Dingen der Vernunft* (ca. 1520), in Sudhoff, *Sämtliche Werke*, 1: 241–83. See Cislo, *Paracelsus's Theory of Embodiment*, esp. chap. 2.

99. Pliny, *Natural History* 8.33.

100. "nicht ungleich einer frauen die in irer monats zeit ist, die auch ein verborgen gift in augen hat"; "also mag sie auch mit irem atem und angrif vil ding vergiften, verderben und kraftlos machen"; "dan der basiliscus wechst und wird geboren aus und von der größten unreinikeit der weiber, aus den menstruis und aus dem blut spermatis." [Pseudo?-]Paracelsus, in Sudhoff, *Sämtliche Werke*, 11: 315.

101. Newman, *Promethean Ambitions*, 203.

102. For the "green lion" in Lullian and Riplean alchemy, see Rampling, "Alchemy of George Ripley," chap. 2, esp. 49–51. For an example of a *menstruum* in Philalethean alchemy, "the menstrual blood of our whore" (the star regulus of antimony), see Newman, "'Decknamen or Pseudochemical Language,'" 181–82.

103. For instance, the English alchemical author George Ripley reinterpreted the famous "green lyon" from the Lullian tradition. Rampling, "Alchemy of George Ripley," chap. 2, esp. 49–51.

104. For more on alchemical bloods, see Nummedal, "Corruption, Generation, and the Problem of Menstrua."

105. "Ehr habe wol ein Exempel von einer rosen geben, Wen ein menes Person des gleichen eine menstruosa ein Jed. Eine rosen abbrochen, und die Irgents an ein fenster stochen od. bej einand. setzen, So verdorrete des mans rose, und des weibs wurde am dritten tage faul."

"Gütliche Aussage des Philipp Sömmering," 10. July, 1574, NLA WO, 1 Alt 9, Nr. 308, fol. 105r.

106. "sie habe Ihr von Ihren Menstruis in warmen wein sippen gegeben, daß habe sie Ihr ins Hauß geschickt, darum daß sie Ihr gram gewesen. Wan daß einer einehme, der gehe [illegible] und Krancke." "Verhör der Anna Maria Ziegler," 14. July, 1574, NLA WO, 1 Alt 9, Nr. 314, fol. 22.

107. "Es were einmahl dem Theophrasto ein frawen hubdt ins Laboratorium bracht, Da were die gleser zersprungen." "Gütliche Aussage des Philipp Sömmering," 10. July, 1574, NLA WO, 1 Alt 9, Nr. 308, fol. 105r.

108. "habe sie Ihme gesagt, ehr solte nicht zu seiner frawen wegen des *Menstruij* gehen." "Gütliche Aussage des Philipp Sömmering," 10. July, 1574, NLA WO, 1 Alt 9, Nr. 308, fol. 131r.

109. "Sie habe Philippen gesagt, der muste sich der Weiber enthalten, sonst konte ehr in d alchimisterij nicht ausrichten." "Verhör der Anna Maria Ziegler," 14. July, 1574, NLA WO, 1 Alt 9, Nr. 314, fol. 11r.

110. Newman, *Promethean Ambitions*, 165.

111. Ibid., 195.

112. "Es ist auch zu wissen, das also menschen mögen geboren werden one natürliche veter und mutter. das ist sie werden nit von weiblichem leib auf natürliche weis wie andere kinder geboren, sonder durch kunst und eines erfarnen spagirici geschiklikeit mage in mensch wachsen und geboren werden, wie hernach wird angezeigt." [Pseudo?-]Paracelsus, *De natura rerum neun Bücher*, in Sudhoff, *Sämtliche Werke*, 11: 318–19.

113. Newman, "Homunculus and His Forebears," 329 (Newman's translation).

114. Ibid.

115. Newman, *Promethean Ambitions*, 217.

CHAPTER 5

1. "Sie habe gesagt, der Chrufurst hett ein haut sonderlich balsamen lassen, darin sie 12. wochen gelegt und erzogen. Die haut und ein adler stein habe Herzog Georg des alter Churfurstn vater fur ein schatz hinter sich erlassen. F. Anna habe gesagt, es sei ein klein haudtlein gewesen in einn Mosch gelegen aus einer weibs leib." "Gütliche Aussage des Philipp Sömmering," 9. July, 1574, NLA WO, 1 Alt 9, Nr. 308, fol. 6iv.

2. On associations between birth in the caul and special spiritual characteristics, see Ginzburg, *Night Battles*, 15–16, 61.

3. Most of these details come from two different transcripts of Heinrich Schombach's response to a point of interrogation: "59. Frau Anna's Birth and Purity [Fraw Anen geburt und Reinigkheit]" According to the first transcript, which is more detailed, Heinrich reported: "Das habe sein fraw selbst gesagt, wie sie zur welt kommen, Doctor Beckel habe sie außgeschnitten, und sie sey nur 3. Vierthel Jahrs in Mutter leib gelegen, wie Ihrer Mutter brud erstochen worden, were die Schwanger Mutter heraußlauffen, und aufm eise ersaufft. Da habe sie der Doctor außgeschnitten, der Doctor solte eine Haut gehabt haben, wan man die geschmiret, were sie dicker worden, darin sie gelegt, und volllents mit Tinctur erzogen. Sie habe nicht volkommliche beine, gar subtile, sei also erzogen, alß ob sie in Mutter leib erzogen, dadurch solte sie sond[er]lich gereinget sein. Wan sie sich umbwende, so knocken sie, so schwach sei sie

auf den beinen. . . . Wisse nicht davon, die haut, darin sie gelegt, solt aus einen Weibe genhomen sein, Eine Maria genat solt auch sechs wochen in der Haudt gelegen sein. Die haudt sole aus Polen gekommen sein. Und were ein Turckisch safft od. Tinctur gewesen, so D. Peter gehabt, und sie damit auferzogen. Ein Tinctur sol es gewesen sein. D. Peter Beber von Asch[reu???] soltes zubereitet habe." "Gütliche Aussage und peinliches Verhör des Heinrichs Schomachs," 5. July, 1575, NLA WO, 1 Alt 9, Nr. 313, fols. 25v-26r. According to the second transcript, Heinrich "Sagt sein weib hette gesagt, wie sie zur welt kommen und das sie nur 3 vierthel in Mutter leibe gelegt, und wie irer Mutter Bruder erstochen hette sich ihr Mutter so sehr erschrocken und unterm Eis ersauffen, do hette einer Doctor Peter Ire Mutter aufgeschnitten und sie aus Mutter leibe genomen und sie in eine Handt milche er gebalsemet und geschmiret mit tinctur aufenzogen und dardurch sonderlichen gereinget worden, Sagt auch sie habe nicht vollenkommen gebein gar subtile sey also erzogen als ob sie in Mutter leibe erwachseen vordurch sie sonderlich gereinigt sein soll, sie sey gar schwach uff Iren beinen, wen sie sich in bette umbgewendet hetten, Ir die beine geknacket." "Aussagen Heinrich Schombach gegen Anna," 5. July, 1574, NLA WO, 1 Alt 9, Nr. 314, fols. 68v-69r.

4. "Sie hett auch gesagt, der Graff hett Ihr eine Physognomniam geben." "Gütliche Aussage des Philipp Sömmering," 9. July, 1574, NLA WO, 1 Alt 9, Nr. 308, fol. 62r.

5. On the body as a site of knowledge and expertise, see Park, *Secrets of Women*; Smith, *Body of the Artisan*.

6. "Fraw Anna habe gesagt, dass sie 22. wochen in Mutter leibe gelegen und darnach mit tinctur vollents auferzogen sei und kein Menstruum habe." "Gütliche Aussage des Philipp Sömmering," 9. July, 1574, NLA WO, 1 Alt 9, Nr. 308, fol. 61r. See also "Verhör und Aussagen Ph. Sömmerings," 9. July, 1574, NLA WO, 1 Alt 9, Nr. 309, fols. 19v-20r ["educatio Anna"] for an additional copy of this testimony.

7. "Ehr habe den fluß an Ihr nicht gespurt, allein wan sie sich gezürnen, hette sich der fluß geferbet wie Ziegel ferbe. . . . Die leute so die tucher gewaschen, habens Ihm nicht berichten, dass alte Weib auch nicht, Ehr habs alle sein leblang nicht erfaren. Sie habe Ihme etwas abgeben, dass ehr sie nicht viel fragen durfen." "Gütliche Aussage und peinliches Verhör des Heinrichs Schomachs," 5. July, 1574, NLA WO, 1 Alt 9, Nr. 313, fol. 26.

8. "Ihre weibliche Zeit habe sie nicht alle Monate gehabt, sonderlich wen sie sich erschrocken." "Verhör der Anna Maria Ziegler," 8. July, 1574, NLA WO, 1 Alt 9, Nr. 314, fol. 14r.

9. Green, "Flowers, Poisons, and Men," 55; Stolberg, "Menstruation and Sexual Difference," 91–92; McClive, *Menstruation and Procreation*.

10. "[F. Annen Reinigkeit] Fraw Anna habe sich fur rein ausgeben, das sie keine menstrua hette, und wen man sich zu ihr hielte, so konten die, so von andern Menstruasisch weibern inficirt, von Ihr gereiniget werden." "Gütliche Aussage des Philipp Sömmering," 9. July, 1574, NLA WO, 1 Alt 9, Nr. 308, fol. 51r-v.

11. "Ehr habe offtmals ihr blut . . . eingetrunken, weil sie mit tinctur solte erzogen sein." "Gütliche Aussage des Philipp Sömmering," 9. July, 1574, NLA WO, 1 Alt 9, Nr. 308, fol. 60v.

12. "Sie habe Philippen nichts beigebracht. Ehr Philip habe von sich selbst wol Ihres bluts getruncken. . . . Das blut, so Philip truncken, habe sie aus der ader gelassen." "Verhör der Anna Maria Ziegler," 13. July, 1574, NLA WO, 1 Alt 9, Nr. 314, fol. 19.

13. "Der Churfurste Sachsen habe auch getruncken, doch unwissend." "Verhör der Anna Maria Ziegler," 13. July, 1574, NLA WO, 1 Alt 9, Nr. 314, fol. 19.

14. "Wisse wol, dass f. annen blut. solt sehr gudt seine, der Graf trunck das, wen ehrs bekommen kont. Ehr wisse aber keinen, d Ihren bluts gossen, alß der Graf und der Konig von Denenmark auß grosser lieb, hetts in einem becher in roten wein genhommen. Weiß nicht,

dass es d Churfurst zu Sachsen gedruncken hette." "Gütliche Aussage und peinliches Verhör des Heinrichs Schomachs," 7. July, 1574, morning, NLA WO, 1 Alt 9, Nr. 313, fols. 67r-v.

15. On the cleansing powers attributed to Christ's blood in late medieval Germany, see Bynum, *Wonderful Blood*.

16. "Philip habe es alles gesagt, und sie die fraw auch, dass sie den fluß nich habe und and[er]e viele tungende und das sie den Engeln gleich were." "Gütliche Aussage und peinliches Verhör des Heinrichs Schomachs, 5. July, 1574, NLA WO, 1 Alt 9, Nr. 313, fol. 25v. "Philip hette es alles gesagt und sie auch das sie den fluß nit hette und anderer mehr tugender und das sie den Engeln gleich sein solle, wie er auch jetz under noch nit anders vermeinet sey es andern so sey er betrogen." "Aussagen Heinrich Schombach gegen Anna," 5. July, 1574, NLA WO, 1 Alt 9, Nr. 314, fols. 68v-69r.

17. "Was fraw Annen reinigkeit anlangt habe ehr sie wol viel gelobt, und der Mutter Gottes vergliechen, wie sie sich dan also selbst angeben. . . . Das habe ehr albereit gesagt, dass ehr Fraw Annen hoch gelobt, und sie der Junckfraw marien gleich gehalten." "Philipp Sömmering's testimony," 9. July, 1574, NLA WO, 1 Alt 9, Nr. 308, fols. 60v-61v.

18. Walsham and Marshall, *Angels in the Early Modern World*, esp. 1–41 and 64–82.

19. Sella, "Northern Italian Confraternities," 602; Ellington, *From Sacred Body to Angelic Soul*.

20. On these debates, see Sella, "Northern Italian Confraternities," 602–4; Brandenbarg, "St. Anne and Her Family," 107; Kreitzer, *Reforming Mary*, 14–16; Ellington, *From Sacred Body to Angelic Soul*.

21. Wood, "Doctor's Dilemma," 719.

22. Ibid., 721.

23. Ibid., 721–25.

24. As one historian has pointed out, "The reformers had no desire to dishonour Mary. They merely wished to moderate her veneration in accordance with their interpretations of Scripture. There was therefore a certain ambiguity in the Protestant position on Mary: reformers condemned her cult, but could not afford to denigrate the Mother of God entirely." Heal, *Cult of the Virgin Mary*, 63. On continuities between medieval and sixteenth-century Protestant attitudes to Mary, see also Kreitzer, *Reforming Mary*, 133–35. It is important to note that Kreitzer has identified a shift in attitudes toward a more critical stance in the mid-sixteenth century.

25. For Luther and other reformers' views on the Immaculate Conception, see Kreitzer, *Reforming Mary*, 123–25.

26. Heal, *Cult of the Virgin Mary*, 64–115 and "Marian Devotion and Confessional Identity."

27. For an overview of Julius's implementation of the Reformation in Braunschweig-Lüneburg, see Mager, "Die Einfürhung der Reformation in Braunschweig-Wolfenbüttel." See also Reller, *Vorreformatorische und Reformatorische Kirchenverfassung*; and Mager, *Die Konkordienformel*.

28. Heal, *Cult of the Virgin Mary*, 23–63; Kreitzer, *Reforming Mary*, 133–42.

29. Although, as Kreitzer has noted, later reformers began to exhibit a more critical stance toward the Virgin. Kreitzer, *Reforming Mary*, 134–35.

30. Heal, *Cult of the Virgin Mary*, 109. On the Protestant "praise of motherhood" more generally, see Rublack, "Pregnancy, Childbirth, and the Female Body," esp. 87–88.

31. Anna described Count Carl as her alchemical tutor in the book of recipes, and, in conversations with her fellow alchemists and with Duke Julius, as Paracelsus's son. See Chapter 4.

32. Most recently, Dane T. Daniel and Amy Cislo have emphasized the connections between Paracelsus's theology, natural philosophy, and medicine. See Cislo, *Paracelsus's Theory of Embodiment*; Daniel, "Invisible Wombs."

33. On Paracelsus's theology and its relationship to his understanding of nature, see Webster, *Paracelsus*.

34. Amy Cislo makes this point convincingly in *Paracelsus's Theory of Embodiment*.

35. Gause, *Paracelsus*, 31–32.

36. "So ist sie die ersste creatur gewesen, ehe das himmel und erdten beschaffen sein worden." Biegger, *"De Invokatione Beatae Mariae Virginis,"* 159.

37. For a discussion of Paracelsus's Mariological texts, including *Libellus de virgine sancta theotoca* (1524), *Von der Geburt Mariae und Christi* (ca. 1525), "Vorredt in das Salue regina" (1520s), *Liber de sancta trinitate* (1524), and *De invocatione beatae mariae virginis* (ca. 1527), see Biegger, *"De Invokatione Beatae Mariae Virginis"* and Gause, *Paracelsus*, 1–73.

38. Anna's refusal to comport herself according to new standards of Lutheran womanhood was certainly an important part of her dispute with Duchess Hedwig, who fully embraced her role as Lutheran *Hausmutter* not only for the princely household but for the entire Duchy of Braunschweig. See Chapter 2.

39. I am grateful to Amy Remensnyder for this observation.

40. "Alß die Grafin diese vermeinten grafen Mutter vom Theophrasto geschwangert worden, were Theophrastus nach des Kaysers hofgezogen und gesagt, dass ein solch kindt von der gräfin wurde geboren werden, welcher Theophrastum Hermetem und alle Philosophen ubertreffen wurde. Und wie d. graff geboren were ein weissen adler das aus von dem himmel gefallen. Und wie sie darnach herauß geritten, hetten sie den adler an der erden liegen sehen, und der hett sich wieder auf [gem] himmel erhoben." "Verhör und Aussagen Philipp Sömmerings," 9. July, 1574, NLA WO, 1 Alt 9, Nr. 309, fol. 20v. For a second transcript of Philipp's testimony on this point, see "Gütliche Aussage des Philipp Sömmering," 9. July, 1574, NLA WO, 1 Alt 9, Nr. 308, fol. 62r-v.

41. "Alß der graff nun erwachsen und seines vaters Theophrastij bucher bekommen, het ehr darin gefunden, dass sein vater mit höchster kunst darnach getrachtet, das ein Megdlein mucht geborn werden, die des Monatlichen flußes entfreiet, und mit der hube [?] der graff kunder zeugen wollen, die biß Immer Jungsten tage leben sollten." "Gütliche Aussage des Philipp Sömmering," 9. July, 1574, NLA WO, 1 Alt 9, Nr. 308, fol. 62v.

42. "Der Graf hett wollen fraw annen haben, Ihren manne Heinrichen Schombach seine Schwester wied geben und also umb beuten, und Ihme noch [zwei?] tonnen goldes zugeben. Und Heinrich hettes gewilligt, und gesagt, Wen ehr das guld bekheme, wolt ehr sie denn Grafen wol ubergeben. Welches ehr fraw Anna zuglaubt." "Gütliche Aussage des Philipp Sömmering," 9. July, 1574, NLA WO, 1 Alt 9, Nr. 308, fols. 65r.

43. "Seine Fraw hatte auch furgeben, das der Graffe alle monat kund mit Ir zeugen konte und wen ein kundt 6 wochen od weniger in Mutter leibe gelegen, solte volkomlich sein und durch die tinctur vom Graff weiter erzogen werden." "Aussagen Heinrich Schombach gegen Anna," 6. July, 1574, NLA WO, 1 Alt 9, Nr. 314, fols. 73v-74v.

44. "1 Kind solte 25. Wochen in Mutter leibe sein, darnach wolte ehr sie mit tinctur aufziehen, das hett ehr aus der Kunst geredt." "Verhör der Anna Maria Ziegler," 8. July, 1574, NLA WO, 1 Alt 9, Nr. 314, fol. 14v.

45. "Die Kinder solte nicht kranck werden, Ihr geordnte zeit erleben, und sterben wan Godt will." "Verhör der Anna Maria Ziegler," 8. July, 1574, NLA WO, 1 Alt 9, Nr. 314, fols. 14v-15.

46. "[D]ie kinder sollen auch eher erwachssen sein den ander, die feminae solten mit dem fluß nicht gehafft sein, wen sie vom sechs Jahrn weren solten sie sich mit einand nehmen khommen, und wen sich der fall mit adam nich hette zugetragen, konten sie woll ewig leben, aber demnach wurden sie so alt werden wie Methusalem und ander patriarchen." "Aussagen Heinrich Schombach gegen Anna," 6. July, 1574, NLA WO, 1 Alt 9, Nr. 314, fol. 73v-74v.

47. Martin Luther, "Lectures on Genesis," as cited in Karant-Nunn and Wiesner, *Luther on Women*, 177.

48. Martin Luther, "Sermon for Christmas," 1522, as cited in ibid., 50.

49. Because menstrual blood and breast milk were often thought to be the same substance, perhaps the fact that Anna did not menstruate suggested that she could not produce breast milk either.

50. "Der Graf habe gesagt, wie ehr ein Weib nheme, so [wurde] ehr dadurch ein newe Welt anrichten. Das wurde mit der tinctur geschehen, das habe sie oft gesagt." "Verhör der Anna Maria Ziegler," 8. July, 1574, NLA WO, 1 Alt 9, Nr. 314, fol. 15.

51. "von solcher Kundern eine newe welt werden." "Aussagen Heinrich Schombach gegen Anna," 6. July, 1574, NLA WO, 1 Alt 9, Nr. 314, fol. 74v.

52. "Sie habe ein ganz kunstbuch gehabt, das habe Theophrastus <gemacht und> geschrieben [ge......] sie habe es vom Graf bekommen, der Graf von seiner Mutter, weiß nicht wieviel hundert stucken darin gewesen. Man [nennt] es die Cabala." "Verhör der Anna Maria Ziegler," 8. July, 1574, NLA WO, 1 Alt 9, Nr. 314, fol. 4v.

53. "es sei auch von der alchimisterey und von die kunst seien fur kleider machen gewesen, fur [Schafsterben], Pferde sterben und anders, vom einen die Pferde [bezaubern] were." "Verhör der Anna Maria Ziegler," 8. July, 1574, NLA WO, 1 Alt 9, Nr. 314, fol. 4v.

54. Recipe books were often passed down in families. See Rankin, *Panaceia's Daughters*; Leong, "Collecting Knowledge for the Family."

55. "Fraw Anna habe Ihme auch ein Buch von den Engeln gezeigt, Wie die Engele ein [in?] Paradeiß Godt angebetten, und das in rhadt Gottes geschlossen, dass Godt denn Menschen fallen lassen wollte, und also durch den graffen und fraw annen die welt wieder anrichten, und den Menschen helffen." "Gütliche Aussage des Philipp Sömmering," 9. July, 1574, NLA WO, 1 Alt 9, Nr. 308, fol. 65v.

56. Barnes, *Prophecy and Gnosis*; Webster, *Paracelsus*; Leppin, *Antichrist und Jüngster Tag*.

57. Webster, *Paracelsus*, 218–24; Green, *Printing and Prophecy*.

58. Barnes, *Prophecy and Gnosis*, 65–66.

59. This "apocalyptic consciousness" would begin to fracture in the early seventeenth century as Lutheran leaders sought to rein in what Barnes calls an increasingly "speculative adventurousness," but it was in full flower when Anna related her Paracelsian prophecy in Wolfenbüttel. Barnes, *Prophecy and Gnosis*, 7–8.

60. Ibid., 94.

61. Lerner, "Refreshment of the Saints."

62. Ibid.; Reeves, *Influence of Prophecy* and *Joachim of Fiore*; McGinn, *Apocalypticism in the Western Tradition*.

63. Lerner, "Refreshment of the Saints," esp. 129–32 on Villanova and Rupescissa; and Calvet, *Les Oeuvres alchimiques*; DeVun, *Prophecy, Alchemy, and the End of Time*.

64. Webster, *Paracelsus*, 222–24; Klaassen, *Living at the End of the Ages*.

65. Webster, *Paracelsus*, 217; Pagel, "Paracelsian Elias Artista"; Breger, "Elias Artista"; Goldammer, "Paracelsische Eschatologie."

66. Webster, *Paracelsus*, 33, 210–12.

67. Harkness, *John Dee's Conversations with Angels*; Kuntz, *Guillaume Postel*.

68. Kuntz, *Guillaume Postel*, esp. 144–68.

69. Harkness, *John Dee's Conversations with Angels*, 144ff.

70. Barnes, *Prophecy and Gnosis*, 60–61.

71. Walsham and Marshall, *Angels in the Early Modern World*, 2.

72. Ibid., 14.

73. Harkness, *John Dee's Conversations with Angels*.

74. Beyer, "Lutheran Popular Prophets." See also Beyer, "Lutherische Propheten in Deutschland" and "Lübeck Prophet."

75. Crisciani, "*Opus* and *Sermo*," 22, 21.

76. Luther, "Of the Resurrection," in *Table-Talk of Martin Luther*, Book DCCLX, 826. For the Latin original, see Luther, *D. Martin Luther's Werke*, Abt. 2, vol. 1, Tischreden, 1149.

77. See Calvet, *Les Oeuvres alchimiques*; Matus, "Alchemy and Christianity in the Middle Ages" and "Reconsidering Roger Bacon's Apocalypticism"; DeVun, *Prophecy, Alchemy, and the End of Time*.

78. DeVun, *Prophecy, Alchemy, and the End of Time*.

79. For a good point of entry into the vast literature on Newton, see Iliffe and Smith, *Cambridge Companion to Newton*.

80. See, for example, the proliferation of prophecy in the wake of the "great flood" (*Sintflut*) that failed to happen in 1524. Zambelli, *"Astrologi hallucinati"*; Barnes, *Prophecy and Gnosis*.

81. Barnes, *Prophecy and Gnosis*, 261.

CHAPTER 6

1. See "Anna Maria Brieff mit eigener Hand an Illm. Julium geschrieben," 1573, NLA WO, 1 Alt 9, Nr. 307, fols. 71r-78r and Philipp Sömmering's reports to Duke Julius on their alchemical projects between August and October, 1573, NLA WO, 1 Alt 9, Nr. 307, fols. 71–86v.

2. "Fraw Anna habe in einen kleinen Speißkammerlein zu Goschlar gearbeitet, und fur gegeben, sie hett Mercurium soluirt. Es stehe noch in dem Speißkammerlein in einen Kleinen Kolbischen." "Gütliche Aussage des Philipp Sömmering," 10. July, 1574, NLA WO, 1 Alt 9, Nr. 308, fol. 83r-v.

3. "Ehr hett den Laboranten aber mit [?] Goschlar genhommen. Franz hett Ihme sollen die tinctur in Goschlar machen, so het ehr das seine auch darzu thun, und es in sechs wochen fertig machen." "Gütliche Aussage des Philipp Sömmering," 9. July, 1574, NLA WO, 1 Alt 9, Nr. 308, fol. 48v-49r.

4. "schreckliche Zeichen am Himmel, auch gespenst gesehen, und grewliche treume gehabt, welche sie da selbst außgedeutet, das Ihnen groß unglук bevorstende und sie alle semptlich schedtlich umbkommen wurden." Regarding the ghost of the woman, Anna said, "Sie habe gedacht, es sei d teuffel. . . . Die leute in d herbergn haben gesagt, es Pflege da so zugehen und viele leut umbzukommen." "Verhör der Anna Maria Ziegler," 8. July, 1574, NLA WO, 1 Alt 9, Nr. 311, fols. 2r-v.

5. Nummedal, *Alchemy and Authority*.

6. Jobst Kettwig to Duke Julius, 28. November, 1573, NLA WO, 1 Alt 9, Nr. 307, fols. 182–87.

7. Nummedal, *Alchemy and Authority.*

8. Jobst Kettwig to Duke Julius, 28. November, 1573, NLA WO, 1 Alt 9, Nr. 307, fol. 182v.

9. "solches wollte Ich nicht alhie aus Haß oder Neidt," Jobst Kettwig to Duke Julius, 28. November, 1573, NLA WO, 1 Alt 9, Nr. 307, fol. 185; "darumb habe Ich aus erkendtlicher scheldigkeit, der Eyde und pflicher, die Ich meinem gnedigen fursten und Herrn, gelobet und geschworen hab," fol. 182v.

10. According to Rhamm, Kettwig and Schulvermann together had robbed a Lübeck merchant named Hans Kapell in 1569. See Rhamm, *Die betrüglichen Goldmacher*, 10.

11. "Paß und Vollmacht für Jobst Kettwig und Hans Schulvermann; Privilegierung fremder Metallsucher 1572 (Druck)," NLA WO, 1 Alt 9, Nr. 331. For the terms of Kettwig's employment at Julius's court, see Rhamm, *Die betrüglichen Goldmacher,* 32 and n63. According to Rhamm (29–34), Kettwig seemed to have a habit of using his affiliation with Julius's court to secure credit at inns while traveling for ducal business and then ducking out on the bill. When Julius began to receive a number of complaints from aggrieved innkeepers, he began to question his agent's integrity.

12. Although Kettwig had been loosely affiliated with Philipp Sömmering through Schulvermann, the fact that Sömmering turned Kettwig in suggests that Julius's two advisers were more rivals than collaborators. Once Kettwig was in custody in Braunschweig, Philipp sent word to Denmark about Kettwig's arrest. He also notified the Duke of Pomerania that Kettwig, with Schulvermann, was involved in the 1569 attack on a Lübeck merchant, Hans Kappell, in his duchy. Perhaps he was trying to usher Kettwig out of Braunschweig-Wolfenbüttel and on to swift justice elsewhere, but this is Rhamm's view and supports the narrative that Philipp was guilty and trying to cover it up. Rhamm, *Die betrüglichen Goldmacher*, 10 and 34. See "Verhör und Aussagen von Sylvester Schulvermann," November 1574, NLO WO, 1 Alt 9, Nr. 315, fols. 12r-50v; "Gefangenhaltung und Flucht des Jobst Kettwig aus Braunschweig und die sich daran anschließenden Auseinandersetzungen mit der Stadt; Herzog Julius bittet um die Verhaftung des Jobst Kettwig und des Sylvester Schulvermann durch Braunschweig," NLA WO, 1 Alt 9, Nr. 320; "Verfolgung, Gefangennahme und Auslieferung des flüchtigen Jobst Kettwig zu Cöln an der Spree und des Sylvester Schulvermann zu Ulm; gütliche Aussage des Sylvester Schulvermann," NLA WO, 1 Alt 9, Nr. 322.

13. "Dem allen nach ist hirmit an [. . .] wolr. mein hoch fleisseiges und emplige anruffen flehen und bitten . . . vor I.f.g . selbst augens und Ohren lassen, schrifftlich, oder mundlichen vorbringen . . . das er sein falsch angebens oder clage ohne weitern vortzugk, alhier zu rechte gegen mir vornehm anfange und ausspure, das doch meine unschuldt und die gottliche warheit, an tagk komme." Jobst Kettwig to Duke Julius, 28. November, 1573, NLA WO, 1 Alt 9, Nr. 307, fol. 186v.

14. "der Durchleuchtiger Hochgeborner Fürst und Herr, Herr Julius zu Braunschweig und Luneburg M. G. f. und Herr, durch allerley falschen bericht, und lugen darzu ist bewogen, das auf J. f. g. bevelich, Ich benebenstandern, unschuldig, gefenglicherr eingezugen werden"; "die Gott vergessene [mene]digeTrawe und erlose, Viertheill henckige und brennessige schelmen Huren und buben, Philip Schreiber, oder Sommerinck von Dambach geborn, Anna Maria, und Irer Menner ein Heinrich Schombach genant sein, dessen allen ein ursache." Jobst Kettwig to Duke Julius, 28. November, 1573, NLA WO, 1 Alt 9, Nr. 307, fol. 182r.

15. "Darumb das der vordampte Hohstraff wirdige unerhorte greuliche Teuffliche laster, Hurerëÿ, flasch [*sic*] lugen und betruch nicht aus licht, und vor M. g. F. und Herrn kommen solten, Jedoch Gott der allmechtige, der aleine ein gereicher Her Ist, und alhe ding biß zu

seiner zeit schicket. Gibt es gnediglich, das die Gottlosen Teufflischen vorlogenen Personen, In Irem Gottlosen veßent und lastern, offenbar zu schanden mussen werden." Jobst Kettwig to Duke Julius, 28. November, 1573, NLA WO, 1 Alt 9, Nr. 307, fol. 182r.

16. Mentzingen was north of Stuttgart (now part of Kraichtal); but Kettwig also said that he visited the "Freyherrn von Etzingen, die in Osterreich und Steyermerck Ire herrschaft haben." Jobst Kettwig to Duke Julius, 28. November, 1573, NLA WO, 1 Alt 9, Nr. 307, fol. 183r.

17. Puchner, "Baumgartner, David" in *Neue Deutsche Biographie* 1 (1953): 663, http://www.deutsche-biographie.de/ppn130353159.html. See also Ortloff, *Geschichte der Grumbachischen Händel*, 2: 329ff., 3: 446ff., 4: 158.

18. Groebner, *Who Are You?*

19. Kettwig's purported attempt to track down Paracelsus's acquaintances—more than three decades after the physician's death—is particularly intriguing.

20. "von anfang der welt, nirgendts gewesen und sein werden." Jobst Kettwig to Duke Julius, 28. November, 1573, NLA WO, 1 Alt 9, Nr. 307, fol. 183v.

21. "Dan auch alle die Grauen zu Ottingen, sich solcher unerhorten graulichen betruglichen lügen zum hogesen vorwunderen und beclagen, Das Ir grauliche reputationes, Nhamen Titul, Handt, und Sigel, zu solchem falschen betrüge sein mißbraucht wurden." Jobst Kettwig to Duke Julius, 28. November, 1573, NLA WO, 1 Alt 9, Nr. 307, fol. 183v.

22. Jobst Kettwig to Duke Julius, 28. November, 1573, NLA WO, 1 Alt 9, Nr. 307, fols. 183v-184r.

23. "greulicher lugen und falscher betrug sein nie erhort oder erfinden wurden, als die Anna Maria und der Pfaff Philip Schreiber M. g. f. und herrn anbracht und vorzutragen haben." Jobst Kettwig to Duke Julius, 28. November, 1573, NLA WO, 1 Alt 9, Nr. 307, fol. 184r.

24. "Das wan die Pfaff und die Huere, mit Irer Alchimei gelt machen, So will Ich es under der Trepfen finden, Jha under den Schuesolen ausnhemen." Jobst Kettwig to Duke Julius, 28. November, 1573, NLA WO, 1 Alt 9, Nr. 307, fol. 184r.

25. According to Rhamm, Peter Dussel stabbed the Frohndiener, then fled with Kettwig. Rhamm, *Die betrüglichen Goldmacher*, 41–42.

26. These claims come from Rhamm (*Die betrüglichen Goldmacher*, 37–38), who cites "Gesandschaft des Herzogs an den Rath zu Braunschweig vom 14. Dezember 1573," Fasc. II. On "sächsischer Frist," or see "Sachsenfrist," in *Pierer's Universal-Lexikon* (Altenburg, 1862), 14: 731, which defines it as, "ein Zeitraum von sechs Wochen und drei Tagen."

27. "Steckbrief gegen Jobst Ketwich wegen Landfriedensbruchs," 1574, NLA BU [Bückeburg], L 1, Nr. 8982. See also Rhamm, *Die betrüglichen Goldmacher*, 87n75 and Fasc. X. Schulvermann was captured in Ulm on January 5, 1574 (although he was not delivered to Wolfenbüttel until the following September because he was not well enough to travel).

28. Kettwig was captured in Halberstadt at the end of September, and eventually condemned alongside Anna and the others. On Schulvermann, see Rhamm, *Die betrüglichen Goldmacher*, 91n102/Fasc. V; on Kettwig, see Rhamm, 91n103, who cites Kettwig's Urgicht from 18. November 1574 and Verhör from 12. November, NLA WO, 1 Alt 9, Nr. 311, fols. 23–35; Rhamm, 100n118. According to the Brandenburger Schöffen (legal experts) Jobst Kettwig was to be "geschleift," broken on the wheel, quartered, with pieces hanged for five crimes, Silvester Schulvermann, the wheel and quartered for eight crimes.

29. Kettwig's letter claims that the would-be Carol Graf von Oettingen was in fact "ein Freÿherr von Etzigen Carolus genant, die bei Keÿser Carull und Jtzigen Konig von Hispanien

viel umbgangen, auch Kaiser Carull zu buecher, von sonderlichen materien zugeschrieben so lange gelt und Edelsteine gemacht, das grosser schulden halben," Jobst Kettwig to Duke Julius, 28. November, 1573, NLA WO, 1 Alt 9, Nr. 307, fol. 183v. Kettwig was likely thinking of the Freiherrn von Mentzingen, whose seat was just north of Stuttgart and just over one hundred miles from Oettingen.

30. "auch kein Graue zu Ottingen, Sondern ein Freÿherr von Etzigen [Mentzingen?] Carolus genant, die bei Keÿser Carull und Jtzigen Konig von Hispanien viel umbgangen, auch Kaiser Carull zu buecher, von sonderlichen materien zugeschrieben so lange golt und Edelsteine gemacht, das grosser schulden halben, die Creditorn Inen die Herrschafft Magenbruck eingenhomen haben." Jobst Kettwig to Duke Julius, 28. November, 1573, NLA WO, 1 Alt 9, Nr. 307, fols. 183v-184r.

31. Nummedal, *Alchemy and Authority*.

32. Rublack, *Crimes of Women*, esp. chap. 4.

33. "ich mus bekennen, das wir damit unrecht gehandelt, ist es aber vorsetzlich aber wissentlich geschenn so sey das leyden und sterben meynes lieben hernn Jhesu Christo an mir und uns allen vorloren." Anna Zieglerin to Duke Julius in Heinrichstadt, 9. January, 1574, NLA WO, 1 Alt 9, Nr. 307, fol. 75r.

34. "bitte der wegen umb gottes willen E. f. g., wolden mich doch aus solchem verdacht schleissen, und durch gottes wilen nicht so verweissen. . . . gott erkenne alle herzenn, und richte sie nach ihrem verdienst und seynem gottlichen erkentnus." Anna Zieglerin to Duke Julius in Heinrichstadt, 9. January, 1574, NLA WO, 1 Alt 9, Nr. 307, fol. 75r.

35. She also reminded Johann Friedrich II that she had married her husband, Heinrich, at the then duke's command and in exchange for a promise of protection while she resided in his lands. Anna Maria Zieglerin to Johann Friedrich II, 19. December, 1573, NLA WO, 1 Alt 9, Nr. 307, fols. 194–196.

36. "So ist mir . . . in dis langwirige ehelend kommen, dermassen ergangen das Ich offtmahls gewunscht, das Ich nicht geboren." Anna Maria Zieglerin to Johann Friedrich II, 19. December, 1573, NLA WO, 1 Alt 9, Nr. 307, fol. 193.

37. "wir wehren vorlangst wid[er] Gott und alle billigkeit auf die fleischbank geopfert, wie sie dan nach teglich wuten und toben." Anna Maria Zieglerin to Johann Friedrich II, 19. December, 1573, NLA WO, 1 Alt 9, Nr. 307, fols. 193–94.

38. "unser Gott vor uns sorget, und vaterliche gottliche wacht auf uns helt, und ab dem Teuffel bey aufhebung seiner verdamnus aufferlege, uns zu plagen und quelen so kont ers nicht erger darthun." Anna Maria Zieglerin to Johann Friedrich II, 19. December, 1573, NLA WO, 1 Alt 9, Nr. 307, fol. 194.

39. "und dieselbig E. f. g. und M. g. f. und H. zugeschickt, hat S. f. g. mich sehen und lesen lassen, wie erschrocklich greulich und unerhort aber der Siebenfachte Radtbruchswirdiger Schelm [i.e., Jobst Kettwig] mich angreifft." Philipp Sömmering to Duke Julius, 27. November, 1573, NLA WO, 1 Alt 9, Nr. 307, fols. 179r-180v.

40. "und dieselbig E. f. g. und M. g. f. und H. zugeschickt, hat S. f. g. mich sehen und lesen lassen, wie erschrocklich greulich und unerhort aber der Siebenfachte Radtbruchswirdiger Schelm mich angreifft." Philipp Sömmering to Duke Julius, 27. November, 1573, NLA WO, 1 Alt 9, Nr. 307, fols. 179r-180v.

41. "My Dear Lord Christ gives me comfort, for he says, 'Blessed are you when people revile you and persecute you and utter all kinds of evil against you falsely on my account,' Matthew 5. And Psalm 26 [*sic*] says, 'Transgression speaks to the wicked deep in their hearts; there is no fear of God before their eyes. For they flatter themselves in their own eyes that their

iniquity cannot be found out and hated.' Against such a godless knave, rogue, thief and miscreant, I can pray with a good heart from Psalm 35, 'O Lord, do not be far from me! Wake up! Bestir yourself for my defense, for my cause, my God and my Lord!' . . . Let all those who rejoice at my calamity be put to shame and confusion; let those who exalt themselves against me be clothed with shame and dishonor. Let those who desire my vindication shout for joy and be glad, and say evermore, 'Great is the Lord, who delights in the welfare of his servant.' " (mein lieber Herr Christes trost gibt, do ehr Spricht selig seidt Ir so euch die Leute schmehen umb meinetwillen, und reden allereley übels von euch, so sie daran lügen, Matt: 5 und der 26 [sic] Psalm: Sagt, Es Ist vom Grundt meines Herzen gesprochen von der Gottloßen wesen, das kein Gotes furchte bei Inen ist, sie schmücken sich unter einander selber das Sie Ire bose sachen fordern, und andere verunglimpfen, Wid solchen gottloßen buben, Schelmen, dieb und bosewicht, Kan Ich mit Gutem Hertzen aus dem 35 Psalm beten, Herr sey nicht fern von mir, erwecke dich und wache auf Zu meinem recht und zu meiner Sache, mein Gott und Herr, Sie mussen sich Schemen, und zu schanden werden alle die sich meines Ubels frewen, sie mussen mit scham und schandt gekleidet werden, die sich wider mich ruhmen, Ruhmen und frewen mussen sich, die mir gönnen das Ich Recht behalte, Wer Herr musse Hochgelobet sein der seinem Knecht wol will.) Philipp Sömmering to Duke Julius, 27. November, 1573, NLA WO, 1 Alt 9, Nr. 307, fols. 179r-v. Translations of Matthew 5, Psalms 36 and 35 are from the NRSV, http://www.Biblegateway.com. Sömmering's German text is nearly identical to the Luther Bible, with some minor variation, probably due to faulty memory.

42. "ein Gottloser unchrist, vielmahl meinaidig, Ehrenvergessener vorzweiffelter Ertz Schelm, verrether, Boswicht, Sehe- und Straßreuber." Philipp Sömmering to Duke Julius, 27. November, 1573, NLA WO, 1 Alt 9, Nr. 307, fol. 179v. Schulvermann was no better, according to Philipp: "He is a rogue above all other rogues except for his companion [Kettwig]; they thank all who house and shelter them and are good to them with treachery, defilement, and nasty behavior [verreterey, schenderej, und schelmerej]." (Summa er ist ein Schelm aller Schelmen sonst seinen gesellen, die da pflegen all denen die si haußen, heimen, und Inen hutes erzeigen mit verretherey, Schenderej, Schlmerej zu danken.) Philipp Sömmering to Duke Julius, 27. November, 1573, NLA WO, 1 Alt 9, Nr. 307, fol. 180r. "Ich gewiß bei M. G. F. und Herr, muß mir vor Gott und der Welt auf sein furstliches gewissen Zeugnus geben, das Ich S. f. g. mein lebtag umb nicht belegen noch betrogen hab." Philipp Sömmering to Duke Julius, 27. November, 1573, NLA WO, 1 Alt 9, Nr. 307, fols. 180r-v.

43. "Ich gewiß bei M. G. F. und Herr, muß mir vor Gott und der Welt auf sein furstliches gewissen Zeugnus geben, das Ich S. f. g. mein lebtag umb nicht belegen noch betrogen hab." Philipp Sömmering to Duke Julius, 27. November, 1573, NLA WO, 1 Alt 9, Nr. 307, fols. 180r-v.

44. See Chapter 1, above.

45. Philipp reminded Julius of how several of his old enemies from the 1560s (including Elector August of Saxony and Weidemann, as well as more minor figures with whom Philipp had sparred over theological issues at Julius's court, such as Nikolaus Selnecker and Lukas Thangel) had relentlessly tried to destroy Philipp's reputation in both his Thuringian homeland and Braunschweig-Wolfenbüttel.

46. According to Philipp, Julius had contacted both the former duke Johann Friedrich II and his successor, Johann Wilhelm of Saxony, to investigate these charges, and was satisfied that the stories about Philipp Sömmering were rooted in old theological battles from the 1560s that were no longer relevant to his utility at the Wolfenbüttel court in the 1570s.

47. See Chapter 3, above.

48. "Wie wir uns uf E. f. g. furstlichs und christlichs gewißen selbst referiren, die vor Got, und dem Welt uns zeugnis geben mußen, das sie uchristlichs zeubreisch nichts auf uns bemerckt haben." Philipp Sömmering to Duke Julius, 8. January, 1574, NLA WO, 1 Alt 9, Nr. 307, fol. 207r.

49. Biagioli, *Galileo Courtier*.

50. "Ich bin ein Christ, hab' von Jugend auf, Ehr Tugend und Gottseligkeit und freier Künste mich beflissen, als ich auch, zu meinen Jhare khomen mich vernemlich od Studium Theologia begeben, hab auch wiewohl unwürdig, 15 Jahr lang das ministerium, Gottlob, in guetem Thun und großem Nutz verwaltet, bei In Meinem Amt, Lehr und Leben unsträflich und in solchen ansehen, bei meinen vicinis gewesen, daß sie so schier bei mir, als dem Superintendenten zu Gotha, [hat] gesucht haben. Darneben hab Ich sonderlich Lust gehabt, an dem schone herlichen geschopff der Metallen und Mineralien bin dadurch in der schonen lustgarten der philosophorum kommen, da ich Gottlob ein wenig vom Saft des himmlischen Tau's und dem Balsam des grünen Erdreichs geschmecket habe, daher ich mich nicht schäme, einen Philosophen zu bekennen, scheue auch keine Facultät, noch hohe Schul, in tota Germania, mit den Allergelehrtesten von der Natur und ihren wunderlichen unbegreiflichen Werken zu disputiren, daraus die Werkstatt des Schöpfers der Natur zu kennen ist. Da nun dieser Profession halber ich sollt' vor einen Zauberer oder Schwarzkünstler gehalten werden, wäre es bei Verstendigen keiner Antwort werth, denn, Gottlob, solcher Leut' viel in der Welt seindt." Philipp Sömmering to Duke Julius, 8. January, 1574, NLA WO, 1 Alt 9, Nr. 307, fols. 207v-8r.

51. On Leonhard Thurneisser, see Spitzer, *. . . und die Spree führt Gold*; Burghartz, "Die Renaissance des Infamen?"; Moran, "Art and Artisanship."

52. "Soll man nun sagen: woher kann er diese Kunst, der Teufel muß ihn lehren, er muß ein Zauberer und Teufelsbändiger sein, So versehe ich mich billig, es werde einem bald das Maul gestopft werden. Ei, warum wolllt' man mich denn für einen Zauberer halten, da ich noch lange nicht so seltsame Dinge vorgeben als derselbige." Philipp Sömmering to Duke Julius, 8. January, 1574, NLA WO, 1 Alt 9, Nr. 307, fol. 208r.

53. "Hält auch der Kurfürst denselben billig in Schutz ja braucht ihn für einen Diener, warum soll dan E. f. G. nicht desgleichen auch zu thun Macht haben?" Philipp Sömmering to Duke Julius, 8. January, 1574, NLA WO, 1 Alt 9, Nr. 307, fol. 208r.

54. L.T.Z.T. [Leonhardt Thurneysser zum Thurn], *Warhafftige Abconterfeyung*.

55. For a description of the details of Lippold's execution, see Ackermann, *Münzmeister Lipold* 62–63. For Grumbach's execution, see Dülmen, *Theatre of Horror*, 93–94.

56. Ackermann discusses early modern chroniclers' accounts of the execution in *Münzmeister Lippold*, 67–71.

57. "Wie hie Kurtz gmelt sind etlich gschicht, Drum traw kein Christ keim Jude nicht." L.T.Z.T., *Warhafftige Abconterfeyung*.

58. "Luppoldt Judenn Uhrgicht, 1573," in Ackermann, *Münzmeister Lippold*, 93 and 97.

59. Johann Georg, Markgraf zu Brandenburg, to Julius, 4. February, 1574, NLA WO, 1 Alt 9, Nr. 307, fols. 90–91.

60. See Chapter 2.

61. Johann Georg, Markgraf zu Brandenburg, to Julius, 4. February, 1574, NLA WO, 1 Alt 9, Nr. 307, fol. 90r.

62. "Dan unser Her Vater seliger gedechtnus (welches wir nicht Ohne sonder betruebnus unsers hertzen kegen E. L. Melden) wolte auch keiner warnung statt geben. Sondern wahr von

dergleichen bösen leuten, durch Ihre Zeuberische künste, wie es hernacher woll an Tagk kohmen, auch also beteubet und eingenohmen." Johann Georg, Markgraf zu Brandenburg, to Julius, 4. February, 1574, NLA WO, 1 Alt 9, Nr. 307, fol. 90v.

63. "Darumb bitten wir Nochmaln, und widerumb gantz guethertziges Und getrewen fleisses, E. L. Wollen sich vor Deren leuthe zeuberischen und betrieglichen hendeln Und vorblendung, sol fursehen." Johann Georg, Markgraf zu Brandenburg, to Julius, 4. February, 1574, NLA WO, 1 Alt 9, Nr. 307, fol. 90v.

64. Philipp Sömmering to Duke Julius, 5. March, 1574, NLA WO, 1 Alt 9, Nr. 307, fol. 212r, and again on fols. 210–213.

65. "Alles Mein thun und laßen ist dahin gerichet . . . das ich, . . . Je und bei efg Regiment gern wolt sei, Wie Joßef In Egypt, Gottes ehr und Reich zu ferderst helffen fordern in Kirchen und schulen, und den Efg Cammerguet nicht mit aufsetz und beschwerung der armen unterthanen, Sondern mit guten gewißen und rechtem Tittel zu vermehren helfen, und neben andern efg trewen Rethen darum mit zu sein, das efg mit allen Ihren angewerten Hern, freunden, Nachtbarn, lehe leuten, und Unterthanen in guetem fried, Ruhe und einigkeyt leben." Philipp Sömmering to Duke Julius, 5. March, 1574, NLA WO, 1 Alt 9, Nr. 307, fol. 211r.

66. Philipp Sömmering to Duke Julius, 5. March, 1574, NLA WO, 1 Alt 9, Nr. 307, fol. 212r, and again on fol. 213r.

67. "Dermals ich auch noch einmhal wil gebeten haben, Nicht ab. derhalben, als befahre ich nicht etwa, der hochgedachter Churf. zu Brandenburg etwas unehrlichs unchristlcihs, od zeuberisch uf mich wird darthue, od beweisen khonnen." Philipp Sömmering to Duke Julius, 5. March, 1574, NLA WO, 1 Alt 9, Nr. 307, fol. 212r, and again on fol. 213r.

68. Philipp Sömmering to Duke Julius, 5. March, 1574, NLA WO, 1 Alt 9, Nr. 307, fol. 212r, and again on fol. 212v.

69. "Das angezogen exempel von dem juden ist jhe schrecklich und aus solcher das verfluchten schelmen ubelthat solten alle Christen lernen, sich vor den feinden Christi Jesu vorzußehen. . . . Mugen derweg efg gleichwol auch an ditz exempel gedencken, und nicht allen frembden, wo sie het kommen, so beld und so viel vertrawen." Philipp Sömmering to Duke Julius, 5. March, 1574, NLA WO, 1 Alt 9, Nr. 307, fol. 212r, and again on fol. 212v.

70. Philipp Sömmering to Duke Julius, 5. March, 1574, NLA WO, 1 Alt 9, Nr. 307, fol. 212r, and again on fol. 212v.

71. Trinity Sunday is the first Sunday after Pentecost. "Darumb will ich unterthenig verhoffen, auch mich gentz l[crease] verlaßen, wo nicht E. f. g. mit auch uf Ostern verleuben wolten, das ich doch uf kueftig Trinitatis wils gott, meinen gnedigen abscheid von e. f.g . bekomen werde." Philipp Sömmering to Duke Julius, 5. March, 1574, NLA WO, 1 Alt 9, Nr. 307, fol. 212r, and again on fol. 213v.

72. Rhamm suggests that Julius went to Berlin for carnival, but gives conflicting dates for the 1574 Easter cycle; see Rhamm, *Die betrüglichen Goldmacher,* 89n84 and 89n87. In 1574, however, Easter would have fallen in mid-April.

73. Rhamm notes that Sylvester Schulvermann encountered one such satirical ballad in Leipzig in the fall of 1573 while he was en route to electoral Saxony. The Braunschweig ballad was recorded in a manuscript, "Pasquill auf Hzg JULIUS von BRAUNSCHWEIG und die betrügerischen Alchimisten an seinem Hofe, 1574," now housed in the Herzog August Bibliothek, Handschriften Cod. Guelf 44.6 Extravagantes, 15r-24v and published as "Ein neue Lied angefangen im 1574 Jahr," in Rhamm, *Die betrüglichen Goldmacher,* 111–23.

74. "Herzog Julio zu Wulfenbuttel / ein groß Lob hat vor dieser Zeit, / nun ist es Alles umgewandt, / sein Lob gebauet auf ein Sand." Rhamm, *Die betrüglichen Goldmacher,* 111.

75. "Seine Leute wären Heinz, Philipp, Anna, / Schalksnarr, Bube, Hur' Contobrina, / mit denen er sich verbunden hätt / zusammen mit geschworen ein' Eid. / Das Sacrament sie nehmen darauf, / einen Beichvater sie nehmen zu Hauf, /Land, Leut, Ehr, Gut, Leib, Leben und Blut / der Fürste bei ihnen wagen thut." Rhamm, *Die betrüglichen Goldmacher,* 115.

76. "Ich bin ein feiner Archemist, / Philosophus Theophrasist / ein feiner kühner Ebentheurer, / Schelmstücke sein mir nicht seltsam heuer. / Die Wahre Kunst der Alchemijen / ist Stehlen, Lügen und Trügen." Rhamm, *Die betrüglichen Goldmacher,* 113.

77. "Mit Zauberei er auch ging um, / der Teufel mit seinen Klauen krumm / ihn haben wollt zweimal vernimm / aus seinem Bett gar ungestrum." Rhamm, *Die betrüglichen Goldmacher,* 118.

78. Rhamm, *Die betrüglichen Goldmacher,* 115–17.

79. "Anna Maria wollte schwanger sein, / ein Kissen legte sie auf die Bein / und treibt es nun lange Zeit, / zuletzt verschwundt das Kriesem leit." Rhamm, *Die betrüglichen Goldmacher,* 120.

80. "Er sprach, Anna Maira zart / hätt gehabt einen Ritter gutter Art / in *philosohia filia* / fromm, jung, schöng, teusch, ohn' Alles ja./ . . . Sie geboren adelig, / im Mutterleib ganz säuberlich / gelegen allein 20 Wochen / darnach heraus an das Licht gebrochen. / Sie kann große Kunst in Hemd machen / mit Charaktieren zu allen Sachen / wie dem Herzog Julio / solch Hemd mach einen Planeto." Rhamm, *Die betrüglichen Goldmacher,* 118.

81. Rhamm claimed that it was incorrect to link Zieglerin's and her colleagues' case to witchcraft because it involved neither a pact nor sexual intercourse with the devil, both of which became increasingly central to witch hunts in the sixteenth century. In a narrow sense, of course, Rhamm was right; Zieglerin was accused of being a sorceress (*Zauberin*), not a witch (*Hexe*), and although she did confess that the devil led her astray, she never admitted to entering into an alliance with him. "Aber mit dem Hexenwesen ist dieser Proceß . . . mit Unrecht in Verbindung gesetzt"; Rhamm, *Hexenglaube und Hexenprocesse,* 74. On the pact with the devil, see Levack, *Witch-Hunt,* 37ff.

82. Hanegraaff, *Esotericism and the Academy.*

83. Kieckhefer, *Magic in the Middle Ages*; Zambelli, *White Magic, Black Magic*; Walker, *Spiritual and Demonic Magic.*

84. Some Catholics critical of alchemy, however, worried that when alchemists inevitably became frustrated at their failures, they might turn to demons for help, intentionally or even accidentally. Alchemical theory, however, did not depend on demons for assistance in manipulating nature. Baldwin, "Alchemy and the Society of Jesus"; Garber, "Transitioning from Transubstantiation to Transmutation."

85. On the witch hunts in the German lands, see Robisheaux, "German Witch Trials." On the witch hunts more generally, see Levack, *Witch Hunt.*

86. Mager, "Elisabeth von Brandenburg—Sidonie von Sachsen"; Lilienthal, *Die Fürstin und die Macht*; Kühn, *Eine "unverstorbene Witwe."* A number of nineteenth-century historians also wrote about Sidonie, including Havemann, "Sidonia, Herzogin zu Braunschweig-Lüneburg"; Weber, *Sidonie, Herzogin von Braunschweig*; Merkel, "Die Irrungen."

87. Merkel, "Die Irrungen," 29–30.

88. Lilienthal, *Die Fürstin und die Macht,* 232n712, citing Havemann, *Sidonie,* 283–303.

89. Merkel, "Die Irrungen," 30.

90. Anna von Rheden, widow of Simon von Rheden and mother of Curt von Rheden, appears in some sources as "die Simonsche." Katharina Dux, born von Dassel, widow of the

Großvogt vom Rübenberge Curt Warnickes, then married to the Oberamtmann von Wolfenbüttel and Hauptmann zu Calvörde Erich Dux, sometimes appears in sources as "Die Warnische," in reference to her first husband. Finally, Margarethe Knigge (sometimes "Die Kniggesche" or "Die Kniggsche") was born Schwarz, from Pattensen.

91. Merkel, "Die Irrungen," 34; Lilienthal, *Die Fürstin und die Macht,* 232ff.

92. Merkel, "Die Irrungen," 67ff.

93. Lilienthal, *Die Fürstin und die Macht,* 237–38.

94. Merckel, "Die Irrungen," 46, 51 on offering refuge.

95. Merkel, "Die Irrungen," 40–41.

96. Katharina Dux, for example, was married to Erich Dux, who had been in Julius's service since 1571. "Bestallungen und Suppliken der Brüder Erich, Wilhelm, Georg und Franz Dux," NLA HA, Cal. Br. 15, Nr. 1028.

97. "Dem Fürsten er wünschet alles Heil, / Daß er sein Herz nicht habe feil, / Die Buben gehören an Galgen doch / ade bis ich mir singe noch. "Ein neue Lied angefangen im 1574 Jahr," in Rhamm, *Die betrüglichen Goldmacher,* 122.

CHAPTER 7

1. According to Albert Rhamm, Philipp appears to have been in Julius's good graces still in April, but was arrested in early June. "Scheint ihm gleich ein ernstlicher Zweifel an der Treue seines Schützlings noch fern gelegen zu haben—denn bis in den April 1574 hinein hat er den Rath desselben bei manchen vertraulichen Sachen in Anspruch genommen." Rhamm, *Die betrüglichen Goldmacher,* 40.

2. "Was Philipp anfangs und zum Eingang in den gute berichtet und aufgesagt," n.d. [June 1574], NLA WO, 1 Alt 9, Nr. 311, fols. 14r-17v and "Gütliche Aussage des Philipp Sömmering," 9. [July] 1574, NLA WO, 1 Alt 9, Nr. 308, fol. 47r.

3. "Verhör und Aussagen von Balthasar Ferber, Franz Brun, Anna Wegener, Heinz Hessen, Bernd Hessen, Matz Rotermund, Heinrich Roßwurm, Sylvester Schulvermann," 15. June, 1574 to 12. January 1575, NLA WO, 1 Alt 9, Nr. 315.

4. Nummedal, *Alchemy and Authority.*

5. "Gütliche Aussage des Philipp Sömmering," 9. July, 1574, NLA WO, 1 Alt 9, Nr. 308, fols. 30r-74v.

6. Werner Schultheiß, "Ebner, Erasmus," in *Neue Deutsche Biographie* 4 (1959): 263f; Adolf Brecher, "Ebner, Erasmus," in *Allgemeine Deutsche Biographie* 5 (1877): 591f. https://www.deutsche-biographie.de/sfz12356.html.

7. For example, "Was Heinrich Schombach anfangs und zum Eingang in den gute bekant und ausgesagt," 5. July, 1574, NLA WO, 1 Alt 9, Nr. 313, fols. 3r-13v; "Verhör der Anna Maria Ziegler," 8. July, 1574, NLA WO, 1 Alt 9, Nr. 314, fols. 2r-17r; "Gütliche Aussage des Philipp Sömmering," fols. 30r-74v.

8. *Des . . . Caroli des V. vnd des Heiligen Römischen Reichs Peinlich Gerichts Ordnung.* See also Rhamm, *Die betrüglichen Goldmacher,* 89n90.

9. According to Rhamm, Mutzeltin pointed out that the *Hüttenwärter* Gregor Greif from Goslar had already lodged a complaint with the Kammergericht that the ducal Oberzehntner Christoph Sander "gegen abgedrungener Urphede bein Reichskammergericht zu Speyer." Rhamm, *Die betrüglichen Goldmacher,* 90n91.

10. Strauss, *Law, Resistance, and the State,* esp. chap. 5.

11. Heinemann, *Geschichte von Braunschweig und Hannover,* 2: 469.

12. On the social and professional roles of executioners in early modern Germany, see Harrington, *Faithful Executioner.*

13. For example, Julius consulted Elector Johann Georg about ordinances on forests/ wood, mills, and the church between 1571 and 1575. See "Briefwechsel des Herzogs Julius mit Kurfürst Johann Georg v. Brandenburg betr. die Mitteilung der Holz-, Forst- u. a. Ordnungen. Darin: Abschrift der brandenburg. Mühlenordnung v. 24. Jan. 1571 und der Holzordnung v. 4. Jan. 1571," NLA WO, 1 Alt 22, Nr. 98; and "Briefwechsel des Herzogs Julius mit Kurfürst Johann Georg v. Brandenburg betr. die Lieferung von 600 Exemplaren der brandenburg. Kirchenordnung, 1575," NLA WO, 1 Alt 22, Nr. 99.

14. "[wir] zweifeln nicht, E. L. werden von ihnen allerlei große Bubenstücke, damit sie die Herren und ander' Leute aufgesatzt [sic], erfahren. Wir schicken auch E. L. zu der Behuf auf derselben freundlichs Suchen unsern Schafrichter, der wird (damit sich E. L. an der Geistlichkeit nich vergreifen) dem Pfaffen wissen die Weihe abzunehmen und ihm und seiner gesellschaft, was hinter ihnen stecket, abzufragen. Wir bitten aber freundlich E. L. wollten denselben nit lange aufhalten, denn wir viel böse Buben sitzen haben, zu denen wir seiner auch bedürfen." Johann Georg von Brandenburg to Julius, 17. June, 1574, NLA WO, 1 Alt 9, Nr. 307, fol. 222r.

15. "Sonst begehren wir von des Pfaffen Goldkunst für uns nichts zu erfragen, denn uns bei Leben und Regierung unsers Herrn Vaters seliger Gedächnus solcher Abenteurer mehr vorkommen, die von Gold- und Silbermachen große Grumpen vorgeben und darauf nichts mehr, denn was im Rauch weggangen, haben zu Wege gebracht. Es ist aber gut, daß einsmals Einer derselben also gestraft wird, daß die Andern die Fürsten und Herrn mit solchen Künsten aufzusetzen und zu betrügen desto mehr Scheu haben müssen." Johann Georg von Brandenburg to Julius, 17. June, 1574, NLA WO, 1 Alt 9, Nr. 307, fols. 222r-v.

16. Duke Julius exchanged several letters with Elector Johann Georg between February and July 1574. See NLA WO, 1 Alt 9, Nr. 307, fols. 90r-119v and 120–122a, 145a. Julius asked Johann Georg for the Grumbach excerpts in a letter dated 28. June, 1574, fol. 120r.

17. "Also uns auch E. L. ersucht, E. L. die auf den allhier gerechtfertigen teuflichen Juden gestellten Fragstücke und Bekenntnus zukommen zu lassen, had unser Fiskal dieselben, weil er beschwerlich krank liegt, nich können aussuchen; E. L. haben aber hierbei die Rechtsfragen, darinnen seine Aussagen zusammengezogen und was darauf geurthelt worden." Kurfürst Johann Georg to Herzog Julius, 17. June, 1574, NLA WO, 1 Alt 9, Nr. 307; Rhamm, *Die betrüglichen Goldmacher,* 90n92.

18. "Auff E. L. Schreiben, haben wir Inn den Grumpachischen und andern domals gerechtfertigten urgichten, auffsuchen lassen, finden von Schilheintzen und den Goldtmachern nichts mehr, dan hiebei liegend ist. Ob nun solches E. L. Hertzogk Julio zuschicken wollen, stellen mich zu E. L. Frendtlichen gefallen." August of Saxony to Johann Georg, accompanied by a brief excerpt about Heinrich Schombach from Grumbach's trial, enclosed with a letter from Johann Georg of Brandenburg to Julius, arrived in Wolfenbüttel 22. July, 1574, NLA WO, 1 Alt 9, Nr. 307, fols. 117r-119v [quote from fol. 118r].

19. "Helt er gentzlich darfur, der Hertzogk sey betzaubert gewesen." "Aus Hansen Beyers Peinlichen Aussage, Folio 28," excerpt from the trials connected to Grumbach sent by Elector August of Saxony to Elector Johann Georg Brandenburg," fol. 199.

20. Julius to Johann Georg, draft copy, 23. July, 1574, NLA WO, 1 Alt 9, Nr. 307, fol. 145a.

21. This is, of course, a long-standing issue for historians who use trial records as evidence. See, for example, Davis, *Fiction in the Archives*.

22. "Philip habe mit den Mullen zuthun gehabt. es sei ein ding des Xennex und Theophrasti, das nheme man aus den mulle und henge es fur die Pestilenz an der Halß." "Anna Zieglerins Verhör," 8. July, 1574, NLA WO, 1 Alt 9, Nr. 314, fol. 10v.

23. Paracelsus, *Zwey Bücher Theophrasti Paracelsi . . . von der Pestilentz*, Book 2, Chapter (sig Fvii). See also Paracelsus, *De la peste*, 143–45, although it is unlikely that Anna Zieglerlin would have been familiar with this text, because it was in French. I am grateful to Peter Forshaw for his consultation on these texts.

24. Baldwin, "Toads and Plague."

25. "Die Mulle thu man in eine topf, lasse sie bl[eiben] 3. Wochen, gebet harten weinessig darauf, davon kome schleim, davon mache man küchlein und Henge an den Hals. Also habe sie zu Gota damit gethan. Die Mulle kriech auf der erden, und nehren sich, wie ander thiere. Die Scheferr Pflegen die Mulle zufolg. In der Fosen kriech die dinge auf dem felde, seien schwarz und gelb. Sie wisse nicht, wer Philippe die Worme bracht." "Anna Zieglerins Verhör," 8. July, 1574, NLA WO, 1 Alt 9, Nr. 314, fol. 10v.

26. "Schlangen schmaltz/ und auch das flaisch ein grosse Curation inn sich helt/nicht allein in frischen wunden/sonder auch inn allen vergifften bissen." Paracelsus, "Von vilerley gifftigen Thiern wie man jhnen das gifft nemen vnd tödten sol," in *Philippi Theophrasti Paracelsi von Hohenheim etliche Tractetlein zur Archidoxa gehörig*.

27. "Also wird auch von den Moltwurmen/unnd Salamander/wie vil haben die Alchimisten auff jn versucht/biß sie auß ihn erfunden haben gut Goldt/-unnd ein Tinctur zumachen/ auff die Metallen/Wiewol er inn Medicine nichts nutz ist/so ist er doch sonst hoch zu loben/ umb der nutzbarkeit wegen/so die Alchimisten auß ihm erfunden haben/wie gemelt." Paracelsus, "Von vilerley gifftigen Thiern wie man jhnen das gifft nemen vnd tödten sol." *Moltwurm* is another word for *Molche*, *Olm*, or *Salamander*. Or perhaps she knew of Paracelsus's essay on elementals, *Ex libro de nymphis, sylvanis, pygmaeis et salamandris*.

28. Paracelsus, *Sieben defensiones*, in Sigerist et al., *Four Treatises*.

29. Serpell, "Guardian Spirits or Demonic Pets," 171. Paracelsus also associated fire elementals, "vulcans," with witches, the devil, and danger. Paracelsus, "A Book on Nymphs, Sylphs, Pygmies, etc.," in Sigerist et al., *Four Treatises*, 240.

30. "VON GIFFT. Ehr habe auch nit gewist, wie Philip durch die Forst[kechter?] Mulchen fangen, anhero bringen lassen, und die gifft davon bereiten, damit hetten sie furgehabt, der Herzoginnen alhero, Ihnen der Marggrafin zu Brandenburg auf Custrin hochloblicher gedechtnus und der Herzogin von Musterberg alß Principalien, so sich gegen sie gesetzt bej dem Herzogen alhier, auch dem Furstlichen Rhaten, welche Ihnen wieder[?] gewesen, zuvergeben, und wen der Herzog Ihnen nicht mehr Ihres willens mit sitzten, sonder ungnadig werden wollen, were es auch wol letztlich an s.f.g. selbst kommen." "Gütliche Aussage und peinliches Verhör des Heinrich Schombach," 5. July, 1574, NLA WO, 1 Alt 9, Nr. 313, fols. 11v-12r.

31. "Philip habe mit den Mullen zuthun gehabt. es sei ein ding des Xennex und Theophrasti, das nheme man aus den mulle und henge es fur die Pestilenz an der Halß. Die Mulle thu man in eine topf, lasse sie bl[eiben] 3. Wochen, gebet harten weinessig darauf, davon kome schleim, davon mache man küchlein und Henge an den Hals. Also habe sie zu Gota damit gethan. Die Mulle kriech auf der erden, und nehren sich, wie ander thiere. Die Scheferr Pflegen

die Mulle zufolg. In der Fosen kriech die dinge auf dem felde, seien schwarz und gelb. Sie wisse nicht, wer Philippe die Worme bracht, Sie habs Philippe ge[lernet], und dem den Proceß gewiesen. Sie habs Philippe heissen [haussen?] in den ofen setzen, Nun macht sie es in der erdt gestanden." "Anna Zieglerins Verhör," 8. July 1574, NLA WO, 1 Alt 9, Nr. 314, fol. 10v.

32. "Sage sie will alles bekennen, wie sie es fur dem Richterstule Christi gestehen will, und bitt sie zu verschonen, und nicht an Ihr schuldig zu werden. Da man sie nicht wil bej der warheit bleiben lassen, so wil sie die Rhäte alle fur dem Richterstuhl Gottes anclagen." "Anna Zieglerins Verhör," 13. July, 1574, NLA WO, 1 Alt 9, Nr. 314, fol. 18r.

33. "Von dem Mullen wisset sie kein gifft zumachen." "Anna Zieglerins Verhör," 13. July, 1574, NLA WO, 1 Alt 9, Nr. 314, fol. 19r.

34. "zwischen solcher Pein und dem Todt kein Unterschied, möchten sie wohl verschont sein, wenn sie sich selbst helfen möchte." "Anna Zieglerins Verhör," 14. July, 1574, NLA WO, 1 Alt 9, Nr. 314, fol. 19v.

35. "daß thu man auch in des Teuffels nhamen und des so beschedigen werde soll. Es sei auf die Herzoginnen gedacht, aber nich in weg gesetzt." "Anna Zieglerins Verhör," 14. July, 1574, NLA WO, 1 Alt 9, Nr. 314, fol. 20v.

36. "in aller Teuffel nhamen, daß all so daruber gingen, muchten verkomment und [v]erlahmen." "Anna Zieglerins Verhör," 14. July, 1574, NLA WO, 1 Alt 9, Nr. 314, fols. 20v–21r.

37. "Wenn sie in d. Kelch gewesen, sei sie Krank worden." "Anna Zieglerins Verhör," 14. July, 1574, NLA WO, 1 Alt 9, Nr. 314, fol. 22r.

38. For this section of the interrogation, see "Kroten gift" and "Guß," "Anna Zieglerins Verhör," 14. July, 1574, NLA WO, 1 Alt 9, Nr. 314, fols. 20r–22r.

39. "Guß," "Anna Zieglerins Verhör," 14. July, 1574, NLA WO, 1 Alt 9, Nr. 314, fols. 24v–25r.

40. "Mulchen Pulver habe sie von Philippen holn lassen. Mulchen Pulver musse man mit essig außziehen wen mas an den Halß hengen will." "Guß," "Anna Zieglerins Verhör," 14. July, 1574, NLA WO, 1 Alt 9, Nr. 314, fols. 22r and 24v–25r.

41. "Philippen fraw," "Philtrum Illimo. Essen kochen," and "Praeparatio Philtrij," "Anna Zieglerins Verhör," 14. July, 1574, NLA WO, 1 Alt 9, Nr. 314, fols. 22r and 24v–25r.

42. "Das Xenex von Theophrasti habe sie daraußzihen wollen. Xenextum heisse vergifft. Alles das man [benenne?], darin sei keine gift. Ale gifftige thiren lasse man faul werden, und mache safft davon und dann damit konne man wol boses aufrichten." "Die Mulchen," "Anna Zieglerins Verhör," 14. July, 1574, NLA WO, 1 Alt 9, Nr. 314, fol. 25v.

43. "Sie habe mit den Mulchen nie was gethan, allein das sie das Xenextum heraußgezogen. Wan man gifft daraus machen wolte, musse man sie putrifieren und in die erde setzen. Sie habe in Philippen Laboratorio Mulchen stehen sehen, und den Laboraten gefragt, was die da machen, da habe der geantworten, es [man]gelte Ihme an fewer, der Laborant were noch nicht fertig. Darumb bleiben die so lange bestehen. Sonst wolte ehr sie im ofen be[weren?], und hett sie darnach in die [Acker?] geschutten. Mit der Putrefaction helte sie man konte allen Menschen damit vergeben. Sie wisse aber keinen Proceß davon." "Die Mulchen" and "Mulle," "Anna Zieglerins Verhör," 14. July, 1574, NLA WO, 1 Alt 9, Nr. 314, fols. 25v–26v.

44. "Luppoldt Judenn Uhrgicht, 1573," as quoted in Ackermann, *Münzmeister Lippold*, 93–98. Lippold's story clearly held interest in nineteenth-century Germany.

45. "sey allerley darinnen geschieben Guths und böses." "Luppoldt Judenn Uhrgicht, 1573," in Ackermann, *Münzmeister Lippold*, 93.

46. "Das Buch, darin Ihre Kunste, den habe Philip nachgetrachtet, do habe sie es in d. Kuchen zu Goschlar verbrant in beisein etzlich Weiber." "Anna Zieglerins Verhör," 16. July, 1574, NLA WO, 1 Alt 9, Nr. 314, fol. 30r.

47. "Luppoldt Judenn Uhrgicht, 1573," in Ackermann, *Münzmeister Lippold*, 97.

48. Jütte, *Strait Gate*, 153–63; Bächtold-Stäubli et al., *Handwörterbuch des deutschen Aberglaubens*, s.v. "Tür"; Rhamm, *Hexen*, 14.

49. "Luppoldt Judenn Uhrgicht, 1573," in Ackermann, *Münzmeister Lippold*, 94.

50. "Guß", "Anna Zieglerins Verhör," 14. July, 1574, NLA WO, 1 Alt 9, Nr. 314, fols. 23–24.

51. "Philtrum Illimo. Essen Kochen" and "Praeparatio philtrij," "Anna Zieglerins Verhör," 14. July, NLA WO, 1 Alt 9, Nr. 314, fols. 24–25.

52. "Die Zarn [?] Dietriche haben die Schlosser zu Schmalcalden Ihrem manne geschencket. Sie habe Philippe ein mal aus seiner laden 100. Thaler gestolen, die habe Ihr d. Schlosser mussen aufmachen, Ihr man habe Ihms wissens Philippen auch 30. Thal gestolen Ihr man habe die [schlussel] allen in seiner laden gehabt Sein darzu gemacht, dß sie Keller und anders in husen damit schliessen wollen." "Anna Zieglerins Verhör," 8. July, 1574, NLA WO, 1 Alt 9, Nr. 314, fol.10r.

53. Jütte, *Strait Gate*, 153–63.

54. That is, *Piper longum*, or Indian long pepper (Pipli), the fruit of which is similar to black pepper.

55. "sagt weil er Müntzer gewesen, und mit Silber umbgegangen, hette er den aus der altenn Apoteckenn, unter dem Schein, als wollte er erzlich goldt damit schmidig machen erlangett." "Luppoldt Judenn Uhrgicht, 1573," in Ackermann, *Münzmeister Lippold*, 97.

56. On toad poison, see Bächtold-Stäubli et al., *Handwörterbuch des deutschen Aberglaubens*, s.v. "Kröte." Toads also had a role in the alchemical tradition (most notably in the spectacular Ripley Scrolls), but it does not appear that Julius's Hofgericht was aware of this connection.

57. On Dr. Kommer, see Chapter 3, above.

58. Johann Georg von Brandenburg to Julius, 17. June, 1574, NLA WO, 1 Alt 9, Nr. 307, fols. 222–24.

59. "Zettel an Wittenberg und Magdeburg, 18. December 1574," NLA WO, 1 Alt 9, Nr. 325, fol. 11. Rhamm notes that "the Urtheile selbst fehlen" (Rhamm, *Die betrüglichen Goldmacher*, 100n118). All that Rhamm could find (and all that remains today) is a summary report in Wolf Ebert's hand of the punishments that the Brandenburger Schöffen recommended for Sömmering and the others: "Aufzug oder Inbegriff der Braundenburgischen Rechts-Urtheile, mit was für Strafen die zu Wolfenbüttel gefangenen bösen *Laboranten* Herzogs Julii wegen ihrer vielen Mißhandtlungen zu belegen, so wol was ein jede ihrer Missethaten in specie verdienet habe als auch, wie ein jeder wegen aller seiner Verbrechen, zusamen genommen, zu bestrafen sey. Die Verurtheilte sind folgende: etc." NLA WO, 1 Alt 9, Nr. 311, fols. 23–35; also summarized in Rhamm, as above.

60. Adalbert Erler, Ekkehard Kaufmann, and Wolfgang Stammler, *Handwörterbuch zur deutschen Rechtsgeschichte* (Berlin: E. Schmidt, 1971–1998), s.v. "Schöffenstuhl."

61. "Der Schöppen zu Brandenburg und Magdeburg Urtheilen gegen Philipp Suommering. 1575 Orig. Etc.," NLA WO, 1 Alt 9, Nr. 312, fols. 36–64. Note that this is the *Schöffensprüchen* from Magdeburg and Brandenburg regarding Sömmering only, however. Rhamm transcribed them, noting, "Von den Schöffensprüchen haben sich nur diejenigen erhalten, welche Philipp Sömmering betreffen" (they had been in private hands and only recently came

back into the Akten, Fasc. VIII). Rhamm also writes that he contacted archives in Magdeburg and Brandenburg about these documents, who said that the legal decisions (*Rechtssprüche*) regarding Sömmering and his companions were no longer extant. Rhamm, *Die betrüglichen Goldmacher,* 92n109.

62. Rhamm, *Die betrüglichen Goldmacher,* 92n109.

63. NLA WO, 1 Alt 9, Nr. 311, fols. 23r-35v; Nr. 312, fols. 33–64; and Nr. 325.

64. The original *Schöffensprüche* regarding Zieglerin are missing, for example; only Ebert's summary of the Brandenburg recommendations is extant. The records are even sometimes a bit contradictory. Julius's court sent to the *Schöffenstühle* a list of confessions to eleven crimes, but on the Day of Justice, Anna confirmed her confessions to only one.

65. "Auszug der brandenburgischen Schöppensprüche gegen Ph. Sömmering, Heinrich Schombach, Anna Maria Ziegler, Jobst Kettwig, Sylvester Schulvermann, Dr. Georg Kommer, Bernd Hueffner [Hübener], Hans Hoyer," n.d. [late 1574/early 1575], NLA WO, 1 Alt 9, Nr. 311, fols. 23r-35v. See also Rhamm's transcription, *Die betrüglichen Goldmacher,* 100n118, although beware that it is not exact.

66. "Wissentlicher Betrug der Alchemie halben und daß sie bekennt, daß Ihr auch Philippen Proceß unbeständig, und daß ihr auch gleich gewiß, daß sie doch Gottes Segen in ihrem gottlosen Leben dabei nit hoffen könne soll." The recommended punishment was "Staupen [flogging] und Landesverweisung [exile]." "Auszug der brandenburgischen Schöppensprüche gegen Ph. Sömmering, Heinrich Schombach, Anna Maria Ziegler, Jobst Kettwig, Sylvester Schulvermann, Dr. Georg Kommer, Bernd Hueffner [Hübener], Hans Hoyer," n.d. [late 1574/early 1575], NLA WO, 1 Alt 9, Nr. 311, fols. 24ff, summarized in Rhamm, *Die betrüglichen Goldmacher,* 100n118.

67. Mutter Eyle/Eile Nöding is a shadowy figure in the sources surrounding Anna Zieglerin. Anna most likely spoke about her verbally, which explains why little trace of her in writing remains. She was supposedly an expert in "cabala" and medicine: "Mutter Eile kenne die Cabala, die habe sie aus Ihre Mutter Hause brachte . . . Mutter Eile konne auch arzneij." "Verhör der Anna Maria Ziegler," 8. July, 1574, NLA WO, 1 Alt 9, Nr. 314, fol. 11r. See also Duke Julius's inquiry with Duke Johann Wilhelm von Sachsen about this "Wundertäterin Edyle Noding, genannt 'Mutter Eyle,'" sometime in 1571. NLA WO, 1 Alt 9, Nr. 306, fols. 17–19.

68. Heinrich Schombach confessed that the alchemists convinced Julius to pay 3,000 gulden for a copy of this valuable book, *Testamentum Hermetis,* which Philipp had recommended that the duke acquire. Instead, they paid Ludwig Hahne, Julius's court pastor, a lower price for the book, and split the profits among them. Rhamm, *Die betrüglichen Goldmacher,* 24–25. Anna Zieglerin and Ludwig Hahne were questioned about this episode in a confrontation on 27. and 30. July, 1574, NLA WO, 1 Alt 9, Nr. 307, fols. 37–66.

69. On the murder of the messenger (Lackei), see "Verhör der Anna Maria Ziegler," 8. July, 1574–26. January, 1574, NLA WO, 1 Alt 9, Nr. 314, fols. 11v-12r, 18r, 25v, 35v, 40r, 42r.

70. On the attempt to poison Philipp's wife, see "Verhör der Anna Maria Ziegler," 14. July, 1574, NLA WO, 1 Alt 9, Nr. 314, fol. 22r (testimony taken directly after a torture session). On Anna's involvement in poisoning Dussel and Finning, which Anna attributed to Philipp's wife, see her testimony on 14. July, 1574 "Verhör der Anna Maria Ziegler," NLA WO, 1 Alt 9, Nr. 314, fol. 23r: "Sie habe keinen gift beibracht allein Philipps frawen." Two days later (July 16), Anna confessed that Philipp gave her salamander powder, which she put into Dussel's eggs: "Philip habe Ihr Mulch Pulver gebe, das habe sie Dussel zu [ange]bott [treulich] in Eiern eingegeben." "Verhör der Anna Maria Ziegler," 14. July, 1574, NLA WO, 1 Alt 9, Nr. 314, fols.

28v-29r: In his summary report on this point, Ebert wrote Sömmering, not Finning, but the fact that Zieglerin was never accused of poisoning Philipp, as well as the frequent pairing of Dussel and Finning in other sources, suggests that this was a mistake.

71. "Anna Marien mishandlungen zusammen sollen vor dem endtlichen toden ein Umführung an die Gerichtsstätte, 6 Zangengriffe & Feuer." "Auszug der brandenburgischen Schöppensprüche gegen Ph. Sömmering, Heinrich Schombach, Anna Maria Ziegler, Jobst Kettwig, Sylvester Schulvermann, Dr. Georg Kommer, Bernd Hueffner, Hans Hoyer," NLA WO, 1 Alt 9, Nr. 311, fol. 25v.

72. On the standard form of the "Endlicher Rechtstag," as outlined in the *Constitutio Criminalis Carolina*, see Ignor, *Geschichte des Strafprozesses*, 61. See also Dezza, *Geschichte des Strafprozessrechts*; and Schroeder, *Die Peinliche Gerichtsordnung Kaiser Karls V.*, Art. 78, "Endtlicher Rechtstag."

73. Rhamm made the most of this episode; indeed, he embellished it. Rhamm added colorful details that do not appear in the transcript of the *Rechtstag*, including, for example, the description of the condemned being seated "auf den mit rothem Tuch überzogenen Blutstuhl, löst ihnen die Bande und halt ihnen die Klagepunkte nochmals summarisch vor" or that "Sömmering und die Zieglerin, von den Nachwehen der Folter und dem langwierigen Gefängniß entkräftet, auf einer Bahre herausgetragen [wurden]." Rhamm, *Die betrüglichen Goldmacher*, 55ff.

74. For the formal invitation to attend the *Rechtstag*, see NLA WO 1, Alt 9, Nr. 325, fols. 1ff.; for a transcript of the proceedings, see fols. 34r-55v. For a complete list of attendees, see fol. 34r; and Rhamm, *Die betrüglichen Goldmacher*, 98–99n113.

75. Possibly, Rhamm suggested, because Julius could not serve as both plaintiff and judge at once. Rhamm, *Die betrüglichen Goldmacher*, 55.

76. Unfortunately, the transcript states only that the recommendations were read out, and does not quote the lost documents themselves.

77. "ungerne vernehme das der teuffel die armen Sund alß verfuhret" and "den Geistlichen Standt in solchen hohen richtig bludtsachen zuerschonen, Sie khonnen aber wol erachten, das solche ubel that ohne straffe nit khonne hingehen, und beten Sie entschuldigt zunemmen." NLA WO, 1 Alt 9, Nr. 325, fols. 35v-36r.

78. NLA WO, 1 Alt 9, Nr. 325, fols. 36r-v.

79. "4. Feb. [1575] In dem Lusthause in beisein hirneben gesetzter Landtstande," NLA WO, 1 Alt 9, Nr. 325, fols. 37r-4rv.

80. "und sagt noch jtzo vor diesen Hern und Junckern das er diese bekhandtnus vor der Welt bestehen khan und wirdet d. Rath zu Braunschweig dargeg nichts erheblichs einbring., will darauf sterben." "4. Feb. [1575] In dem Lusthause in beisein hirneben gesetzter Landtstande," NLA WO, 1 Alt 9, Nr. 325, fol. 53r.

81. "Ehr bekhennet und gestehet es noch das ese mit der gifft so er Peter Dussel und Vinning gegeben geschen sey." "In dem Lusthause in beisein hirneben gesetzter Landtstande," NLA WO, 1 Alt 9, Nr. 325, fol. 47v.

82. "Nun bete er undthenig und dienstlich die Hern wolen seinen unfall darzu er durch die gifftig schlangen und zeuberej komen, er weg und umb Gotts willen vor Ihne geg Illm bitten, das sfg Ine nit nach verdiest sond gnaden straffen und das leben schencken und mit Ime dispensiren, well er mit sfg richtiger sachen helber, die er nit notig zu repetiren halte, geredet, und nach dem der liebe Godt Ine begrabet und sovil erfaren, das er den Lapide Philosophicum [*sic*] zurichten khonne Wie er bej Johans Friderich auch Ins werckh richten wollen,

und bittet nochmahls wo es unmuglich, Ine das leben zuschencken, und muste er sagen wogel-
ert leut als Adam von Bodenstein und andern Theophrastischen die uff Ine bezeugen sollen,
das er etwas zu wegen bring khan, und will s. F. G. nit eine 2 od 3 thunne golt sond etzliche
Million goles zu wegen bringen und soll das landt aller steur und schatzungen frej werden.
Hette auch noch einen bericht von einen Bergwerckh gethan davon sfg auch einen dobbelten
nutz bekomen und erlangen khonten, und stehet seine hofnunge, sie werden vor Ihne bitten.
Illij Hen: Julius Er solten vor Ihm kheinen fueßfall thun, er solte umb vergebung seiner Sunde
bitten. *Philip*: Sagt er bethe gnade und khein recht." "4. Feb. [1575]. In dem Lusthause in
beisein hirneben gesetzter Landtstande," NLA WO, 1 Alt 9, Nr. 325, fols. 51v-52r.

    83. This heart-wrenching detail is documented in a petition from Kloster Marienthal to
Duke Julius, asking that Philipp's children (all but one of whom appear to have been living in
the cloister that fall, most likely in the wake of their mother's death that summer) not be held
financially responsible for their father's crimes. "Auf dem General-Consistorio zu Marienthal
versandete Consistoriale bitten für des gefange sitzenden Philipp Therocycli nunmundige
Kinder bey Herzog Julio, dieselber ihres Vaters Missethat nicht entgelten zu lassen, sondern
ihren an einen Orte notdürftig Unterhalt zu geben." 12. October, 1574, NLA WO, 1 Alt 9, Nr.
322, fols. 10–12.

    84. "Sie bittet Lauter umb Gots willen, man wolle Ihr alles vergeben was Sie gegen Gott
und die welt gesundiget hette, und bittet einer vorbitt gegen Illm und die Hertzogin." "In dem
Lusthause in beisein hirneben gesetzter Landtstande," 4. February [1575], NLA WO, 1 Alt 9,
Nr. 325, fol. 53v-54r.

    85. "Verstehen wir keiner anderen Meinung geschehen, als daß wir uns bald von Kind
auf sollen lernen erinnern justitiam, d. i. Recht und Gerechtigkeit, in allen Handeln zu üben,
die Frommen zu schützen und dem Bösen zu wehren, auf daß also Gottesfurcht, ehrbar christ-
lich Leben, gutter Handel und Wandel im Land gepflanzt, hegt und gehandhabt werden
möchte, wie solches die heilige Schrift aller christlichen Obrigkeit zu thun gebeut." "In dem
Lusthause in beisein hirneben gesetzter Landtstande," 4. February [1575], NLA WO, 1 Alt 9,
Nr. 325, fols. 72ff. Also Rhamm, *Die betrüglichen Goldmacher*, 99n115.

    86. Arnold, bei der Wieden, and Gleixner, *Herzog Heinrich Julius zu Braunschweig und
Lüneburg*; Brüdermann, "Heinrich Julius, Herzog zu Braunschweig-Lüneburg," 324ff.; Albrecht
Eckhardt, "Heinrich Julius," in *Neue Deutsche Biographie* 8 (1969): 352–54; Ferdinand Spehr,
"Heinrich Julius, Herzog von Braunschweig-Wolfenbüttel," in *Allgemeine Deutsche Biographie*
11 (1880): 500–505. On the witch hunts under Duke Heinrich Julius, see Soldan, *Geschichte der
Hexenprozesse* 2: 88–89.

    87. Greenblatt, *Renaissance Self-Fashioning*.

    88. Dülmen, *Theatre of Horror*; Schuster, *Verbrecher, Opfer, Heilige*.

CONCLUSION

    1. Note that this account does not mention the execution of Bernd Hübener, who alleg-
edly was planning to help them escape, or the Braunschweig bailiff Hans Hoyer, both of whom
were also condemned in the "Auszug der brandenburgischen Schöppensprüche gegen Ph. Sömmering, Heinrich Schombach, Anna Maria Ziegler, Jobst Kettwig, Sylvester Schulvermann, Dr.
Georg Kommer, Bernd Hueffner [Hübner], Hans Hoyer," n.d. [late 1574/early 1575], NLA
WO, 1 Alt 9, Nr. 311, fols. 23r-35v. "Anno 1575 hat H. Julius am 7. Tage des Hornungs/zu
Wolffenbüttel nachfolgende Personen (die jm und seinem hertzlieben Gemahl/nach Leib und

Leben gestanden/mit der Alchimisterey grossen betrug getrieben/etlichen Leuten mit Gifft vergeben/und auch sonsten viel böser thaten begangen) gebürlicher weise hinrichten und tödten lassen. Der schiele Heintze ward entheuptet/und darnach geviertheilt. Sein Weib Anna Maria/ward also ein Zeuderin [sic] verbrand. D. Kummer ward entheuptet/Magister Philips mit heissen Zangen 5. Mal angriffen/und darnach geuiertheilt/Sylvester Schulver und Ketwich/ wurden geredert/geuiertheiltund die stücke an Kniegalgen auffgehangen." Bünting, *Braunschweigische und Lunebürgische Chronica*, 1: 150. Bünting was also the author of *Itinerarium sacrae scripturae* (1581), which included several well-known maps, including a clover-leaf map of the world with Jerusalem at the center and a map of Europe in the shape of a virgin.

2. Grunow, *Die Spur führt nach Wolfenbüttel*; Lehrmann, *Goldmacher, Gelehrte und Ganoven*; Hodemacher, *Der Fall Sömmering*. Zieglerin also appears in two of Susanne Gantert's historical novels: *Fürstenlied* and *Das Mädchenreigen*.

3. See Paul Zimmermann, "Rehtmeyer, Philipp Julius," in *Allgemeine Deutsche Biographie* 27 (1888): 604–6, https://www.deutsche-biographie.de/gnd120539101.html#adbcontent; Rehtmeyer, *Braunschweig-Lüneburgische Chronica*, 2: 1015–17.

4. "Anna Maria ward verbrant; . . . also nahm diese Rotte ein Ende. Es hatten diese böse Leutenicht allein grossen Betrug mit der Alchymisterey getrieben, sondern auch ausser obige That, da sie der Herzogin nach Leib und Leben gestanden, auch der Stadt Braunschweig Secretarium, Jacob Tonning genant, mit Gift vergeben, und sonst viel böses begangen." This passage also included a footnote to "Algermann in vita Duc. Julii." Rehtmeyer, *Braunschweig-Lüneburgische Chronica*, 2: 1016–17. Note that Julius has dropped out as a target, leaving only Hedwig in danger "life and limb."

5. On this genre, see Dixon, "Sense of the Past."

6. On Beckmann (1739–1811), see Carl Graf von Klinckowstroem, "Beckmann, Johann," in *Neue Deutsche Biographie* 1 (1953): 727–28, https://www.deutsche-biographie.de/gnd 118654624.html#ndbcontent.

7. "Wie sehr sich Herzog Julius, der sonst große Verdienste um sein Land hatte, sich von Godmachern täuschen ließ, beweiset die Geschichte bey Rehtmeier S. 1016. worüber ich vom H. Obercommiss. Ribbentrop einen alten geschriebenen Bericht erhalten habe, den man nicht ohne Erstaunen lessen kan. Noch zeigt man am Schlosse zu Wolfenbüttel den eisernen Stuhl, worauf den 5. Fbr. 1575. Die Betriegerinn Anna Maria Zieglerinn genant Schlüter Ilsche, verbrant worden." Beckmann, *Beyträge zur Geschichte der Erfindungen*, 3: 404.

8. Ribbentrop, *Beschreibung der Stadt Braunschweig*; *Sammlung der Landtagsabschiede, Fürstlichen Reversalen und anderer Urkunden*; and *Vollständige Geschichte und Beschreibung der Stadt Braunschweig*. On Ribbentrop, see Ribbentrop, Philipp Christian, Indexeintrag: *Deutsche Biographie*, https://www.deutsche-biographie.de/gnd116503653.html.

9. *Schlüter = Schliesser*, or a person responsible for locking up.

10. Algermann, "Leben, Wander und tödtlicher Abgang des Herrn Juliussen."

11. "Unter diesen Umständen gehörte eine genaue Vergleichung aller zehn Handschriften dazu, um einen Text, wie der hier mitgetheilte ist, herzustellen." Algermann, "Leben, Wander und tödtlicher Abgang des Herrn Juliussen," preface.

12. Which he had already discussed at great length. See Chapter 2.

13. Algermann, "Leben, Wander und tödtlicher Abgang des Herrn Juliussen," 32.

14. Ibid., 35.

15. "Anne Marie Ziegler, des Schumpacht (oder, wie Andere Schreiben, Schombach) Weib, war eine äußerst listige Betrügerinn und ein Beweiß, daß die Laster und Thorheiten der Menschen stets dieselben bleiben. Sie gab vor, engelrein zu seyn und nur mit der Mutter Gottes

verglichen werden zu können. Ein Mann, der ihrer Liebe genosse, nahme Theil an dieser Reinheit, und wurde hundert Jahr alter als ein gewöhnlicher Mensch. Sie sei mit einem Grafen von Oettingen vermahlt, der ein Sohn des Theophrastus Paracelsus sey, und sie ware bestimmt, mit ihrem Gemahl Kinder zu zeugen, aus denen eine neue Welt der Unschuld und Reinheit hervorgehen werde. Kurz, ein Cagliostro hätte night großprahlender reden können, als sie, und sie wurde geglaubt, so gut als dieser, bis sie ihren Tod in den Flammen fand." Friedrich Karl von Strombeck, footnote, in Algermann, "Leben, Wander und tödtlicher Abgang des Herrn Juliussen," 33.

16. Citing Johannes Beckmann, for example, Johann Friedrich Gmelin noted in his own 1797 *History of Chemistry* that Duke Julius executed the *"Betrügerin* Anna Maria Zieglerin, known as Schlüter Ilsche, in an iron chair": "doch wurde auch er [Herzog Julius] so ergrimmt, daß er 1575 eine solche Betrügerin Anna Maria Zieglerin, genannt Schlüter Ilsche, in einem eisernen Stuhle verbrennen lies"; Gmelin, *Geschichte der Chemie* 1: 260–61. Nearly a century later, Hermann Kopp did the same. His 1886 *History of Alchemy* included a section on the Wolfenbüttel alchemists, adding that "Schombach's wife, Anne Marie Zieglerin, [was] pinched with tongs and burned to death in an iron chair." (Schombach's Frau, die Anne Marie Zieglerin mit Zangen gezwickt und in einem eisern Stuhle verbrannt [wurde].) Kopp, *Die Alchemie in älterer und neuerer Zeit,* 1: 173. Kopp claimed that one could see it "until a few decades ago" hanging in the vault of Wolfenbüttel Palace. He dismissed Schlüter-Ilsche, however, as folk legend: "Under the name Schlüter-Liese oder Schlüter-Ilsche (the meaning of neither of these names has been explained with any certainty) the alchemist who ended this way lived on in the mouths of the folk as a witch and poisoner." (Unter dem Namen der Schlüter-Liese oder Schlüter-Ilsche [die Bedeutung keines dieser Namen ist mit einiger Wahrschinlichkeit erklärt] hat die Alchemistin, die in solcher Weise geendet, also Hexe und Giftmischerin in dem Munde des Volkes fortgelebt.) Kopp, *Die Alchemie in älterer und neuerer Zeit,* 1: 173.

17. Lentz, *Bücher der Geschichten der Lande Braunschweig und Hannover,* 148; Görges, *Vaterländische Geschichten,* 297ff.; Vehse, *Geschichte der deutschen Höfe seit der Reformation,* 144.

18. Lentz, *Bücher der Geschichten der Lande Braunschweig und Hannover,* 148; Vehse, *Geschichte der deutschen Höfe seit der Reformation,* 144.

19. Görges, *Vaterländische Geschichten,* 298.

20. Vehse corrected an error in a previous volume with a note that quoted from a letter from local historian Karl Heinrich August Steinmann, which mentioned (among other things) *"der Sage nach noch jetzt im alten Wolfenbüttler Schlosse spoken gehende famose 'Schlüter-Liese'—'Anna marie Schulfermanns'."* Vehse, *Geschichte der deutschen Höfe seit der Reformation,* 30er Band, Fünfte Abtheilung: Sachsen, Dritter Theil (Hamburg: Hoffmann und Campe, 1854), xi-xii. See also Lentz, *Bücher der Geschichten der Lande Braunschweig und Hannover;* Görges, *Vaterländische Geschichten.*

21. The identity of "Fr. Brandes" is unclear, and it seems likely that he was related to the Braunschweig landscape painter Heinrich Georg Brandes (1803–1868).

22. There was one prominent exception to this pattern: the Hungarian rebel and "peasant king" György Dózsa, who was executed in 1514 while seated in a red-hot iron chair with a metal crown on his head. Dózsa's execution was commemorated in the early sixteenth century in two printed images: *Die auffrur so geschehen ist im Vngerlandt,* 4; and Taurinus, *Stauromachia.* Dózsa had an interesting afterlife, too, as, among other things, the subject of an eponymous 1867 opera by Ferenc Erkel.

23. "lange, gespenstische Gestalt"; "eine weißgekleidete, tief verschleierte Frau"; "So sehr Herzog Julius auch durch die Wissenschaft und durch angeborene Gerzhaftigkeit mit Muth

versehen war—er bebte dennoch unwillkürlich, als er diese bleiche, zwar schöne aber unheimlich Gesicht erblickte Die Züge der Frau waren wie aus Marmor gehauen, keine Muskel zuckte und nur die Augen drehten sich wie schwarze Ruader in den von dichten Brauen überschatteten Höhlen." Hiltl, "Der Teufelsdoktor von Wolfenbüttel," 7, 35, and 39.

24. "'Das war ein ganz böses Geschuaftchen!' rief die bleiche Frau, ihre schleier abwerfend und mit einem in Wasser getauchten Schwamm die Farbe von ihrem Gesichte entfernend, welche ihr jene Geisterblässe verliehen hatte." Hiltl, "Der Teufelsdoktor von Wolfenbüttel," 60.

25. "Der Herzog war . . . ein Kind seiner Zeit. Die seltsame Idee vom Stein der Weisen erfüllte damals die Köpfe. Das Verlangen [for everlasting youth and riches] war wie eine Krankheit über Viele gekommen, mit der die düstere Manie der Hexerei Hand in Hand ging." Hiltl, "Der Teufelsdoktor von Wolfenbüttel," 71.

26. "Für die Schulvermann ward eine besonders raffinirte und schreckliche Strafe von den grausamen Richtern ersonnen, deren Spruch der Herzog nicht ändern durfte. Sie wurde auf einen eisernen Stuhl gesetzt gefesselt und dann, also sitzend, lebendig verbrannt. Sämmtliche Stücke diese Stuhles sind heute noch auf dem Schlosse zu sehen." Hiltl, "Der Teufelsdoktor von Wolfenbüttel," 126. Confusingly, Hiltl links this execution to "die Schilvermann" rather than Zieglerin, demonstrating that even in 1873, when he published his novella, confusion about Anna Zieglerin remained. Zieglerin's only role in the novella is to prove that Philipp's elixir is effective; on the stormy night when the group attacked Hedwig to steal the ducal treasure, she is present, but in no way as a leader of the activity. Hiltl attributes a significant plot point, however—unlocking the doors to Hedwig's chambers—to another female character, "die Schulvermann, die Schliesserin," repeating the confusion about Zieglerin/Schulvermann that first emerged in Rehtmeyer's 1722 history. It is Schulvermann, not Zieglerin, whom he places in the iron chair at the execution scene.

27. Beckmann, "Therocyclus in Wolfenbüttel," 564.

28. "das historische Bild des interessantes Fürsten an verwischten Stellen der Wahrheit gemäß zu retouchiren,—sollte auch etwas mehr Schatten in das Bild kommen." Beckmann, "Therocyclus in Wolfenbüttel," 551.

29. "zwar hauptsächlich nach alten Manuscripten, die sich im Besitze des Verfassers befinden und manches Neue enthalten." Beckmann, "Therocyclus in Wolfenbüttel," 553.

30. "ungenannten entrüsteten Zeitgenossen und Unterthanen des Herzogs," "Bericht von Anna Zieglerin," Beckmann, "Therocyclus in Wolfenbüttel," 557.

31. Beckmann, "Therocyclus in Wolfenbüttel," 557.

32. "mit diesem Herren [Carl] wird der Herzog schändlich betrogen, und genarrt. . . . Die Hure giebt dem Frommen Herzog teuflische und unglaubliche Dinge vor, und bezaubert den Herzog alles zu glauben, um ihres Willens zu folgen. Es ist zu schreiben und zu erzehlen unmöglich." Beckmann, "Therocyclus in Wolfenbüttel," 560.

33. Beckmann, "Therocyclus in Wolfenbüttel," 564.

34. Although Rhamm identified himself as "A. Rhamm, Amtsrichter" on the title page of his books, his full name was Friedrich Franz Karl Albert von Rhamm.

35. "Was uns bisher von Philipp Sömmerings abenteuerlichen Umtrieben am Hofe zu Wolfenbüttel bekannt geworden ist, beschränkt sich im Wesentlichen auf die Mittheilungen, die Franz Algermann in seinen Aufzeichnungen über das Leben des Herzogs Julius von Braunschweig hinterlassen hat." Rhamm, *Die betrüglichen Goldmacher*, III.

36. "geht hervor, daß auch Algermann nicht völlig als sicherer Gewährsmann anzusehen ist, daß seine Erzählung hier übertreibt, dort den Sachverhalt nicht erschöpfend wiedergiebt." Rhamm, *Die betrüglichen Goldmacher*, IV.

37. "Aber diese Thatsache is nicht richtig," Rhamm declared. "Frau Anne had auf dem Schloß weder Dienst noch Aufenthalt gefunden." Rhamm, *Die betrüglichen Goldmacher,* 109n44.

38. Ibid., 109n43.

39. "Von den Schöffensprüchen haben sich nur diejenigen erhalten, welche Philipp Söm-mering betreffen (in neuerer Zeit aus Privat-besitz zu den Akten—Fasc. VIII—gekommen). Auf Nachfrage bei den betreffendend Stätten, dem königl. Staats-Archiv zu Magdeburg und dem königl. Amstgericht Brandenburg ist mir die Mittheilung geworden, daß dort Rechtsprüche, Sömmering und seine Genossen anlagend, sich überhaupt nich mehr vorfinden." Rhamm, 92–98n109.

40. "Wenigstens ließe es sich dann erklären, wie so schnell nach der That eine Legende sich bilden konnte und wie beispielsweise schon Algermann, der ja bereits 1575 nach Wolfenbüt-tel in höfischen Dienst, kam, den Giftguß Bartol Taubes zu einem nächtlichen Überfall umgestalten mochte, bei welchem die Verschworenen mit Mordwaffen ins Gemach der Herzogin zu dringen suchen, die Fürstin zu erdolchen und die Schatzkammern zu plündern. Diese Erzählung ist neben einer Reihe anderer Ungenauigkeiten von allen späteren Ncahrichten wiederholt, bis in der neuesten Zeit auch die dichterische Erfindung des dankbaren Stoffes sich bemächtigte und in Georg Hiltl's Novelle 'Der Teufelsdoktor von Wolfenbüttel' . . . zu Sömmering und seiner Rotten noch ein romantisches Liebespärchen hinzugesellte, nach dessen Urbilde in den vergblten Aktenstößen allerdings vergeblich gesucht wird." Rhamm, *Die betrüglichen Gold-macher,* 67.

41. "An der Stätte der dargestellten Ereignisse ist die Erinnerung an dieselben nicht gänzlich erloschen. Bis vor wenigen Jahrzehnten hint der Stuhl, auf welchem die Zieglerin den Feuertod erlitten, von einem Gewölbe des Schlosses, dem alten Richtplatz gegenüber, in Ketten herab und unter dem Namen der 'Schlüter-Liese' lebt Frau Anne, als Hexe und Giftmischerin, noch heute im Munde des Volks, wenngleich um ihre Gestalt, ihre Verbrechen, ihre Gefährten Wahrheit und Dichtung in buntem Gemisch sich verwoben haben." Rhamm, *Die betrüglichen Goldmacher,* 67.

42. Hodemacher, *Der Fall Sömmering*; Lehrmann, *Goldmacher, Gelehrte und Ganoven*; Grunow, *Die Spur führt nach Wolfenbüttel.*

43. "Sie trieb mit derlei Versprechungen nicht minder, wie durch ihre persönliche Liebenswürdigkeit den armen Philipp immer tiefer in ihre Netze, bis er sich in ihrem Garne gefangen sah und, zu spat, als betrogenen Betrüger erkannte." (She lured poor Philipp [Sömmering] deeper and deeper into her web, no less with her promises than with her personal charm, until he saw that he was caught in her threads and recognized, too late, that he was a cheat who had been cheated.) Rhamm, *Die betrüglichen Goldmacher,* 13.

BIBLIOGRAPHY

MANUSCRIPT COLLECTIONS

HStA Dresden. Hauptstaatsarchiv Dresden, Geheimer Rat (Gehemes Archiv), Loc. 7262/11. (Urgicht etlicher Personen, so Herzog Julius zu Braunschweig und Lüneburg älterer Linie, Wolfenbüttel, erstlich für Seine Fürstliche Gnaden getreue Diener und Räte gehalten, hernach aber aus Befindung ihrer bösen taten ernstlich strafen und hinrichten lassen, deren Namen auf folgendem Blatt verzeichnet zu befinden, von Seiner Fürstlichen Gnaden selbst überschickt.)

LATh-HStA Weimar. Landesarchiv Thüringen–Hauptstaatsarchiv Weimar, Ernestinisches Gesamtarchiv, Reg. P (Grumbachsche Händel und die daraus entstandenen Gothaischen Händel).

NLA HA. Niedersächsisches Landesarchiv Hannover, Cal. Br. 15, Nr. 1028.

NLA WO. Niedersächsisches Landesarchiv Wolfenbüttel, 1 Alt 9 (Acta publica des Herzogs Julius), Nr. 306–36.

StAC. Staatsarchiv Coburg, Altbestände: Die älteren Behörden des Fürstentums und Herzogtums Sachsen-Coburg, Lokat A (Akten über persönliche Verhältnisse der herzoglichen Familie, Korrespondenzen, Hofbauwesen, Hofhandwerk und Hofpersonal aus der Zeit von 1373 bis 1955).

PRINTED PRIMARY SOURCES

Algermann, Franz. "Leben, Wandel und tödtlicher Abgang des Herrn Juliussen, Herzogen zu Braunschweig." In *Feier des gedächtnisses der vormahligen hochschule Julia Carolina zu Helmstedt, veranstaltet zu monate Mai des jahres 1822; hinzugefügt ist die lebensbeschreibung des herzogs Julius von Braunschweig von Franz Algermann,* edited by Friedrich Karl von Strombeck. Helmstedt: C. G. Fleckeisen, 1822.

Aristotle. *Generation of Animals.* Translated by A. L. Peck. Loeb Classical Library 366. Cambridge, MA: Harvard University Press, 1942.

August, Churfürst zu Sachsen. *Künstlich Obstgarten-Büchlein.* Berlin: Georg Runge, 1619.

Aureum Vellus, Oder Güldin Schatz und Kunstkammer. Rorschach am Bodensee, 1599.

Bock, Hieronymus. *De Stirpium.* Strasbourg, Vuendelinus Rihelius, 1552.

Brunschwig, Hieronymus. *Kleines Distillierbuch.* Strasbourg, 1500.

———. *Liber de arte distillandi de simplicibus . . .* Strasbourg, 1500.

———. *Liber der arte distulandi simplicia . . .* Strasbourg, 1509.

Bünting, Heinrich. *Braunschweigische und Lüneburgische Chronica.* Magdeburg, 1596.

De Alchimia. Nuremberg: J. Petreius, 1541.

Dee, John. *Monas Hieroglyphica.* Antwerp: G. Silvius, 1564.

Della Porta, Giambattista. *Phytognomonica.* Frankfurt: Wechel & Fischer, 1591.

Des Allerduchleutigsten, Großmechtigsten, Unüberwindlichsten Keyser Caroli des V. vnd des Heiligen Römischen Reichs Peinlich Gerichts Ordnung, auff den Reichßtagen zu Augßpurg vnd Regenspurg, in Jahren 30. vnd 32. Gehalten, auffgericht, vnd beschlossen. Vnd jetzo von . . . Herrn Julio, hertzogen zu Braunschweig vnd Lüneburg etc. in jrer F. G. Lande, im Jahr 1.5.7.0. den 4. tag des Monats Februarij, angenommen vnd publiciret. Wolffenbüttel: Horn, 1570.

Die auffrur so geschehen ist im Vngerlandt/ mit den Creuetzern/ Vnnd auch darbey wie man der Creuetzer Haubtman hat gefangen vnnd getœdt. Nuremberg: Wolfgang Huber, 1514.

Galen. *On Semen.* Edition, translation, and commentary by Phillip De Lacy. Berlin: Akademie Verlag, 1992.

Gessner, Conrad. *De Raris et Admirandis Herbis.* Zurich: Andreas Gesnerus & Iacobs Gesnerus, fratres, 1555.

Gratarolo, Guglielmo, ed. *Veræ alchimiae artisque metallica.* Basel: Henricus Petrus & Petrus Perna, 1561.

Jacob, Cyriacus, ed. *De alchimia opuscula complura veterum philosophorum, quorum catalogum sequens pagella indicabit.* 2 vols. Frankfurt am Main, 1550.

Jacob, Johann Christoph. *Die Fruchtbare Boriza, oder das heilsame Mond-Kraut: Mit vielen chymischen und lunarischen Früchten abgebildet.* Brieg: Jacob, 1681.

Karant-Nunn, Susan C., and Merry E. Wiesner. *Luther on Women: A Sourcebook.* Cambridge; New York: Cambridge University Press, 2003.

Languet, Hubert. *Historische beschreibung der ergangenen Execution, wider des Heil. Röm. Reichs auffrhrürische Echter, unnd derselben Receptatorn, sampt einen kurtzen bericht, wie die Stad Gotha eingenomen, vnd die Festunge Grimmenstein zerschleifft worden ist, im Jar nach Christi geburt 1567, den 13. Monatstag Aprilis.* [N.p.], 1569.

L.T.Z.T. [Thurneysser zum Thurn, Leonhardt]. *Warhafftige Abconterfeyung oder gestalt/ des angesichts Leupolt Jüden/ sampt fürbildung der Execution/ welche an ihme/ seiner wolverdienten grausamen und unbmenschlichen thaten halben.* [N.p.], [ca. 1573].

Lullus, Raimundus. *De Secretis Naturae sive Quinta Essentia.* Strassbourg: Beck, 1541.

———. *Künstliche Eröffnung aller Verborgenheyten vn[d] Geheymnussen der Natur durch woelche die war Kunst der Artzney vnd Alchimey so mit mancherley Saffte[N] Düffte Gewaechsen . . .* Augsburg, 1532.

———. *Testamentum duobus libris universam artem chymican complectens.* Cologne: Johann Byrckmann, 1566.

Luther, Martin. *D. Martin Luther's Werke: Kritische Gesamtausgabe.* Weimar: H. Böhlaus Nachfolger, 1912–21.

———. *The Table-Talk of Martin Luther.* Translated by William Hazlitt. London: George Bell and Sons, 1878.

Maier, Michael. *Atalanta fugiens.* Oppenheim: Johann Theodor de Bry, 1618.

Paracelsus. *Archidoxa Ex Theophrastia. Sampt den Büchern Præparationum, De Tinctura Physicorum, De renovatione et restauratione vitæ, und De Vita Longa, alle Teutsch . . . inn Druck gleben. Von D. Iohanne Alberto Wimpineo.* Munich: bey Adam Berg, 1570.

———. *Archidoxa Philippi Theophrasti Paracelsi Bombast. Von Heymligkeyten der Natur, Zehen Bücher. Item, I. De Tinctura Physicorum. II. De Occulta Philosophia.* Strassburg: T. Rihel, 1570.

———. *Bücher und Schriften, des edlen, hochgelehrten und bewehrten Philosophi unnd Medici, Philippi Theophrasti Bombast von Hohenheim, Paracelsus genannt.* Edited by Johannes Huser. 10 vols. Basel: Conrad Waldkirch, 1589–91.

———. *De la peste et de ses causes et ses accidents.* Translated by Pierre Hassard. Avers, 1570.

———. *De Lapide Philosophorum, Drey Tractat. I. Manuale De Lapide Medicinali. II. De Tinctura Phyicorum. III. De Tinctura Planetarum . . .* Strassburg: bey Niclauss Wyriot, 1572.

———. *Des Hoch Gelerten unnd Weit Berümpten Herren, D. Theophrasti Paracelsi Büchlin von der Tinctura Physica.* Basel: Peter Perna, 1570.

———. *De signis Zodiaci et ejus mysteriis.* In Paracelsus, *De Spiritibus Planetarum sive Metallorum*, fol. 12r-p4r. Basel: [Peter Perna], 1571.

———. *Etliche Tractetlein zur Archidoxa.* Munich, 1570.

———. *Ex libro de nymphis, sylvanis, pygmaeis et salamandris.* Nissae Silesiorum: Joannes Cruciger, 1566.

———. *Philippi Theophrasti Paracelsi von Hohenhaim etliche Tractetlein zur Archidoxa gehörig: 1. Von dem Magneten vnnd seiner wunderbarlichen tugend in allerley kranckheiten . . . zugebrauchen. 2. De occulta Philosophia, darinnen tractirt wird De Consecrationibus. De Coniurationibus. De Caracteribus. Von allerley erscheinungen im schlaff. Von den jrdischen Geistern oder Schroetlein. Von der Imagination. Von den verborgnen Schätzen. Wie der mensch vom Teuffel besessen wird. Wie man den bösen Geist von den besessenen leuten außtreiben sol. Von dem Vngewitter. 3. Die recht weiß zu Administrirn die Medicin von Theophrasti aigner hand gezogen. 4. Von vilerley gifftigen Thiern wie man jhnen das gifft nemen vnd tödten sol.* Munich: Berg, 1570.

———. *Sieben defensiones.* In *Four Treatises of Theophrastus Von Hohenheim: Called Paracelsus*, edited by Henry E. Sigerist, Clarice Lilian Shelley Temkin, George Rosen, and Gregory Zilboorg, 1–41. Baltimore: Johns Hopkins University Press, 1941.

———. *Zwey Bücher Theophrasti Paracelsi . . . von der Pestilentz und ihren zufallen.* Edited by Adam von Bodenstein. Strassbourg, 1559.

Pliny. *Natural History.* Translated by H. S. Jones and H. Rackham. 10 vols. Loeb Classical Library. Cambridge, MA: Harvard University Press, 1940.

Rosarium philosophorum. Secunda pars alchimiae de lapide philosophico vero modo praeparando, continens exactam eius scientiae progressionem. Cum figuris [. . .] Francoforti: ex officina Cyriaci Iacobi, 1550.

Ruland, Martin. *Lexicon alchemiae, siue Dictonarium alchemisticum, cum obscuriorum verborum, & rerum hermeticarum, tum Theophrast-Paracelsicarum phrasium, planam explicationem continens.* Frankfurt am Main, 1612.

Ryff, Walther Hermann. *Das new groß Distillierbuch.* Frankfurt: Christian Egenolff, 1545.

———. *Des aller fürtrefflichsten, höchsten unnd adelichsten Gschöpffs aller Creaturen.* Strassbourg: Beck, 1541.

Schroeder, Friedrich-Christian, ed. *Die Peinliche Gerichtsordnung Kaiser Karls V. und des Heiligen Römischen Reichs von 1532.* Stuttgart: Reclam, 2000.

Solea, Nicolaus. *Ein Büchlien von dem Bergwergk: wie man dasselbige nach der Rutten vnnd Witterung bawen sol . . .* Brieg in Schlesien, 1600.

Sudhoff, Karl, ed. *Paracelsus, Sämtliche Werke. I. Abteilung. Medizinische naturwissenschaftliche und philosophische Schriften.* Pt. I, 14 vols. Munich: R. Oldenbourg, 1929–33.

Taurinus, Stephanus. *Stauromachia.* Vienna, 1519.

Trismosin, Salomon, et al. *Aureum Vellus, oder, Guldin Schatz und Kunstkammer darinnen der aller fürnemisten, fürtreffenlichsten, ausserlesenesten, herrlichisten und bewährteste Auctorum*

Schrifften und Bucher . . . von . . . Salomone Trissimosino . . . und Medici Theophrasti Paracelsi . . . in sonderbare underschiedliche Tractätlein disponiert und in das Teutsch gebracht. 3 vols. Getruckt zu Rorschach am Bodensee: getruckt in dess F. Gottshauss S. Gallen Reichshoff, 1598.

Ulsted, Philipp. *Coelum philosophorum, Heimlichkeit der Naturen genannt.* Straßburg, 1527.

PRINTED SECONDARY SOURCES

Ackermann, Aaron. *Münzmeister Lippold: Ein Beitrag zur Kultur- und Sittengeschichte des Mittelalters, nach urkundlichen Quellen bearbeitet.* Frankfurt am Main: J. Kauffmann, 1910.

Akkerman, Nadine, and Birgit Houben. *The Politics of Female Households: Ladies-in-Waiting Across Early Modern Europe.* Leiden; Boston: Brill, 2014.

Angstmann, Else. *Der Henker in der Volksmeinung: Seine Namen und sein Vorkommen in der mündlichen Volksüberlieferung.* Bonn: Fritz Klopp, 1928.

Arnold, Werner, Brage bei der Wieden, and Ulrike Gleixner, eds. *Herzog Heinrich Julius zu Braunschweig und Lüneburg (1564–1613): Politiker und Gelehrter mit europäischem Profil.* Braunschweig: Appelhans Verlag, 2016.

Ash, Eric H. "Introduction: Expertise and the Early Modern State." *Osiris* 25, no. 1 (2010): 1–24. DOI:10.1086/657254.

———. *Power, Knowledge, and Expertise in Elizabethan England.* Baltimore: Johns Hopkins University Press, 2004.

Bächtold-Stäubli, Hanns, Eduard Hoffmann-Krayer, and Gerhard Lüdtke. *Handwörterbuch des deutschen Aberglaubens: Herausgegeben unter besonderer Mitwirkung.* 10 vols. Berlin; Leipzig: W. de Gruyter, 1927–42.

Baldwin, Martha R. "Alchemy and the Society of Jesus in the Seventeenth Century: Strange Bedfellows?" *Isis* 40 (1993): 41–64.

———. "Toads and Plague: Amulet Therapy in Seventeenth-Century Medicine." *Bulletin of the History of Medicine* 67 (1993): 227–47.

Barnes, Robin Bruce. *Prophecy and Gnosis: Apocalypticism in the Wake of the Lutheran Reformation.* Stanford: Stanford University Press, 1988.

Beck, August. *Johann Friedrich der Mittlere, Herzog zu Sachsen: Ein Beitrag zur Geschichte des sechzehnten Jahrhunderts.* 2 vols. Weimar: Böhlau, 1858.

Beckmann, A. "Therocyclus in Wolfenbüttel, 1568–1575." *Zeitschrift für deutsche Kulturgeschichte, Neue Folge,* September 1857, 551–65.

Beckmann, Johann. *Beyträge zur Geschichte der Erfindungen.* Vol. 3, pt. 3. Leipzig: P.G. Kummer, 1792.

Benzenhöfer, Udo. *Paracelsus.* Reinbek bei Hamburg: Rowohlt Taschenbuch Verlag, 1997.

Beyer, Jürgen. "A Lübeck Prophet in Local and Lutheran Context." In *Popular Religion in Germany and Central Europe, 1400–1800,* edited by R. W. Scribner and Trevor Johnson, 166–82. London: Palgrave Macmillan, 1996.

———. "Lutheran Popular Prophets in the Sixteenth and Seventeenth Centuries: The Performance of Untrained Speakers." *Arv. Nordic Yearbook of Folklore* 51 (1995): 63–86.

———. "Lutherische Propheten in Deutschland und Skandinavien im 16. und 17. Jahrhundert: Entstehung und Ausbreitung eines Kulturmusters zwischen Mündlichkeit und Schriftlichkeit." In *Europe in Scandinavia: Kulturelle und soziale Dialoge in der frühen Neuzeit,* edited by Robert Bohn, 35–55. Frankfurt am Main: Peter Lang International Academic Publishers, 1994.

Biagioli, Mario. *Galileo Courtier: The Practice of Science in the Culture of Absolutism.* Chicago: University of Chicago Press, 1994.

Biegger, Katharina. *'De Invokatione Beatae Mariae Virginis': Paracelsus und die Marienverehrung.* Stuttgart: Franz Steiner Verlag, 1990.

Blanckmeister, Franz. *Die Kirchenbücher im Königreich Sachsen.* Leipzig: Johann Ambrosius Barth, 1901.

Blaschke, Karlheinz. *Moritz von Sachsen: Ein Reformationsfürst der zweiten Generation.* Göttingen: Muster-Schmidt, 1983.

———. *Politische Geschichte Sachsens und Thüringens.* Munich: Haus der Bayerischen Geschichte, 1991.

Blaschke, Karlheinz, Uwe John, Reiner Gross, and Holger Starke. *Geschichte der Stadt Dresden.* Stuttgart: Theiss, 2005.

Brady, Thomas A. *German Histories in the Age of Reformations, 1400–1650.* Cambridge: Cambridge University Press, 2009.

Brandenbarg, Ton. "St. Anne and Her Family: The Veneration of St. Anne in Connection with Concepts of Marriage and the Family in Early Modern Europe." In *Saints and She-Devils: Images of Women in the 15th and 16th Centuries,* edited by Lène Dresen-Coenders, 101–26. London: Rubicon Press, 1987.

Breger, Herbert. "Elias Artista—a Precursor of the Messiah in Natural Science." In *Nineteen Eighty-Four: Science Between Utopia and Dystopia,* edited by Everett Mendelsohn and Helga Novotny, 49–72. Dordrecht; Boston: D. Reidell Publishing, 1984.

Brown, Richard D. "Microhistory and the Post-Modern Challenge." *Journal of the Early Republic* 23, no. 1 (2003): 1–20.

Brüdermann, Stefan. "Heinrich Julius, Herzog zu Braunschweig-Lüneburg (Wolfenbüttel)." In *Braunschweigisches Biographisches Lexikon—8. bis 18. Jahrhundert,* edited by Horst-Rüdiger Jarck and Dieter Lent, 324–35. Braunschweig: Appelhans Verlag, 2006.

Buck, Meike. "Angeblicher Hexenstuhl." In *Tatort Geschichte: 120 Spurensuche im Braunschweigischen Landesmuseum,* edited by Meike Buck, Hans-Jörgen Derda, and Heike Pöppelmann, 120–21. Braunschweig: Michael Imhof Verlag, 2011.

Burghartz, Susanne. "Die Renaissance des Infamen? Leonhard Thurneysser zwischen Geschichte und Gegengeschichte." In *Faltenwürfe der Geschichte: Entdecken, entziffern, erzählen,* edited by Sandra Maß and Xenia von Tippelskirch, 335–50. Frankfurt am Main: Campus Verlag, 2014.

Bynum, Caroline Walker. *Fragmentation and Redemption: Essays on Gender and the Human Body in Medieval Religion.* New York; Cambridge, MA: Zone Books; Distributed by MIT Press, 1991.

———. *Holy Feast and Holy Fast: The Religious Significance of Food to Medieval Women.* Berkeley: University of California Press, 1987.

———. *Wonderful Blood: Theology and Practice in Late Medieval Northern Germany and Beyond.* Philadelphia: University of Pennsylvania Press, 2007.

Cadden, Joan. *The Meanings of Sex Difference in the Middle Ages: Medicine, Science, and Culture.* Cambridge; New York: Cambridge University Press, 1993.

Calvet, Antoine. *Les oeuvres alchimiques attribuées à Arnaud de Villeneuve: Grand oeuvre, médecine et prophétie au Moyen-Âge.* Paris: Archè Milan, 2011.

Capra, Carlo. "Governance." In *The Oxford Handbook of Early Modern European History, 1350–1750*, vol. 2, *Cultures and Power*, edited by Hamish Scott, 478–511. New York: Oxford University Press, 2015. DOI:10.1093/oxfordhb/9780199597260.013.18.

Cislo, Amy Eisen. "Paracelsus's Conception of Seeds: Rethinking Paracelsus's Ideas of Body and Matter." *Ambix* 55, no. 3 (2008): 274–82.

———. *Paracelsus's Theory of Embodiment: Conception and Gestation in Early Modern Europe.* London; Brookfield, VT: Pickering & Chatto, 2010.

Classen, Albrecht, ed. *Paracelsus im Kontext der Wissenschaften seiner Zeit: Kultur- und mentalitätsgeschichtliche Annäherungen.* Berlin; New York: Walter de Gruyter, 2010.

Cohen, Elizabeth S. "Honor and Gender in the Streets of Early Modern Rome." *Journal of Interdisciplinary History* 22, no. 4 (1992): 597–625.

Crisciani, Chiara. "*Opus* and *Sermo*: The Relationship Between Alchemy and Prophecy (12th–14th Centuries)." *Early Science and Medicine* 13 (2008): 4–14.

Crowther-Heyck, Kathleen. "'Be Fruitful and Multiply': Genesis and Generation in Reformation Germany." *Renaissance Quarterly* 55, no. 3 (2002): 904–35.

Daniel, Dane Thor. "Invisible Wombs: Rethinking Paracelsus's Concept of Body and Matter." *Ambix* 53, no. 2 (2006): 129–42.

———. "'Invisible Wombs' and Paracelsus's *Astronomia Magna* (1537/38): Bible-Based Science and the Religious Roots of the Scientific Revolution." PhD diss., Indiana University, 2003.

Darmstaedter, Ernst. "Zur Geschichte des Aurum potabile." *Chemiker-Zeitung* 48 (1924): 653–55, 678–80.

Darnton, Robert. *The Great Cat Massacre and Other Episodes in French Cultural History.* New York: Vintage Books, 1985.

Davis, Natalie Zemon. *Fiction in the Archives: Pardon Tales and Their Tellers in Sixteenth-Century France.* Stanford: Stanford University Press, 1987.

———. *The Return of Martin Guerre.* Cambridge, MA: Harvard University Press, 1983.

Debus, Allen G. *The Chemical Philosophy: Paracelsian Science and Medicine in the Sixteenth and Seventeenth Centuries.* New York: Science History Publications, 1977.

DeVun, Leah. "'Human Heaven': John of Rupescissa's Alchemy at the End of the World." In *History in the Comic Mode*, edited by Bruce Holsinger and Rachel Fulton, 251–61. New York: Columbia University Press, 2007.

———. "The Jesus Hermaphrodite: Science and Sex Difference in Premodern Europe." *Journal of the History of Ideas* 69, no. 2 (2008): 193–218.

———. *Prophecy, Alchemy, and the End of Time: John of Rupescissa in the Late Middle Ages.* New York: Columbia University Press, 2009.

Dezza, Ettore. *Geschichte des Strafprozessrechts in der Frühen Neuzeit: Eine Einführung.* Berlin; Heidelberg: Springer-Lehrbuch, 2017. DOI:10.1007/978-3-662-53244-7_5.

Dilg, Peter, and Hartmut Rudolph. *Resultate und Desiderate der Paracelsus-Forschung.* Stuttgart: F. Steiner, 1993.

Dilg-Frank, Rosemarie, ed. *Kreatur und Kosmos: Internationale Beiträge zur Paracelsusforschung.* Stuttgart; New York: Gustav Fischer Verlag, 1991.

Dingel, Irene. "The Culture of Conflict in the Controversies Leading to the Formula of Concord (1548–1580)." In *Lutheran Ecclesiastical Culture, 1550–1675*, edited by Robert Kolb, 15–64. Leiden: Brill, 2008.

Dixon, Scott. "The Sense of the Past in Reformation Germany: Part II." *German History* 30, no. 2 (2012): 175–98.

Duden, Barbara. *The Woman Beneath the Skin: A Doctor's Patients in Eighteenth-Century Germany.* Cambridge, MA: Harvard University Press, 1991.

Duindam, Jeroen Frans Jozef. *Vienna and Versailles: The Courts of Europe's Dynastic Rivals, 1550–1780.* Cambridge: Cambridge University Press, 2007.

Dülmen, Richard van. *Theatre of Horror: Crime and Punishment in Early Modern Germany.* Cambridge: Polity Press, 1990.

Egmond, Florike, and Peter Mason. *The Mammoth and the Mouse: Microhistory and Morphology.* Baltimore: Johns Hopkins University Press, 1997.

Ellington, Donna Spivey. *From Sacred Body to Angelic Soul: Understanding Mary in Late Medieval and Early Modern Europe.* Washington, DC: Catholic University of America Press, 2001.

Erdélyi, Gabriella. "Tales of a Peasant Revolt: Taboos and Memories of 1514 in Hungary." In *Memory Before Modernity: Practices of Memory in Early Modern Europe*, edited by Erika Kuijpers, Judith Pollmann, Johannes Müller, and Jasper van der Steen, 93–109. Leiden: Brill, 2013.

Findlen, Paula. "The Scientist's Body: The Nature of a Woman Philosopher in Enlightenment Italy." In *The Faces of Nature in Enlightenment Europe,* edited by Gianna Pomata and Lorraine Daston, 211–36. Berlin: Berliner Wissenschafts-Verlag, 2003.

———. "The Two Cultures of Scholarship?" *Isis* 96 (2005): 230–37.

Fissell, Mary. "Gender and Generation: Representing Reproduction in Early Modern England." *Gender & History* 7, no. 3 (1995): 433–56.

Forshaw, Peter. "'Alchemy in the Amphitheatre': Some Considerations of the Alchemical Content of the Engravings in Heinrich Khunrath's *Amphitheatre of Eternal Wisdom* (1609)." In *Art and Alchemy,* edited by Jacob Wamberg, 195–220. Copenhagen: Museum Tusculanum Press, 2006.

———. "Vitriolic Reactions: Orthodox Responses to the Alchemical Exegesis of Genesis." In *The Word and the World: Biblical Exegesis and Early Modern Science,* edited by Kevin Killeen and Peter J. Forshaw, 111–36. Basingstoke, UK; New York: Palgrave Macmillan, 2007.

Frietsch, Ute. "Leben und Sterben in der Alchemie: Anna Maria Ziegler und der eiserne Stuhl am Schloss Wolfenbüttel." In *Wolfenbütteler Hefte* 37 (2019), edited by Petra Feuerstein-Herz, forthcoming.

Gantert, Susanne. *Das Mädchenreigen: Historischer Roman.* Meßkirch: Gmeinder-Verlag, 2016.

———. *Fürstenlied: Historischer Kriminalroman.* Meßkirch: Gmeinder-Verlag, 2015.

Garber, Margaret D. "Transitioning from Transubstantiation to Transmutation: Catholic Anxieties over Chymical Matter Theory at the University of Prague." In *Chymists and Chymistry: Studies in the History of Alchemy and Early Modern Chemistry,* edited by Lawrence Principe, 63–76. Sagamore Beach, MA: Science History Publications/USA, 2007.

———. "Untwisting the Greene Lyon's Tale." *Historical Studies in the Natural Sciences* 39, no. 4 (2009): 491–500.

Gause, Ute. *Paracelsus: Genese und Entfaltung seiner fruehen Theologie; Spaetmittelalter und Reformation.* Tübingen: Mohr, 1993.

Ginzburg, Carlo. *The Night Battles: Witchcraft and Agrarian Cults in the Sixteenth and Seventeenth Centuries.* Translated by John Tedeschi and Anne Tedeschi. Baltimore: Johns Hopkins University Press, 1983.

Ginzburg, Carlo, and Anna Davin. "Morelli, Freud, and Sherlock Holmes: Clues and Scientific Method." *History Workshop* 9 (Spring 1980): 5–36.

Gmelin, Johann Friedrich. *Geschichte der Chemie seit dem Wiederaufleben der Wissenschaften bis an das Ende des achzehnten Jahrhundert*. 3 vols. Göttingen: Johann Georg Rosenbusch, 1797–99.

Goldammer, Kurt. "Paracelsische Eschatologie." *Nova Acta Paracelsica* 5 (1948): 45–85.

Goodare, Julian. *The European Witch-Hunt*. London; New York: Routledge, 2016.

Görges, Wilhelm. *Vaterländische Geschichten und Denkwürdigkeiten der Vorzeit: Mit vielen Abbildungen von Städten, Flecken . . . der Lande Braunschweig und Hannover, größtentheils, wie dieselben vor 200 Jahren sich darstellten*. Braunschweig: Meinecke, 1844.

Gowing, Laura. *Domestic Dangers: Women, Words, and Sex in Early Modern London*. Oxford: Clarendon Press, 1998.

Graefe, Christa, ed. *Staatsklugheit und Frömmigkeit: Herzog Julius zu Branschweig-Lüneburg, ein Norddeutscher Landesherrr des 16. Jahrhunderts*. Weinheim: VCH Verlagsgesellschaft, Acta Humaniora, 1989.

Grafton, Anthony. *Forgers and Critics: Creativity and Duplicity in Western Scholarship*. Princeton: Princeton University Press, 1990.

Green, Jonathan. *Printing and Prophecy: Prognostication and Media Change, 1450–1550*. Ann Arbor: University of Michigan Press, 2012.

Green, Monica H.. "Flowers, Poisons, and Men: Menstruation in Medieval Western Europe." In *Menstruation: A Cultural History*, edited by Andrew Shail and Gillian Howie, 51–64. Basingstoke, UK; New York: Palgrave Macmillan, 2005.

Greenblatt, Stephen Jay. *Renaissance Self-Fashioning: From More to Shakespeare*. Chicago: University of Chicago Press, 1980.

Grell, Ole Peter, ed. *Paracelsus: The Man and His Reputation*. Leiden; Boston: Brill, 1998.

Groebner, Valentin. *Who Are You? Identification, Deception, and Surveillance in Early Modern Europe*. New York: Zone Books, 2007.

Grunow, Heinz. *Die Spur führt nach Wolfenbüttel: Bericht über den grössten Kriminalprozess des 16. Jahrhunderts*. Wolfenbüttel: H. Grunow, 1976.

Hanegraaff, Wouter J. *Esotericism and the Academy: Rejected Knowledge in Western Culture*. Cambridge: Cambridge University Press, 2012.

Harkness, Deborah. *John Dee's Conversations with Angels: Cabala, Alchemy, and the End of Nature*. Cambridge: Cambridge University Press, 1999.

Harrington, Joel F., trans. and ed. *The Executioner's Journal: Meister Frantz Schmidt of the Imperial City of Nuremberg*. Charlottesville: University of Virginia Press, 2016.

———. *The Faithful Executioner: Life and Death, Honor and Shame in the Turbulent Sixteenth Century*. New York: Farrar, Straus and Giroux, 2013.

Havemann, Wilhelm. "Sidonia, Herzogin zu Braunschweig-Lüneburg." *Vaterländisches Archiv des Historischen Vereins für Niedersachsen* 11 (1843): 278–303.

Heal, Bridget. *The Cult of the Virgin Mary in Early Modern Germany: Protestant and Catholic Piety, 1500–1648*. Cambridge: Cambridge University Press, 2007.

———. "Marian Devotion and Confessional Identity in Sixteenth-Century Germany." In *The Church and Mary*, edited by R. N. Swanson, 218–27. Rochester, NY: Published for the Ecclesiastical History Society by Boydell Press, 2004.

Hedesan, Georgiana D. *An Alchemical Quest for Universal Knowledge: The "Christian Philosophy" of Jan Baptist Van Helmont (1579–1644)*. London: Routledge, Taylor & Francis Group, 2016.

Heinemann, Otto von. *Geschichte von Braunschweig und Hannover*. 3 vols. Gotha: Friedrich Andreas Perthes, 1882–92.

Hiltl, Georg. *Der Teufelsdoktor von Wolfenbüttel*. In *Historische Novellen*, vol. 1, 1–127. Berlin: Wedekind & Schweiger, 1873.

Hirai, Hiro. *Le concept de semence dans les théories de la matière à la Renaissance de Marsile Ficin à Pierre Gassendi*. Turnhout: Brepols, 2005.

Hirschbiegel, Jan, and Werner Paravicini, eds. *Das Frauenzimmer die Frau bei Hofe in Spätmittelalter und früher Neuzeit: 6. Symposium der Residenzen-Kommission der Akademie der Wissenschaften in Göttingen, veranstaltet in Zusammenarbeit mit dem Deutschen Historischen Institut Paris, dem Sonderforschungsbereich 537 der Technischen Universität Dresden und dem Landsamt für Archäologie des Freistaaates [i.e. Freistaates] Sachsen, Dresden, 26. bis 29. September 1998*. Residenzenforschung 11. Dresden: J. Thorbecke, 2000.

Hodemacher, Jürgen. *Der Fall Sömmering*. Braunschweig: DöringDRUCK, 2015.

Hoheisel, Karl. "Christus und der philosophische Stein: Alchemie als über- und nichtchristlicher Heilsweg." In *Die Alchemie in der europäischen Kultur- und Wissenschaftsgeschichte*, edited by Christoph Meinel, 61–84. Wiesbaden: Harrassowitz, 1986.

Ignor, Alexander. *Geschichte des Strafprozesses in Deutschland: 1532–1846; von der Carolina Karls V. bis zu den Reformen des Vormärz*. Paderborn; Munich: Schöningh, 2002.

Ilić, Luka. *Theologian of Sin and Grace: The Process of Radicalization in the Theology of Matthias Flacius Illyricus*. Göttingen: Vandenhoeck & Ruprecht, 2014.

Iliffe, Rob, and George E Smith. *The Cambridge Companion to Newton*. 2nd ed. Cambridge: Cambridge University Press, 2016.

Janacek, Bruce. *Alchemical Belief: Occultism in the Religious Culture of Early Modern England*. University Park: Pennsylvania State University Press, 2011.

Jütte, Daniel. *Das Zeitalter des Geheimnisses: Juden, Christen und die Ökonomie des Geheimen (1400–1800)*. Göttingen; Oakville, CT: Vandenhoeck & Ruprecht, 2011.

———. *The Strait Gate: Thresholds and Power in Western History*. New Haven: Yale University Press, 2015.

Jütte, Robert. *Paracelsus Heute—im Lichte der Natur*. Heidelberg: Karl F. Haug Verlag, 1994.

Kahn, Didier. *Alchimie et paracelsisme en France à la fin de la Renaissance (1567–1625)*. Geneva: Droz, 2007.

Kämmerer, Ernst Wilhelm. *Das Leib-Seele-Geist-Problem bei Paracelsus und einigen Autoren des 17. Jahrhunderts*. Wiesbaden: F. Steiner, 1971.

Karant-Nunn, Susan C. *Luther's Pastors: The Reformation in the Ernestine Countryside*. Philadelphia: American Philosophical Society, 1979.

Kassell, Lauren. *Medicine and Magic in Elizabethan London: Simon Forman, Astrologer, Alchemist, and Physician*. Oxford: Clarendon Press, 2005.

Kavey, Allison. "Mercury Falling: Gender Malleability and Sexual Fluidity in Early Modern Popular Alchemy." In *Chymists and Chymistry: Studies in the History of Alchemy and Early Modern Chemistry*, edited by Lawrence M. Principe, 125–35. Sagamore Beach, MA: Science History Publications/USA, 2007.

Keller, Katrin. *Hofdamen: Amtsträgerinnen im Wiener Hofstaat des 17. Jahrhunderts*. Vienna: Böhlau, 2005.

———. "Kommunikationsraum altes Reich: Zur Funktionalität der Korrespondenznetze von Fürstinnen im 16. Jahrhundert." *Zeitschrift für Historische Forschung* 31, no. 2 (2004): 205–30. http://www.jstor.org/stable/43570001.

———. *Kurfürstin Anna von Sachsen (1532–1585)*. Regensburg: Pustet, 2010.

———. "Zwischen zwei Residenzen: Der Briefwechsel der Kurfürstin Anna von Sachsen mit Freiin Brigitta Trautson." In *Viatori per urbes castraque: Festschrift für Herwig Ebner zum*

75. Geburtstag, edited by Gerhard Jartiz, Helmut Bräuer, and Käthe Sonnleitner, 365–82. Graz: Institut für Geschichte der Karl-Franzens-Universität, 2003.

Kieckhefer, Richard. *Magic in the Middle Ages*. Cambridge: Cambridge University Press, 1989.

Kirwan, R. M. "Empowerment and Representation at the University in Early Modern Germany: Helmstedt and Würzburg 1576–1634." PhD diss., Trinity College Dublin. Published: Wiesbaden: Harrassowitz, 2009.

Klaassen, Walter. *Living at the End of the Ages: Apocalyptic Expectation in the Radical Reformation*. Lanham, MD; Waterloo, ONT: University Press of America; Institute for Anabaptist and Mennonite Studies Conrad Grebel College, 1992.

Kobuch, Agatha. *Die Geschichte der Familie von Schönberg: Ausstellung der von Schönberg'schen Stiftung, Nossen 2004*. Panitzsch: Von Schönberg'sche Stiftung, 2004.

Kolb, Robert. *Bound Choice, Election, and Wittenberg Theological Method: From Martin Luther to the Formula of Concord Lutheran*. Grand Rapids, MI: Wm. B. Eerdmans, 2005.

Kolbe, Heinz, Wolfram Forche, und Max Humburg. *Die Geschichte der Saline Salzliebenhalle und der alten Salzstadt*. Salzgitter: Archiv der Stadt Salzgitter, 1988.

Kollmann, Nancy. *By Honor Bound: State and Society in Early Modern Russia*. Ithaca: Cornell University Press, 1999.

Kopp, Hermann. *Die Alchemie in älterer und neuerer Zeit*. 2 vols. Heidelberg: Carl Winter's Universitätsbuchhandlung, 1886.

Kraschewski, Hans-Joachim. *Wirtschaftspolitik im deutschen Territorialstaat des 16. Jahrhunderts: Herzog Julius von Braunschweig-Wolfenbüttel*. Cologne; Vienna: Böhlau Verlag, 1978.

Kreitzer, Beth. *Reforming Mary: Changing Images of the Virgin Mary in Lutheran Sermons of the Sixteenth Century*. New York: Oxford University Press, 2004.

Krusch, Bruno. "Die Entwickelung der Herzoglichen Braunschweigischen Centralbehörden, Canzlei, Hofgericht und Consistorium bis zum Jahre 1584." Pt. II. *Zeitschrift des Historischen Vereins für Niedersachsen* 54 (1894): 39–179.

Kühn, Helga-Maria. *Eine "unverstorbene Witwe": Sidonia Herzogin zu Braunschweig-Lüneburg geborene Herzogin zu Sachsen 1518–1575; Ein aus Archivquellen nachgezeichneter Lebensweg*. Hannover: Hahnsche Buchhandlung, 2009.

Kuntz, Marion Leathers. *Guillaume Postel, Prophet of the Restitution of All Things: His Life and Thought*. The Hague; Boston; Hingham, MA: Nijhoff; Kluwer Boston, distributor, 1981.

Lander Johnson, Bonnie, and Eleanor Decamp, eds. *Blood Matters: Studies in European Literature and Thought, 1400–1700*. Philadelphia: University of Pennsylvania Press, 2018.

Lehrmann, Joachim. *Goldmacher, Gelehrte und Ganoven: Die Suche nach dem Stein der Weisen in den Ländern Braunschweig, Hannover, Hildesheim*. Hannover: Lehrmann, 2008.

Lentz, Carl Georg Heinrich. *Bücher der Geschichten der Lande Braunschweig und Hannover*. 2nd ed. Braunschweig: bei Oelune et Müller, 1840.

Leong, Elaine Yuen Tien. "Collecting Knowledge for the Family: Recipes, Gender, and Practical Knowledge in the Early Modern English Household." *Centaurus* 55, no. 2 (2013): 81–103.

———. "Making Medicines in the Early Modern Household." *Bulletin of the History of Medicine* 82, no. 1 (2008): 145–68.

Leppin, Volker. *Antichrist und Jüngster Tag: Das Profil apokalyptischer Flugschriftenpublizistik im deutschen Luthertum 1548–1618*. Gütersloh: Gütersloher Verlagshaus, 1999.

Lerner, Robert. "Refreshment of the Saints: The Time after Antichrist as a Station for Earthly Progress in Medieval Thought." *Tradition* 32 (1976): 97–144.

Levack, Brian P. *The Witch-Hunt in Early Modern Europe*. Abingdon, UK; New York: Routledge, 2016.

Lilienthal, Andrea. *Die Fürstin und die Macht: Welfische Herzoginnen im 16. Jahrhundert; Elisabeth, Sidonia, Sophia*. Hannover: Hahn, 2007.

Lippelt, Christian, and Gerhard Schildt, eds. *Braunschweig-Wolfenbüttel in der frühen Neuzeit: Neue historische Forschungen*. Braunschweig: Appelhans Verlag, 2003.

Maaser, Michael. *Humanismus und Landesherrschaft: Herzog Julius (1528–1589) und die Universität Helmstedt*. Stuttgart: Steiner, 2010.

Mager, Inge. "Die Einführung der Reformation in Braunschweig-Wolfenbüttel und die Gründung der Universität Helmstedt." In *Staatsklugheit und Frömmigkeit: Herzog Julius zu Branschweig-Lüneburg, ein norddeutscher Landesherr des 16. Jahrhunderts*, edited by Christa Graefe, 25–33. Weinheim: VCH Verlagsgesellschaft, Acta Humaniora, 1989.

———. *Die Konkordienformel im Fürstentum Braunschweig-Wolfenbüttel: Entstehungsbeitrag, Rezeption, Geltung*. Göttingen: Vandenhoeck und Ruprecht, 1993.

———. "Elisabeth von Brandenburg—Sidonie von Sachsen: Zwei Frauenschicksale im Kontext der Reformation von Calenberg-Göttingen." In *450 Jahre Reformation im Calenberger Land: Bis zum Jubiläum im Jahr 1992*, 23–32. Laatzen: e.-l. K. Laatzen-Pattensen, 1992.

Magnússon, Sigurður G., and István Szíjártó. *What Is Microhistory? Theory and Practice*. Milton Park, Abingdon, UK: Routledge, 2013.

Margoszy, Daniel. "Certain Fakes and Uncertain Facts: Jan Jonston and the Question of Truth in Religion and Natural History." In *Fakes!? Hoaxes, Counterfeits, and Deception in Early Modern Science*, edited by Marco Beretta and Maria Conforti, 190–225. Sagamore Beach, MA: Science History Publications, 2014.

Martin, John Jeffries. *Myths of Renaissance Individualism*. Basingstoke, UK: Palgrave Macmillan, 2004.

Martinón-Torres, Marcos. "Some Recent Developments in the Historiography of Alchemy." *Ambix* 58, no. 3 (2011): 215–37.

Matus, Zachary A. "Alchemy and Christianity in the Middle Ages." *History Compass* 10 (2012): 934–45.

———. *Franciscans and the Elixir of Life: Religion and Science in the Later Middle Ages*. Philadelphia: University of Pennsylvania Press, 2017.

———. "Reconsidering Roger Bacon's Apocalypticism in Light of His Alchemical and Scientific Thought." *Harvard Theological Review* 105, no. 2 (2012): 189–222. http://www.jstor.org/stable/41474572.

———. "Resurrected Bodies and Roger Bacon's Elixir." *Ambix* 60, no. 4 (2013): 323–40.

McClive, Cathy. *Menstruation and Procreation in Early Modern France*. Farnham, Surrey, UK; Burlington, VT: Ashgate, 2015.

McGinn, Bernard. *Apocalypticism in the Western Tradition*. Aldershot, UK; Brookfield, VT: Variorum, 1994.

———. *Visions of the End: Apocalyptic Traditions in the Middle Ages*. New York: Columbia University Press, 1979.

Merkel, Johannes. "Die Irrungen zwischen Herzog Erich II. und seiner Gemahlin Sidonie." *Zeitschrift des Historischen Vereins für Niedersachsen* 63/64 (1899): 11–101.

———. "Herzog Julius von Braunschweig und Lüneburg (1529–1589)." *Jahrbuch der Gesellschaft für Niedersächsische Kirchengeschichte* 1–2 (1896): 20–44.

Mohrmann, Wolf-Dieter. "Vater-Sohn-Konflikt und Staatsnotwendigkeit: Zur Auseinandersetzung zwischen den Herzögen Heinrich der Jungere und Julius von Braunschweig-Wolfenbüttel." *Braunschweigisches Jahrbuch* 76 (1995): 63–100.

Moran, Bruce. *The Alchemical World of the German Court: Occult Philosophy and Chemical Medicine in the Circle of Moritz of Hessen (1572–1632)*. Stuttgart: Franz Steiner Verlag, 1991.

———. "Art and Artisanship in Early Modern Alchemy." *Getty Research Journal* 5 (2013): 1–14.

———. *Distilling Knowledge: Alchemy, Chemistry, and the Scientific Revolution*. Cambridge, MA: Harvard University Press, 2005.

———, ed. "Focus: Alchemy and the History of Science." *Isis* 102, no. 2 (2011): 300–337.

———. "German Prince-Practitioners: Aspects in the Development of Courtly Science, Technology, and Procedures in the Renaissance." *Technology and Culture* 22, no. 2 (1981): 253–74.

———, ed. *Patronage and Institutions: Science, Technology, and Medicine at the European Court, 1500–1750*. Rochester, NY: Boydell Press, 1991.

Muir, Edward, and Guido Ruggiero, eds. *Microhistory and the Lost Peoples of Europe*. Baltimore: Johns Hopkins University Press, 1991.

Newman, William R. "'Decknamen or Pseudochemical Language'? Eirenaeus Philalethes and Carl Jung." *Revue d'Histoire des Sciences et de Leurs Applications* 49 (1996): 159–88.

———. "The Homunculus and His Forebears: Wonders of Art and Nature." In *Natural Particulars: Nature and the Disciplines in Renaissance Europe*, edited by Anthony Grafton and Nancy Siraisi, 321–45. Cambridge, MA: MIT Press, 1999.

———. *Promethean Ambitions: Alchemy and the Quest to Perfect Nature*. Chicago: University of Chicago Press, 2004.

———, ed. *The Summa Perfectionis of Pseudo-Geber: A Critical Edition, Translation, and Study*. Leiden; New York: E.J. Brill, 1991.

———. "Technology and Alchemical Debate in the Late Middle Ages." *Isis* 80, no. 3 (1989): 423–45.Newman, William R., and Lawrence M. Principe. "Alchemy vs. Chemistry: The Etymological Origins of a Historiographic Mistake." *Early Science and Medicine* 3, no. 1 (1998): 32–65.

Nummedal, Tara. "Alchemical Reproduction and the Career of Anna Maria Zieglerin." *Ambix* 48 (2001): 56–68.

———. "The Alchemist." In *A Companion to the History of Science*, edited by Bernard V. Lightman, 58–70. Chichester, UK; Malden, MA: John Wiley & Sons, 2016.

———. "The Alchemist in His Laboratory." In *Goldenes Wissen: Die Alchemie—Substanzen, Synthesen, Symbolik; Ausstellung der Herzog August Bibliothek Wolfenbüttel (Bibliotheca Augusta: Augusteerhalle, Schatzkammer, Kabinett) vom 31. August 2014 bis zum 22. Februar 2015*, edited by Petra Feuerstein-Herz and Stefan Laube, 121–28. Wolfenbüttel: Herzog August Bibliothek, 2014.

———. *Alchemy and Authority in the Holy Roman Empire*. Chicago: University of Chicago Press, 2007.

———, ed. "Alchemy and Religion in Christian Europe." *Ambix* 60 (4): 311–414.

———. "Anna Zieglerin's Alchemical Revelations." In *Secrets and Knowledge in Medicine and Science*, edited by Elaine Yuen Tien Leong and Alisha Rankin, 125–41. Aldershot, UK; Brookfield, VT: Ashgate, 2011.

———. "Corruption, Generation, and the Problem of Menstrua in Early Modern Alchemy." In *Blood Matters: Studies in European Literature and Thought, 1400–1700*, edited by Bonnie Lander Johnson and Eleanor Decamp, 111–22. Philadelphia: University of Pennsylvania Press, 2018.

———. "Introduction: Alchemy and Religion in Christian Europe." *Ambix* 60 (4): 311–22.

———. "Words and Works in the History of Alchemy." *Isis* 102, no. 2 (2011): 330–37.

Obrist, Barbara. *Les débuts de l'imagerie alchimique (XIVe-XVe siècle).* Paris: Le Sycomore, 1982.

Ortloff, Friedrich. *Geschichte der Grumbachischen Händel.* 4 vols. Jena: Friedrich Frommann, 1868–70.

Pagel, Walter. "The Paracelsian Elias Artista and the Alchemical Tradition." *Medizinhistorisches Journal* 16 (1981): 6–19.

———. *Paracelsus: An Introduction to Philosophical Medicine in the Era of the Renaissance.* Basel: Karger, 1982.

Pal, Carol. *Republic of Women: Rethinking the Republic of Letters in the Seventeenth Century.* Cambridge; New York: Cambridge University Press, 2012.

Park, Katharine. *Secrets of Women: Gender, Generation, and the Origins of Human Dissection.* New York; Cambridge, MA: Zone Books; Distributed by MIT Press, 2006.

Peltonen, Matti. "Clues, Margins, and Monads: The Micro-Macro Link in Historical Research." *History and Theory* 40, no. 3 (2001): 347–59.

Pereira, Michela. *The Alchemical Corpus Attributed to Raymond Lull.* London: Warburg Institute, University of London, 1989.

———. "Alchemy and the Use of Vernacular Languages in the Late Middle Ages." *Speculum* 74, no. 2 (1999): 336–56.

Pomata, Gianna. "*Praxis Historialis*: The Uses of *Historia* in Early Modern Medicine." In *Historia: Empiricism and Erudition in Early Modern Europe*, edited by Gianna Pomata and Nancy G. Siraisi, 105–46. Cambridge, MA: MIT Press, 2005.

Pouchelle, Marie-Christine. *The Body and Surgery in the Middle Ages.* New Brunswick: Rutgers University Press, 1990.

Priesner, Claus, and Karin Figala. *Alchemie: Lexicon einer hermetischen Wissenschaft.* Munich: Beck, 1998.

Principe, Lawrence M. "Alchemy Restored." *Isis* 102, no. 2 (2011): 305–12.

———. "Revealing Analogies: The Descriptive and Deceptive Roles of Sexuality and Gender in Latin Alchemy." In *Hidden Intercourse: Eros and Sexuality in the History of Western Esotericism*, ed. Wouter J. Hanegraaf and Jeffrey J. Kripal, 209–29. Leiden; Boston: Brill, 2008.

———. *The Secrets of Alchemy.* Chicago; London: University of Chicago Press, 2013.

Principe, Lawrence M., and William R. Newman. "Some Problems with the Historiography of Alchemy." In *Secrets of Nature: Astrology and Alchemy in Early Modern Europe*, edited by William R. Newman and Anthony Grafton, 385–431. Cambridge, MA: MIT Press, 2001.

Rampling, J. M. "The Alchemy of George Ripley, 1470–1700." PhD diss., Cambridge University, 2009.

Rankin, Alisha Michelle. "Becoming an Expert Practitioner: Court Experimentalism and the Medical Skills of Anna of Saxony (1532–85)." *Isis* 98 (2007): 23–53.

———. *Panaceia's Daughters: Noblewomen as Healers in Early Modern Germany.* Chicago: University of Chicago Press, 2013.

Ray, Meredith K. *Daughters of Alchemy: Women and Scientific Culture in Early Modern Italy.* Cambridge, MA: Harvard University Press, 2015.

Reeves, Marjorie. *The Influence of Prophecy in the Later Middle Ages: A Study in Joachimism.* Oxford: Clarendon Press, 1969.

———. *Joachim of Fiore and the Prophetic Future: A Medieval Study in Historical Thinking.* New, rev. ed. Stroud: Sutton, 1999.

Rehtmeyer, Philipp Julius, ed. *Braunschweig-Lüneburgische Chronica, Oder: Historische Beschreibung der Herzogen zu Braunschweig und Lüneburg*. Braunschweig: Detleffsen, 1722.

Reller, Horst. "Vorreformatorische und Reformatorische Kirchenverfassung im Fürstentum Braunschweig-Wolfenbüttel." PhD diss., Universität Göttingen. Published: Göttingen: Vandenhoeck & Ruprecht, 1959.

Rhamm, Albert. *Die betrüglichen Goldmacher: Am Hofe des Herzogs Julius von Braunschweig; Nach den Processakten*. Wolfenbüttel: Julius Zwißler, 1883.

———. *Hexenglaube und Hexenprocesse: Vornämlich in den braunschweigischen Landen*. Wolfenbüttel: Zwißler, 1882.

Ribbentrop, Carl Philipp [*sic*]. *Vollständige Geschichte und Beschreibung der Stadt Braunschweig*, 2 vols. Braunschweig, 1796.

Ribbentrop, Philipp Christian. *Beschreibung der Stadt Braunschweig*. 2 vols. Braunschweig: Meyer, 1791–96.

———. *Sammlung der Landtagsabschiede, Fürstlichen Reversalen und anderer Urkunden, die Landschaftliche Verfassung des Herzogthums Braunschweig-Lüneburg, Wolfenbüttelschen Theils Betreffend*. Helmstedt: Fleckeisen, 1793–97.

Robisheaux, Thomas. "The German Witch Trials." In *The Oxford Handbook of Witchcraft in Early Modern Europe and Colonial America*, edited by Brian P. Levack, 179–98. Oxford: Oxford University Press, 2013.

Roper, Lyndal. *The Holy Household: Women and Morals in Reformation Augsburg*. Oxford; New York: Clarendon Press; Oxford University Press, 1989.

Rowland, Ingrid D. *The Scarith of Scornello: A Tale of Renaissance Forgery*. Chicago: University of Chicago Press, 2004.

Rublack, Ulinka. *The Crimes of Women in Early Modern Germany*. Oxford; New York: Clarendon Press, 1999.

———. "Pregnancy, Childbirth, and the Female Body in Early Modern Germany." *Past and Present* 150 (1996): 84–110.

Sabean, David. *Power in the Blood: Popular Culture and Village Discourse in Early Modern Germany*. Cambridge; New York: Cambridge University Press, 1984.

Schattkowsky, Martina, ed. *Witwenschaft in der frühen Neuzeit: Fürstliche und adlige Witwen zwischen Fremd- und Selbstbestimmung*. [Leipzig]: Leipziger Universitätsverlag, 2003.

Scheliga, Thomas. "A Renaissance Garden in Wolfenbüttel, North Germany." *Garden History* 25, no. 1 (1997): 1–27.

Schott, Heinz, and Ilana Zingler, eds. *Paracelsus und seine Internationale Rezeption in der frühen Neuzeit: Beiträge zur Geschichte des Paracelsismus*. Leiden; Boston: Brill, 1998.

Schreiner, Klaus and Gerd Schwerhoff, eds. *Verletzte Ehre: Ehrkonflikte in Gesellschaften des Mittelalters und der Frühen Neuzeit*. Vol. 5. Cologne; Weimar; Wien: Böhlau Verlag, 1995.

Schuster, Peter. *Verbrecher, Opfer, Heilige: Eine Geschichte des Tötens*. Stuttgart: Klett-Cota, 2015.

Schulte, Regina, ed. *The Body of the Queen: Gender and Rule in the Courtly World, 1500–2000*. New York: Berghahn Books, 2006.

Sehling, Emil, and Anneliese Sprengler-Ruppenthal, eds. *Die Evangelischen Kirchenordnungen des XVI. Jahrhunderts 6: Niedersachsen; I: Die Welfischen Lande; 1: Die Fürstentümer Wolfenbüttel und Lüneburg mit den Städten Braunschweig und Lüneburg*. Tübingen: Mohr, 1955.

Sella, Barbara. "Northern Italian Confraternities and the Immaculate Conception in the Fourteenth Century." *Journal of Ecclesiastical History* 49, no. 4 (1998): 599–619.

Serpell, James A. "Guardian Spirits or Demonic Pets: The Concept of the Witch's Familiar in Early Modern England." In *The Animal-Human Boundary: Historical Perspectives*, edited by Angela N. H. Creager, William C. Jordan, and Shelby Cullom Davis Center for Historical Studies, 157–92. Rochester: University of Rochester Press, 2002.

Shackelford, Jole. "Paracelsianism and the Orthodox Lutheran Rejection of Vital Philosophy in Early Seventeenth-Century Denmark." *Early Science and Medicine* 8, no. 3 (2003): 210–52.

———. "Seeds with a Mechanical Purpose: Severinus' Semina and 17th-Century Matter Theory." In *Reading the Book of Nature: The Other Side of the Scientific Revolution*, ed. Allen G. Debus and Michael T. Walton, 15–44. Kirksville, MO: Sixteenth Century Publishers, 1998.

———. "Unification and the Chemistry of the Reformation." In *Infinite Boundaries: Order, Disorder, and Reorder in Early Modern German Culture*, edited by Max Reinhart, 291–312. Kirksville, MO: Sixteenth Century Journal Publishers, 1998.

Shaw, Frank. "Friedrich II as the 'Last Emperor.'" *German History* 19, no. 3 (2001): 321–39.

Sibum, H. Otto. "Reworking the Mechanical Value of Heat: Instruments of Precision and Gestures of Accuracy in Early Victorian England." *Studies in the History and Philosophy of Science* 26, no. 1 (1995): 73–106.

Smith, Pamela H. *The Body of the Artisan: Art and Experience in the Scientific Revolution.* Chicago: University of Chicago Press, 2004.

———. *The Business of Alchemy: Science and Culture in the Holy Roman Empire.* Princeton: Princeton University Press, 1994.

———. "Vermilion, Mercury, Blood, and Lizards: Matter and Meaning in Metalworking." In *Materials and Expertise in Early Modern Europe: Between Market and Laboratory*, edited by Ursula Klein and E. C. Spary, 29–49. Chicago; London: University of Chicago Press, 2009.

Snyder, Jon R. *Dissimulation and the Culture of Secrecy in Early Modern Europe.* Berkeley: University of California Press, 2009.

Soldan, Wilhelm Gottlieb. *Geschichte der Hexenprozesse.* Edited by Heinrich Heppe. Stuttgart: Cotta, 1880.

Spieß, Karl-Heinz. "Witwenversorgung im Hochadel: Rechtlicher Rahmen und praktische Gestaltung um Spätmittelalter und zu Beginn der Frühen Neuzeit." In *Witwenschaft in der Frühen Neuzeit: Fürstliche und adlige Witwen zwischen Fremd- und Selbstbestimmung*, edited by Martina Schattkowsky, 87–114. Leipzig: Leipziger Universitätsverlag, 2003.

Spitzer, Gabriele. *. . . und die Spree führt Gold: Leonhard Thurneysser Zum Thurn; Astrologe, Alchimist, Arzt und Drucker im Berlin des 16. Jahrhunderts.* Berlin: Staatsbibliothek zu Berlin, Preussischer Kulturbesitz, 1996.

Starenko, Peter Elsel. "In Luther's Wake: Duke John Frederick II of Saxony, Angelic Prophecy, and the Gotha Rebellion of 1567." PhD diss., University of California, Berkeley, 2002.

Stavenhagen, Lee. "The Original Text of the Latin Morienus." *Ambix* 17, no. 1 (1970): 1–12.

———, ed. *A Testament of Alchemy; Being the Revelations of Morienus, Ancient Adept and Hermit of Jerusalem, to Khalid Ibn Yazid Ibn Mu'awiyya, King of the Arabs, of the Divine Secrets of the Magisterium and Accomplishment of the Alchemical Art.* Hanover, NH: Published for the Brandeis University Press by the University Press of New England, 1974.

Stolberg, Michael. "Menstruation and Sexual Difference in Early Modern Medicine." In *Menstruation: A Cultural History*, edited by Andrew Shail and Gillian Howie, 90–101. Basingstoke, UK; New York: Palgrave Macmillan, 2005.

Strauss, Gerald. *Law, Resistance, and the State: The Opposition to Roman Law in Reformation Germany*. Princeton: Princeton University Press, 1986.

Struve, Tilman. "Utopie und gesellschaftliche Wirklichkeit: Zur Bedeutung des Friedenskaisers im späten Mittelalter." *Historische Zeitschrift* 225 (1977): 65–95.

Stuart, Kathy. *Defiled Trades and Social Outcasts: Honor and Ritual Pollution in Early Modern Germany*. Cambridge; New York: Cambridge University Press, 1999.

Sudhoff, Karl. *Bibliographica Paracelsica: Besprechung der unter Hohemheims Namen 1527–1893 erschienenen Druckschriften*. Graz: Akademische Druck- und Verlaganstalt, 1958.

Telle, Joachim, ed. *Analecta Paracelsica: Studien zum Nachleben Theophrast von Hohenheims im deutschen Kulturgebiet der Frühen Neuzeit*. Stuttgart: F. Steiner, 1994.

———. "Der *Splendor Solis* in der frühneuzeitlichen Republica Alchemica." *Daphnis* 35 (2006): 421–48.

———, ed. *Parerga Paracelsica: Paracelsus in Vergangenheit und Gegenwart*. Stuttgart: Franz Steiner Verlag, 1991.

———, ed. *Rosarium philosophorum: Ein alchemisches Florilegium des Spätmittelalters; Faksimile der illustrierten Erstausgabe Frankfurt 1550*. Translated by Lutz Claren and Joachim Huber. 2 vols. Weinheim: VCH Verlagsgesellschaft, 1992.

———. *Sol und Luna: Literar- und alchemiegeschichtliche Studien zu einem altdeutschen Bildgedicht; Mit Text- und Bildanhang*. Hürtgenwald: Guido Pressler Verlag, 1980. Tentzel, Wilhelm Ernst. "Vorstellung oder kurtze Beschreibung des hochadelischen uralten Geschlechts, derer von Ziegler und Kliphausen, Vermehrt von F. V. K. A. 1737." *Beyträge zur Historie derer Chur- und Fürstlichen Sächsishen Lande, gesammelt von M. George Christoph Kreysig* 1 (1754): 35–76.

Thöne, Friedrich. "Hans Vredeman de Vries in Wolfenbüttel." *Braunschweigisches Jahrbuch* 41 (1960): 47–68.

———. *Wolfenbüttel: Geist und Glanz einer alten Residenz*. Munich: F. Bruckmann, 1963.

———. "Wolfenbüttel unter Herzog Julius (1568 bis 1589): Topographie und Baugeschichte." Special issue, *Braunschweigisches Jahrbuch* 33 (1952): 1–74.

Tilton, Hereward. "Alchymia Archetypica: Theurgy, Inner Transformation, and the Historiography of Alchemy." In *Transmutatio: La via ermetica alla felicità/The Hermetic Way to Happiness*, edited by Daniela Boccassini e Carlo Testa, 179–215. Special Issue, *Quaderni di Studi Indo-Mediterranei* 5 (2012).

Triebs, Michaela. *Die medizinische Fakultät der Universität Helmstedt (1576–1810): Eine Studie zu ihrer Geschichte unter besonderer Berücksichtigung der Promotions- und Übungsdisputationen*. Wiesbaden: Harrassowitz, 1995.

Vehse, Carl Eduard. *Geschichte der deutschen Höfe seit der Reformation*. Vol. 22, pt. 3: *Geschichte der Höfe des Hauses Braunschweig in Deutschland und England*, pt. 5. Hamburg: Hoffmann und Campe, 1853.

Walker, D. P. *Spiritual and Demonic Magic from Ficino to Campanella*. London: Warburg Institute, University of London, 1958.

Walsham, Alexandra, and Peter Marshall. *Angels in the Early Modern World*. Cambridge: Cambridge University Press, 2006.

Wandel, Lee Palmer, ed. *A Companion to the Eucharist in the Reformation*. Boston: Brill, 2014.

———. *The Eucharist in the Reformation: Incarnation and Liturgy*. Cambridge; New York: Cambridge University Press, 2006.

Weber, Karl von. *Sidonie, Herzogin von Braunschweig: Aus vier Jahrhunderten*. Leipzig: Verlag von Bernhard Tauchnitz, 1858.

Webster, Charles. *Paracelsus: Medicine, Magic, and Mission at the End of Time.* New Haven; London: Yale University Press, 2008.

Weeks, Andrew. *Paracelsus: Speculative Theory and the Crisis of the Early Reformation.* Albany: State University of New York Press, 1997.

Weyer, Jost. *Graf Wolfgang II. von Hohenlohe und die Alchemie: Alchemistische Studien in Schloss Weikersheim, 1587–1620.* Sigmaringen: Thorbecke, 1992.

Winzlmaier, Hans. *Wilhelm von Grumbach und seine mit ihm hingerichteten Gefährten: Kurz Biografien und Hinrichtungsbericht nach einer Handschrift von 1568.* Rimparer Geschichtsblätter 6. Würzburg: noitasprint, 2012.

Wood, Charles T. "The Doctor's Dilemma: Sin, Salvation, and the Menstrual Cycle in Medieval Thought." *Speculum* 56, no. 4 (1981): 710–27.

Wood, Christopher S. "The Credulity Problem." In *Antiquarianism and Intellectual Life in Europe and China, 1500–1800,* ed. Peter N. Miller and François Louis, 149–79. Ann Arbor: University of Michigan Press, 2012.

———. *Forgery, Replica, Fiction: Temporalities of German Renaissance Art.* Chicago: University of Chicago Press, 2008.

Wunder, Heide. *Dynastie und Herrschaftssicherung in der Frühen Neuzeit: Geschlechter und Geschlecht.* Berlin: Duncker & Humblot, 2002.

———. *Er ist die Sonn', Sie ist der Mond: Frauen in der Frühen Neuzeit.* Munich: Beck, 1992.

Yuval, Israel. "Jews and Christians in the Middle Ages: Shared Myths, Common Language." In *Demonizing the Other: Antisemitism, Racism, and Zenophobia,* edited by Robert S. Wistrich, 88–107. Amsterdam: Harwood Academic, 1999.

Zambelli, Paola. *"Astrologi hallucinati": Stars and the End of the World in Luther's Time.* Berlin; New York: W. de Gruyter, 1986.

———. *White Magic, Black Magic in the European Renaissance.* Leiden: Brill, 2007.

Zöller-Stock, Bettina. *Stühle: Sitzmöbel von der Renaissance bis zur Gegenwart aus dem Berliner Kunstgewerbemuseum.* Berlin: Pädagogischer Dienst, Staatliche Museen Preussischer Kulturbesitz, 1991.

Zuber, Mike. "Spiritual Alchemy from the Age of Jacob Boehme to Marry Anne Atwood, 1600–1900." PhD diss., University of Amsterdam, 2017.

INDEX

Note: Elites are listed by their first names.

ACKNOWLEDGMENTS

When I first encountered Anna Zieglerin, I thought she might be an interesting subject for an article, perhaps about gender and alchemy. Like many big projects that start small, however, my research in Anna Zieglerin gradually grew and took on a life of its own. I published my early thoughts about Zieglerin's significance in the history of alchemy in *Ambix* in 2001 and was fortunate to spend 2001–2002 developing a book project around Zieglerin, as the Sidney M. Edelstein International Fellow in the History of the Chemical Sciences and Technologies at the Chemical Heritage Foundation (now the Science History Institute) in Philadelphia. Over the years since, I have juggled other projects, as well as personal and professional responsibilities, but have returned to this project intermittently. In 2008, I was delighted to share my work on Zieglerin at a conference on the topic "Secrets and Knowledge" organized by Alisha Rankin and Elaine Leong at Trinity College, Cambridge. Rankin and Leong, along with the other participants in the conference, helped me begin to see the significance of Ziegerlin's life and work in the history of science and medicine more broadly—insights that I shared in an essay that appeared in *Secrets and Knowledge in Medicine and Science*, edited by Alisha Rankin and Elaine Leong.

Two fellowships were transformative for this project: the American Council of Learned Societies Frederick Burkhardt Residential Fellowship for Recently Tenured Scholars (Huntington Library, 2010–2011) and the John Simon Guggenheim Memorial Foundation Fellowship (2011–2012). I am enormously grateful for the support of these institutions, which allowed me precious time to complete my research, write intensively, and, most of all, reimagine the book as part of something more than the scholarly conversations with which I was most familiar, the history of alchemy and the history of science. I would not have had the ability or courage to venture into early modern religious, gender, and political history without the support of these fellowships, or the intellectual and social companionship of my extraordinary "fellow fellows" at the Huntington Library in 2010–2011, especially David Blight, Jeannine DeLombard, Mary Fuller, Margaret Garber, Daniel Horowitz, Helen Horowitz, Bruce Moran, Marcy Norton, Daniel Richter, Susannah Shaw Romney, and Sean Willentz. Conversations with each of these colleagues shaped this project in important ways.

I have been honored by the opportunity to share my work over the years with colleagues and students in seminars, workshops, and conferences at the New England Renaissance Conference; the Early Modern History Workshop, Early Science Working Group, and the Seminar on Women & Culture in the Early Modern World at Harvard University; the 2008 International Conference on "Chymia: Science and Nature in Early Modern Europe (1450–1750)," El Escorial, Madrid, Spain; the Theorizing Early Modern Studies Research Collaborative and the History of Science and Technology Colloquium at the University of Minnesota; the History Department at Washington University, St. Louis; the History of Science Seminar at UCLA; the Past Tense Seminar at the USC-Huntington Early Modern Studies Institute; the Institute of History & Philology at the Academia Sinica, Taiwan; the Medieval and Early Modern Studies Workshop at Stanford University; the Graduate Student Conference on "The Charlatan in Europe, 1500–1700" and the Davis Center for Historical Studies at Princeton University; the 2014 Blood Conference and 2016 Scientiae Conference at the University of Oxford; the Renaissance Studies Program at Indiana University; the Science Studies Program Colloquium at UC San Diego; the History Department at Tufts University; the Department of the History of Science at the University of Oklahoma; the Glass Department at the Rhode Island School of Design; the Medieval and Early Modern History Seminar at Brown University; the Forschungskolloquium Vormoderne/Historisches Seminar at the University of Lucerne; and the Department of History of Science and Ideas at Uppsala University. Every one of these audiences has challenged me to think broadly, deeply, and carefully, and this book would not be the same without their questions and suggestions.

I owe a special debt to a number of individuals who supported my research and read this book manuscript, in part or in its entirety, along the way. First and foremost, the wonderful staff of the Niedersächsisches Landesarchiv in Wolfenbüttel deserves my deep gratitude for always welcoming me and my queries. Long ago, Peter Starenko shared with me his own research on the "angel seer" Hänschen, modeling the best collegiality and collaborative spirit of our profession. Anne-Charlotte Trepp demonstrated the utmost scholarly generosity and trust very early on by giving me her microfilms of documents in the Niedersächsisches Landesarchiv Wolfenbüttel, and, at the very end of this project, Ute Friesch demonstrated that same generosity and trust by sharing her own research and writing on Zieglerin, particularly regarding the iron chair with which I begin this book. Meike Buck astonished me by alerting me to the existence of this chair in the first place, and kindly welcomed me at the Landesmuseum in Braunschweig in 2012 to view it in situ as part of the museum's exhibit. Conversations at strategic moments with Deborah Cohen and Moshe Sluhovsky helped me understand what was at stake with this book and how to write it, and periodic lunches with my colleague Adam Teller helped me articulate the relationship between the figures of the "court Jew" and the "court alchemist." I am thankful to Elaine Leong and Alisha Rankin for lending their expertise and careful reading to chapters of the book at a crucial moment in the project, and to Ann Durham for her willingness to read my academic writing and to offer her honest feedback on style and accessibility. I am especially grateful to the two scholars who read this book for the University of Pennsylvania Press, as well as to Leah DeVun, Joel Harrington, Hal Cook, and Amy Remensnyder for taking the time to read the entire manuscript once it came together and to discuss it at length. I am deeply indebted to this "dream team" for offering critical interventions on structure, and for pushing me to be bolder with my claims and my voice. Many people helped see this book into print, and I would like to extend my thanks to Hannah Blake and Lily Palladino at the University of Pennsylvania Press, as well as to indexer Nancy Gerth. Most of all, however, I am grateful to Jerry Singerman for seeing the potential of this project very early on, for waiting patiently during the many years of its gestation, and for his expert editorial eye (and good humor!) in its final stages.

Janice Neri was the first person I told about Anna Zieglerin when I encountered this curious alchemist during a very cold trip to Wolfenbüttel in the winter of 1999, and it is impossible to fully acknowledge all of the discussions we had over the years about the shape of this book. When Janice was diagnosed with cancer, our scholarly conversations came to be overshadowed by more pressing concerns, but she continued to ask about Anna Zieglerin as long as she could. I miss her enormously, both as a friend and as a colleague, but I know she would have been pleased to see this book make its way into print.

Finally, my deepest gratitude goes to Mila and to Seth, to whom this book is dedicated. Mila has lived with this book her entire life (whether she realizes it or not), but I am delighted that she now shares at least some of my interest in alchemy and in Germany. For over two decades, Seth has been an unwaveringly supportive partner in life, parenting, work, and scholarship. I am extraordinarily lucky to share my life with such a talented person.

Printed and bound by CPI Group (UK) Ltd, Croydon, CR0 4YY

27/10/2024

14580744-0001